# Mixed-flow Pumps

## Wiley-ASME Press Series

Thermal Spreading and Contact Resistance: Fundamentals and Applications
*Yuri S. Muzychka and M. Michael Yovanovich*

Analysis of ASME Boiler, Pressure Vessel, and Nuclear Components in the Creep Range
*Maan H. Jawad and Robert I. Jetter*

Voltage-Enhanced Processing of Biomass and Biochar
*Gerardo Diaz*

Pressure Oscillation in Biomedical Diagnostics and Therapy
*Ahmed Al-Jumaily and LuluWang*

Robust Control: Youla Parameterization Method
*Farhad Assadian and Kevin R. Mellon*

Metrology and Instrumentation: Practical Applications for Engineering and Manufacturing
*Samir Mekid*

Fabrication of Process Equipment
*Owen Greulich and Maan H. Jawad*

Engineering Practice with Oilfield and Drilling Applications
*DonaldW. Dareing*

Flow-Induced Vibration Handbook for Nuclear and Process Equipment
*Michel J. Pettigrew, Colette E. Taylor, and Nigel J. Fisher*

Vibrations of Linear Piezostructures
*Andrew J. Kurdila and Pablo A. Tarazaga*

Bearing Dynamic Coefficients in Rotordynamics: Computation Methods and Practical Applications
*Lukasz Brenkacz*

Advanced Multifunctional Lightweight Aerostructures: Design, Development, and Implementation
*Kamran Behdinan and Rasool Moradi-Dastjerdi*

Vibration Assisted Machining: Theory, Modelling and Applications
*Li-Rong Zheng, Dr. Wanqun Chen, and Dehong Huo*

Two-Phase Heat Transfer
*Mirza Mohammed Shah*

Computer Vision for Structural Dynamics and Health Monitoring
*Dongming Feng and Maria Q Feng*

Theory of Solid-Propellant Nonsteady Combustion
*Vasily B. Novozhilov and Boris V. Novozhilov*

Introduction to Plastics Engineering
*Vijay K. Stokes*

Fundamentals of Heat Engines: Reciprocating and Gas Turbine Internal Combustion Engines
*Jamil Ghojel*

Offshore Compliant Platforms: Analysis, Design, and Experimental Studies
*Srinivasan Chandrasekara and R. Nagavinothini*

Computer Aided Design and Manufacturing
*Zhuming Bi and Xiaoqin Wang*

Pumps and Compressors
*Marc Borremans*

Corrosion and Materials in Hydrocarbon Production: A Compendium of Operational and Engineering Aspects
*Bijan Kermani and Don Harrop*

Design and Analysis of Centrifugal Compressors
*Rene Van den Braembussche*

Case Studies in Fluid Mechanics with Sensitivities to Governing Variables
*M. Kemal Atesmen*

The Monte Carlo Ray-Trace Method in Radiation Heat Transfer and Applied Optics
*J. Robert Mahan*

Dynamics of Particles and Rigid Bodies: A Self-Learning Approach
*Mohammed F. Daqaq*

Primer on Engineering Standards, Expanded Textbook Edition
*Maan H. Jawad and Owen R. Greulich*

Engineering Optimization: Applications, Methods and Analysis
*R. Russell Rhinehart*

Compact Heat Exchangers: Analysis, Design and Optimization using FEM and CFD Approach
*C. Ranganayakulu and Kankanhalli N. Seetharamu*

Robust Adaptive Control for Fractional-Order Systems with Disturbance and Saturation
*Mou Chen, Shuyi Shao, and Peng Shi*

Robot Manipulator Redundancy Resolution
*Yunong Zhang and Long Jin*

Stress in ASME Pressure Vessels, Boilers, and Nuclear Components
*Maan H. Jawad*

Combined Cooling, Heating, and Power Systems: Modeling, Optimization, and Operation
*Yang Shi, Mingxi Liu, and Fang*

Applications of Mathematical Heat Transfer and Fluid Flow Models in Engineering and Medicine
*Abram S. Dorfman*

Bioprocessing Piping and Equipment Design: A Companion Guide for the ASME BPE Standard
*William M. (Bill) Huitt*

Nonlinear Regression Modeling for Engineering Applications: Modeling, Model Validation, and Enabling Design of Experiments
*R. Russell Rhinehart*

Geothermal Heat Pump and Heat Engine Systems: Theory and Practice
*Andrew D. Chiasson*

Fundamentals of Mechanical Vibrations
*Liang-Wu Cai*

Introduction to Dynamics and Control in Mechanical Engineering Systems
*Cho W.S. To*

Thermal Spreading and Contact Resistance: Fundamentals and Applications
*Yuri S. Muzychka and M. Michael Yovanovich*

Mixed-flow Pumps: Modeling, Simulation, and Measurements
*Wei Li, Leilei Ji, Ramesh Agarwal, Weidong Shi, and Ling Zhou*

This Work is a co-publication between John Wiley & Sons, Inc. and ASME Press.

# Mixed-flow Pumps

## Modeling, Simulation, and Measurements

*Wei Li*
Jiangsu University, China

*Leilei Ji*
Jiangsu University, China

*Ramesh Agarwal*
Washington University in St. Louis, USA

*Weidong Shi*
Nantong University, China

*Ling Zhou*
Jiangsu University, China

ASME PRESS
WILEY

This Work is a co-publication between John Wiley & Sons, Inc. and ASME Press.
Published simultaneously in Canada.



*Library of Congress Cataloging-in-Publication Data:*

Names: Li, Wei (Professor of engineering), author. | Ji, Leilei, author. | Agarwal, R. K. (Ramesh K.), author. | Shi, Weidong (College presidents), author. | Zhou, Ling (Professor), author.
Title: Mixed-flow pumps : modeling, simulation, and measurements / Wei Li, Leilei Ji, Ramesh Agarwal, Weidong Shi, Ling Zhou.
Description: Hoboken, NJ : Wiley, [2024] | Series: Wiley-ASME press series | Includes index.
Identifiers: LCCN 2024009766 (print) | LCCN 2024009767 (ebook) | ISBN 9781119910787 (hardback) | ISBN 9781119910794 (adobe pdf) | ISBN 9781119910374 (epub)
Subjects: LCSH: Centrifugal pumps. | Computational fluid dynamics.
Classification: LCC TJ919 .L38 2024 (print) | LCC TJ919 (ebook) | DDC 621.6/7–dc23/eng/20240319
LC record available at https://lccn.loc.gov/2024009766
LC ebook record available at https://lccn.loc.gov/2024009767

Cover Design: Wiley
Cover Image: Courtesy of Wei Li, Leilei Ji, Ramesh Agarwal, Weidong Shi, Ling Zhou

Set in 9.5/12.5pt STIXTwoText by Straive, Chennai, India

# Contents

# Preface

Pumps are among the most power-consuming general-purpose equipment in energy conversion devices and significantly impact the modern industrial economy. A mixed-flow pump can be considered a type of pump design between centrifugal pump and axial flow pump since it employs the combined effect of centrifugal force and thrust generated by the rotation of the impeller to convey fluid, and the fluid flows axially in and diagonally out through the impeller. It can also be called oblique flow pump with high flow rate, high efficiency, strong anti-cavitation performance, etc. It is widely used for agricultural irrigation, municipal water supply and drainage, water circulation in power industry, naval water jet propulsion, underwater weapon launches, and regional water transfer projects.

Compared to other pump types, the internal flow of mixed-flow pumps is more complex, and the secondary flow and deliquescence are more prominent. There are not only inherent unsteady flow problems caused by the static and dynamic interference but also unsteady problems induced by the wheel edge leakage vortex and its trailing off, rotational stall, and other complex flow phenomena which seriously affect the operational stability and efficiency of the mixed-flow pumps. Therefore, there is a need to explore the spatial and temporal evolution of flow structures and flow dynamics of the internal flow field of a mixed-flow pump as well as to achieve the desired targeted optimized solutions. In addition, the internal vortex energy loss characteristics of mixed-flow pumps, cavitation damage, and other phenomena also need to be studied systematically. Understanding and mastering the physical mechanisms of the internal flow in a mixed-flow pump is a prerequisite for improving the operational stability, reliability, and efficiency of the pump.

In previous studies, the flow field and performance characteristics of a mixed-flow pump were generally determined and analyzed by experimental means; however, the experimental approach is not only expensive, but it is often difficult to observe and obtain all the details of the flow field experimentally due to its complex structure. In recent years, the emergence of computational fluid dynamics has provided an effective tool to study the finer details of the flow structure inside the hydraulic machinery, which is uniquely beneficial in analyzing the internal flow field in a mixed-flow pump at multiple scales for a full range of operating conditions. Currently, there are no reference books providing the computational approach for the study of the flow fields and performance of mixed-flow pumps. Therefore, this book selects a typical model of a guide vane-type mixed-flow pump as the object of study and systematically investigates the complex internal flow structure through numerical simulations and experiments aiming to provide a reference work for industrial practitioners, academics, and students interested in the field of hydraulic machinery.

The book is divided into 12 chapters; the content of each chapter is as follows.

The first chapter provides a brief introduction to the definitions, types, and applications of mixed-flow pumps. The second chapter provides a detailed description of the basic concepts of mixed-flow pumps and the related theories. Chapter 3 focuses on computational fluid dynamics (CFD) simulation technology including geometric modeling, meshing, governing equations of fluid flow, CFD methods classification, turbulence models, solution algorithms, near-wall surface treatment, and boundary conditions. Chapter 4 describes different analysis methods including entropy production analysis, vortex analysis, and wavelet methods. Chapter 5 details the experimental methods, data, and analysis such as pressure pulsation measurements, PIV measurements, and axial trajectory measurements. Chapter 6 covers the application of turbulence models and compares the applicability of several turbulence models in the performance prediction of mixed-flow pumps. Chapter 7 investigates and analyzes the energy characteristics, flow structure, instability characteristics, and dynamic and static interference of the tip leakage flow of the mixed-flow pump. Chapter 8 investigates and analyzes the energy characteristics, flow structure, and the effect of tip clearance on the rotational stall and its propagation characteristics as well as the causes of incipient and deep rotational stall in the mixed-flow pump. Chapter 9 provides several passive suppression techniques to control the rotating stall in the mixed-flow pump. Chapter 10 analyzes the cavitation flow field and cavitation energy characteristics of the mixed-flow pump. Chapter 11 describes a special application of the Wray–Agarwal (WA) one-equation turbulence model to analyze the vortex dynamics characteristics in the tip region of the mixed-flow pump to demonstrate the accuracy and efficiency of the WA model for computing such complex flows compared to the other widely used turbulence models. Chapter 12 investigates the influence of the sediment particles on internal energy dissipation of the mixed-flow pump with different solid-phase volume fractions.

This book has been limited in terms of the depth and breadth of data collection. Furthermore, there could inevitably be some shortcomings and errors in the book. We hope that readers will provide comments and input so that future editions can be improved.

# Acknowledgments

This work was sponsored by the Key International Cooperative Research Program of the National Natural Science Foundation of China (No. 52120105010), the National Key R&D Project (No. 2020YFC1512405) of China, the National Natural Science Foundation of China (No. 52179085 and No. 52309112), the Sixth "333 High Level Talented Person Cultivating Project" of Jiangsu Province, projects of the "Blue Project" in Jiangsu Colleges and Universities, and China Postdoctoral Science Foundation (No. 2022TQ0127 and No. 2023M741414); Open Research Subject of Key Laboratory of Fluid and Power Machinery (Xihua University), Ministry of Education (LTDL-2022010).

The publication of this book was made possible by the help of several colleagues and students at the National Engineering Research Center for Pumps and Systems of Jiangsu University and the CFD Laboratory of Washington University in St. Louis, USA, to whom the authors would like to express their sincere gratitude.

In particular, special thanks go to Mingjiang Liu, Shuo Li, Yi Yang, Shenglei He, and others from Jiangsu University for completing the translation of part of this book from Chinese.

The authors would like to dedicate this book to their respective families for their unwavering support, perseverance, and encouragement during the preparation of this book.

# List of Acronyms

| | |
|---|---|
| BPF | blade-passing frequency |
| BV | bounded vortex |
| CFD | computational fluid dynamics |
| CFL | Courant–Fredrick–Levy |
| DES | detached Eddy simulation |
| DNS | direct numerical simulation |
| FFT | fast Fourier transform |
| GCI | grid convergence index |
| HRN | high Reynolds number |
| LE | leading edge |
| LES | large eddy simulation |
| LNG | liquefied natural gas |
| LRN | low Reynolds number |
| LVC | local vortical cavitation |
| PDE | partial differential equation |
| PIV | particle image velocimetry |
| PS | pressure side |
| PTLV | primary tip leakage vortex |
| PV | passage vortex |
| RANS | Reynolds-averaged Navier-Stokes |
| RHD | right-hand side |
| RSI | rotor-stator interaction |
| RSM | Reynolds stress model |
| SA | Spalart–Allmaras |
| SGS | subgrid-scale |
| SIMPLE | semi-implicit method for pressure-linked equations |
| SS | suction side |
| SST | shear stress transport |
| STLV | secondary tip leakage vortex |
| SV | secondary vortex |
| TCC | tip clearance cavitation |
| TE | trailing edge |
| TLF | tip leakage flow |

| | |
|---|---|
| TLV | tip leakage vortex |
| TLVC | tip leakage vortex core |
| TMR | turbulence modeling resource |
| WA | Wray–Agarwal |
| WTC | wavelet transforms coherence |

# List of Symbols

## Nomenclature

| | |
|---|---|
| $b_2$ | the width of the impeller, mm |
| $C_p$ | pressure pulsation coefficient |
| $D_1$ | the inlet diameter of the impeller, mm |
| $D_2$ | the outlet diameter of the impeller, mm |
| $D_3$ | the inlet diameter of the guide vane, mm |
| $D_4$ | the outlet diameter of the guide vane, mm |
| $D_\omega$ | the cross-diffusion term in turbulence models |
| $G_b$ | generation of turbulence due to buoyancy in turbulence models |
| $G_k$ | generation of turbulent kinetic energy due to the mean velocity gradients in turbulence models |
| $G_\omega$ | production term of the turbulent dissipation rate in turbulence models |
| $h_{\Delta p}$ | head drop loss coefficient |
| $H$ | head, m |
| $H_t$ | theoretical head, m |
| $i, j$ | stands for the $x$, $y$, $z$ direction |
| $k$ | turbulent kinetic energy, $m^2/s^2$ |
| $M_t$ | turbulent Mach number in turbulence models |
| $m_{ji}$ | time-averaged viscous stress tensor |
| $\dot{m}$ | mass discharge from each domain of the pump, kg/s |
| $NPSH$ | net positive suction head, m |
| $NPSH_a$ | net inlet pressure available, m |
| $NPSH_r$ | net inlet pressure required, m |
| $n$ | rated speed of the impeller, r/min |
| $n_s$ | specific speed |
| $p_1, p_2$ | total pressure at the inlet and outlet of each domain, Pa |
| $P_e$ | effective power, W |
| $P_{tol}$ | total input power, W |
| $q$ | heat flux, J/s |
| $Q$ | flow rate, $m^3/h$ |
| $Q_{des}$ | designed flow rate, $m^3/h$ |
| $\dot{Q}$ | energy transfer rate |
| $R$ | the cross-diffusion term in WA model ($= k/\omega$) |
| $R(a,b)$ | the coherence coefficient |

| | |
|---|---|
| $S$ | strain rate, smoothing operator, standard deviation |
| $s$ | specific entropy, J/(kg K) |
| $\dot{S}_D'''$ | local entropy production rate, kW/m$^3$/K$^3$ |
| $\dot{S}_{\overline{D}}'''$ | entropy production rate induced by time-averaged movement, kW/m$^3$/K$^3$ |
| $\dot{S}_{D'}'''$ | entropy production rate induced by velocity fluctuation, kW/m$^3$/K$^3$ |
| $t$ | time, s |
| $T$ | temperature, K |
| $u$ | velocity, m/s |
| $u_j$ | stands for the velocity in different coordinate directions |
| $x$ | coordinate, m |
| $x_j$ | stands for the coordinate directions |
| $Y_M$ | the effect of the expansion of compressible turbulence on the total dissipation rate in turbulence models |
| $Y_k$, $Y_\omega$ | the dissipation terms of $k$ and $\omega$ in turbulence models |
| $y$ | the distance from the wall |
| $Z$ | number of impeller blades |
| $Z_\mathrm{d}$ | number of guide vane blades |
| $\alpha_3$ | average inlet blade angle of guide vane, ° |
| $\alpha_4$ | average outlet blade angle of guide vane, ° |
| $\beta$ | the coefficient of thermal expansion |
| $\beta_1$ | average inlet blade angle of the impeller, ° |
| $\beta_2$ | average outlet blade angle of the impeller, ° |
| $\Gamma_k$, $\Gamma_\omega$ | the coefficients of diffusion term for $k$ and $\omega$ in turbulence models |
| $\delta_{ij}$ | Kronecker delta symbol |
| $\varepsilon$ | turbulent dissipation rate, m$^2$/s$^3$ in turbulence models |
| $\eta$ | efficiency of mixed-flow pump, % |
| $\mu$ | dynamic viscosity, Pa s |
| $\mu_t$ | turbulent viscosity, m$^2$/s |
| $\rho$ | density, kg/m$^3$ |
| $\sigma_k$, $\sigma_\varepsilon$ | turbulent Prandtl numbers for $k$ and $\varepsilon$ in turbulence models |
| $\phi$ | scalar variable |
| $\omega$ | turbulent eddy frequency, s$^{-1}$ in turbulent models |

# 1

# Introduction

## 1.1 What Is a Mixed-flow Pump?

A mixed-flow pump is a centrifugal pump with a mixed-flow impeller [1]. The specific speed ($n_s$) lies between 35 and 80 rpm for low-speed mixed-flow pumps and between 80 and 160 rpm for higher-speed mixed-flow pumps (in special cases, even higher). It has characteristics of both radial flow and axial flow pumps. As liquid flows through the impeller of a mixed-flow pump, the impeller blades push the liquid out away from the pump shaft and to the pump suction at an angle greater than 90°. The impeller of a typical mixed-flow pump and the flow through a mixed-flow pump are shown in Fig. 1.1.

## 1.2 Types of Mixed-flow Pumps

Based on the type of suction chamber, mixed-flow pumps can be divided into two types: volute mixed-flow pumps and guide vane mixed-flow pumps, as shown in Fig. 1.2. The former is close to the design of a centrifugal pump, and the latter is close to the design of an axial flow pump.

At present, majority of mixed-flow pumps are volute mixed-flow pumps which are similar to a single-suction centrifugal pump but are different in the type of impeller: the impeller of a mixed-flow pump of high specific speed is similar to that of an axial flow pump which is open type with adjustable blades; the impeller of a mixed-flow pump of low specific speed, on the other hand, is closed type which is similar to that of a single-suction centrifugal pump, but its flow channel is wider and the blade outlet is inclined.

Compared to the axial flow pump, the guide vane mixed-flow pump has slightly higher efficiency and a relatively flat efficiency characteristic curve. In other words, it can ensure higher efficiency when the water level changes; hence, it is very suitable for farmland drainage and irrigation and saves power, but compared to the volute mixed-flow pump, its diameter is smaller. For the vertical guide vane mixed-flow pump, the impeller is submerged in water during operation, so there is no need for water diversion equipment, and therefore the needed floor area is small. Therefore, in places where the axial flow pump is used (except for the axial flow pump with large adjustable blades), it is advantageous to replace it with an appropriate model of guide vane mixed-flow pump.

Other classifications of mixed-flow pumps are:

1. According to the inspection and disassembly form, they can be divided into the extractable mixed-flow pump and the non-extractable mixed-flow pump.
2. According to the blade regulation type, they can be divided into the fixed mixed-flow pump, the semi-regulated submersible axial flow pump, and the fully regulated mixed-flow pump.

*Mixed-flow Pumps: Modeling, Simulation, and Measurements*, First Edition.
Wei Li, Leilei Ji, Ramesh Agarwal, Weidong Shi, and Ling Zhou.

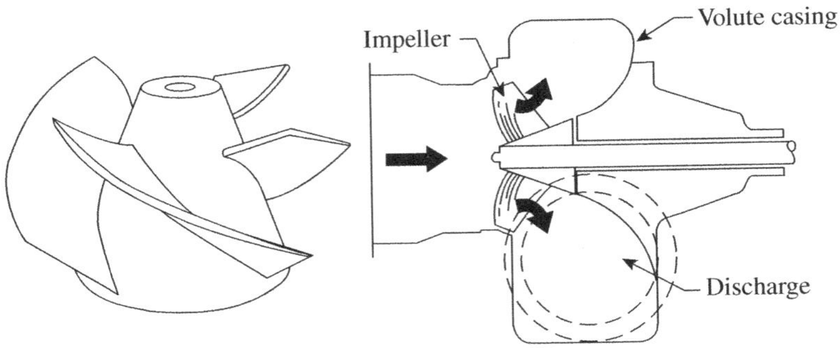


**Figure 1.1** Mixed-flow pump impeller and mixed-flow pump model.

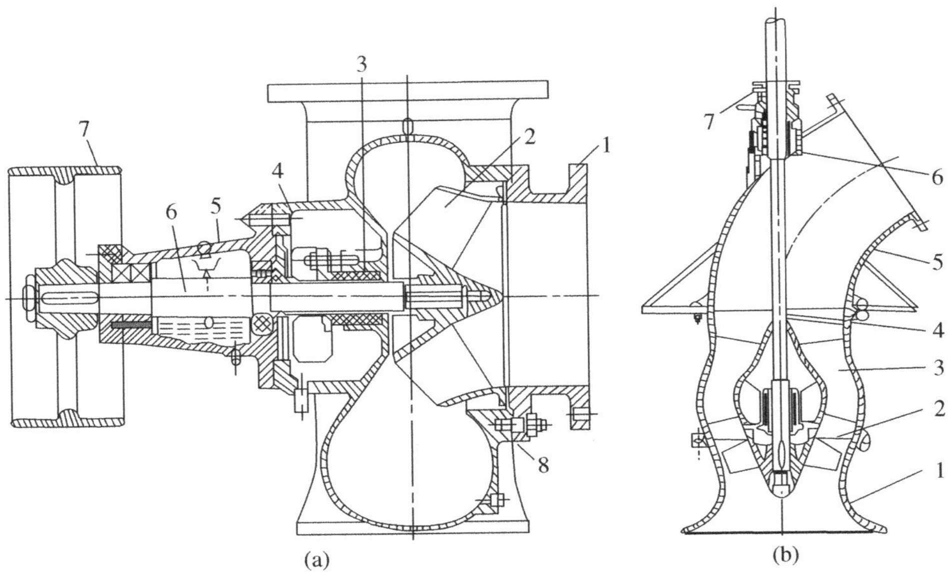


**Figure 1.2** Classification of mixed-flow pumps. (a) Structural diagram of volute mixed-flow pump 1. Pump cover, 2. Impeller, 3. Packing, 4. Pump body, 5. Bearing body, 6. Pump shaft, 7. Pulley, 8. Bolt. (b) Structural diagram of guide vane mixed-flow pump. 1. Suction horn, 2. Impeller, 3. Guide vane, 4. Outlet elbow, 5. Pump shaft, 6. Rubber bearing, 7. Stuffing box.

## 1.3 Agricultural and Industrial Applications of Pumps

Due to the characteristics of moderate head and large flow rate, the mixed-flow pump is widely used in farmland irrigation, flood prevention and drainage, sewage treatment, power station cooling systems, and other applications.

In agricultural production, the main function of the mixed-flow pump is irrigation and drainage. There are vast rural areas in the world, thus a large number of pumps are needed every year. Generally, agricultural pumps account for more than half of the total output of the pumps.

In the mining and metallurgical industries, mixed-flow pumps are also widely used. The mixed-flow pump is used for drainage and water supply in the process of beneficiation, smelting, and rolling in mines.

In the power sector, power stations need a large number of boiler feed pumps, condensate pumps, circulating pumps, and ash pumps, among which mixed-flow pumps account for the majority.

In the shipbuilding industry, many advanced water jet propulsion pumps are of mixed-flow pump types.

The following are examples of large-scale mixed-flow pump station projects in which Chinese companies have been engaged inside China as well as in neighboring countries for development of shipping, flood discharge, and other functions. The three representative projects are briefly described below:

1. Pumping station of Zaohe River in Suqian, Jiangsu province, China [2].
   The first-stage renovation project of the Zaohe River pumping station in the eastern route of the south-to-north water diversion project is located in Zaohe town, Suyu district, Suqian City, Jiangsu Province, China. Its primary task is to pump the diverted water from the Liulaodian pumping station into Luoma Lake, achieving a target water delivery of 175 $m^3/s$ to Luoma Lake and addressing the drainage needs in the regions of Pihong River and Huangdun Lake.
   The Zaohe pumping station, shown in Fig. 1.3 is currently equipped with two sets of 5700HLQ100-4.78 vertical fully adjustable guide vane mixed-flow pumps. The pumps are designed with a net head of 4.78 meters, a design flow rate of 100 $m^3/s$, an impeller diameter of 5.70 meters, a rated speed of 75 r/min, and an adjustable blade angle in the range $+2°$ to $-18°$. They are paired with TL7000-80/7400 vertical synchronous electric motors with a rated capacity of 7000 kW and a total installed capacity of 14 000 kW. The first unit was successfully started on April 8, 2011, at 15:40 in the afternoon.
2. Qushou pumping station of Qinglongshan irrigation area in Heilongjiang Sanjiang Plain, China [3].

**Figure 1.3** Pumping station of Zaohe River in Suqian, Jiangsu province, China. Source: [2]. Jiangsu Aerospace Hydraulic Equipment Co., Ltd. https://www.pumpcj.com/case/95.html. Last accessed 17 January, 2024.

**Figure 1.4** Qushou pumping station of Qinglongshan irrigation area in Heilongjiang Sanjiang Plain, China. Source: [3]. Jiangsu Aerospace Hydraulic Equipment Co., Ltd. https://www.pumpcj.com/case/97.html. Last accessed 17 January, 2024.

The installed flow rate capacity of the Qushou pumping station of Qinglongshan irrigation area is 381 $m^3/s$, and the total installed power capacity is 56 000 kW. It has six sets of 3300HLQ38.1-9.74 fully adjustable mixed-flow pumps to irrigate the largest irrigation area in Heilongjiang province. Furthermore, it is the second largest mixed-flow pump station in China, as shown in Fig. 1.4. This infrastructure plays a crucial role in realizing increased grain production and efficiency, optimizing the regional water resource allocation, and implementing the coordinated scheduling of surface water, groundwater, and rainwater resources for irrigation in the Sanjiang region – the largest granary in the country. It contributes significantly to promoting the coordinated and sustainable development of the economy, society, and ecology in the region.

3. The Belt and Road project of Chongqing Electromechanical Group – the Hyderabad flood control irrigation project in Telangana, India – has been successfully tested recently [4]. The 24 large, closed-volute mixed-flow pumps and 12 large synchronous motors used in the project have all been developed by Chongqing Hydro Turbine Co. Ltd. with independent intellectual property rights. Twenty-four large mixed-flow pumps are installed in this flood control and irrigation project. Each water pump has a flow of 41 $m^3/s$, a lift of 11 m, and a rotational speed of 136.6 r/min. It is the largest closed mixed-flow pump with single unit power of 6500 kW synchronous motor. The energy index and stability index of water pumps and synchronous motors have reached an international advanced level.

## 1.4 Summary

This chapter provides an overview of the main structural forms of the mixed-flow pump, its classifications, and industrial applications. In terms of the rotational speed, the specific speed (ns) ranges between 35 and 80 rpm for low-speed mixed-flow pumps and between 80 and 160 rpm for higher-speed mixed-flow pumps. Considering the structure of the suction chamber, mixed-flow pumps can be categorized as volute mixed-flow pumps and guide vane mixed-flow pumps. Additionally, the broad applications of mixed-flow pumps in agricultural irrigation and other major industrial projects attest to their excellent operational range, performance, and stability.

## References


1 https://www.ksb.com/centrifugal-pump-lexicon/mixed-flow-pump/192030.
2 https://www.pumpcj.com/case/95.html.
3 https://www.pumpcj.com/case/97.html.
4 https://rmh.pdnews.cn/Pc/ArtInfoApi/article?id=7493665.

# 2

# Basic Concepts and Theory of Mixed-flow Pumps

## 2.1 Basic Flow and Performance Parameters

The parameters characterizing the performance of the pump are as follows.

### 2.1.1 Volume Flow $Q$

Flow is the volume (or mass) of liquid delivered in unit time. Volume flow is expressed in $Q$, the unit is $m^3/s$, $m^3/h$, or L/s, etc. The mass flow is expressed as $Q_m$ and the unit is ton/h, kg/s, etc. The relationship between the mass flow and volume flow is

$$Q_m = \rho Q \tag{2.1}$$

where $\rho$ is the density of the liquid in $kg/m^3$; $\rho$ of clean water is generally taken as $1000\,kg/m^3$ at room temperature.

### 2.1.2 Head $H$

Head is the increment of energy per unit weight of liquid pumped by the pump from the pump inlet to the pump outlet. It is the effective energy obtained by 1 kg of liquid through the pump. The unit therefore is (N • m/N) = m, which is the equivalent liquid column height pumped by the pump shaft; the liquid column height is called the pump head $H$ and is conventionally given in meters (m). The pump head can be written as

$$H = E_d - E_S \tag{2.2}$$

where $E_d$ is energy per unit weight of liquid at the pump outlet in meters and $E_s$ is energy per unit weight of liquid at the pump inlet in meters.

The energy per unit weight of the liquid is called head in hydraulics, which is usually composed of the pressure head $\frac{p}{\rho g}$ (m), the velocity head $\frac{v^2}{2g}$ (m), and the position head $z$ (m), i.e.

$$E_d = \frac{p_d}{\rho g} + \frac{v_d^2}{2g} + z_d \tag{2.3}$$

and

$$E_s = \frac{p_s}{\rho g} + \frac{v_s^2}{2g} + z_s \tag{2.4}$$

*Mixed-flow Pumps: Modeling, Simulation, and Measurements*, First Edition.
Wei Li, Leilei Ji, Ramesh Agarwal, Weidong Shi, and Ling Zhou.

Therefore

$$H = \frac{p_d - p_s}{\rho g} + \frac{v_d^2 - v_s^2}{2g} + (z_d - z_s) \tag{2.5}$$

where $p_d$ and $p_s$ are the static pressure of the liquid at the pump outlet and inlet respectively, $v_d$ and $v_s$ are the liquid velocity at the pump outlet and inlet, respectively, and $z_d$ and $z_s$ are the distance from the pump outlet and inlet to a specified measuring datum plane respectively.

The head $H$ of the pump is a key performance parameter of the pump, which is only related to the energy of the liquid at the inlet and outlet flanges of the pump and is not directly related to the type of the pump. However, using the energy equation, the pump head can be expressed by the energy of the liquid in the pump device.

### 2.1.3 Speed *n*

The speed $n$ is the number of revolutions per unit time of the pump shaft, and its unit is revolutions r/min.

### 2.1.4 *NPSH*

The *NPSH*, an abbreviated form for the net positive suction head, is the main parameter indicating the cavitation performance of the pump. The *NPSH* has also been represented by $\Delta h$ in the literature by some scientists.

### 2.1.5 Power and Efficiency

Pump power usually refers to the input power; it is the power transmitted by the prime mover to the pump shaft and is also known as the shaft power expressed by $P$.

The effective power of the pump, also known as the output power is expressed by $P_e$. It is the effective energy obtained by the liquid output from the pump per unit time.

Since the pump head is the effective energy obtained from the pump by the unit mass of liquid output from the pump, the product of head, mass flow, and gravity acceleration is the effective energy obtained from the liquid output from the pump in unit time, which is the effective power of the pump.

$$P_e = HQ_m g = \rho g Q H \text{ (Watt W)} \tag{2.6}$$

or

$$P_e = \frac{\rho g Q H}{1000} = \frac{\gamma Q H}{1000} \text{(kW)} \tag{2.7}$$

where $\rho$ is the density of liquid delivered by the pump in kg/m$^3$, $\gamma = \rho\, g$ is the specific gravity of the liquid delivered by the pump in N/m$^3$, $Q$ is the pump flow in m$^3$/s, $H$ is the pump head in m, and g is the gravitational acceleration in m/s$^2$.

If the unit of specific gravity of the liquid is kg f/m$^3$, then

$$P_e = \frac{\gamma Q H}{100} \text{(kW)} \tag{2.8}$$

The difference in the shaft power $P$ and the effective power $P_e$ is the power loss in the pump, which is used to determine and define the efficiency of the pump. The efficiency of the pump is the ratio of the effective power to the shaft power expressed as

$$\eta = P_e / P \tag{2.9}$$

## 2.2 Typical Type of Flows in the Mixed-flow Pumps

### 2.2.1 Tip Leakage Flow

Since the tip leakage flow (TLF) in turbomachinery has significant impact on the performance and even safety of the machine, it is very important in the study of hydraulic machinery. Since the 1950s, the understanding of the TLF has been one of the major research topics in fluid mechanics of pumps, compressors, and turbines. The TLF model developed by Rains [1] is considered as the first original and seminal contribution which has served as a stepping stone toward the comprehensive understanding of this important flow phenomenon in turbomachines. Using this model, the velocity of TLF at the top of the outlet of the suction surface can be approximately estimated. At the same time, the change in the runner efficiency caused by TLF can be analyzed, but this model cannot calculate the micro-flow structure of the flow field. Later, Chen et al. [2] simplified the TLF model and deduced the trajectory coordinates of the two-dimensional tip leakage vortex (TLV) theoretically. Early experimental research and numerical simulations in the field of gas turbines provided a lot of information and data for exploring and analyzing the cause of formation as well as flow field structure of the flange leakage vortex [3, 4]. All these research efforts have enormously contributed to the present understanding of TLF and TLV.

Compared to the gas turbine, due to the large viscosity of water and more complex flow field in the end wall region, the research progress on TLF of the mixed-flow pump has been relatively slow. Yi et al. [5] employed the Reynolds-Averaged Navier-Stokes (RANS) equations with the SST $k$-$\omega$ turbulence model, revealing the formation mechanism of TLV and its influence on the performance of mixed-flow pump. Liu et al. [6] studied the shape and trajectory of the TLV in the mixed-flow pump, qualitatively and quantitatively analyzed the TLV, and determined that the TLV formed by the mixing of TLF and mainstream is the main reason for the deterioration of flow pattern and performance of the mixed-flow pump. Wu [7, 8], and Miorini et al. [9] used PIV technology to test and measure the flow field structure of TLV in the axial-flow water-jet propeller. It was found that the instantaneous TLV structure was formed by the unsteady vortex propagating to the top area of the blade channel, entraining with the mainstream and then breaking when reaching the pressure surface of adjacent blades. Using the PIV measurements, Masahiro et al. [10] tried to determine the generation mechanism of TLV of the mixed-flow pump at low flow rates and its impact in creating instability in the flow. It was found that the load on the impeller blade inlet rim increased with an increase in the leakage flow, and the TLV developed further with a decrease in the flow rate, forming a shroud of leakage flow.

Since TLF has a direct relationship with the clearance size, many researchers have studied the TLF under different tip clearances. Hah [11] and Li et al. [12] employed the LES to reveal the unsteady flow properties of TLF and TLV and analyzed the structure of TLF for five different tip clearances. Li et al. [13] studied the leakage flow for different tip clearances by performing numerical calculations and analyzed the formation and development process as well as the losses due to TLF and TLV for different tip clearances, and found that the strength of TLV increases with increase in tip clearance resulting in increase in losses. Zhang et al. [14] conducted the numerical simulation of a mixed-flow pump with low specific speed and analyzed its internal flow field for different tip clearances and found that the larger the tip clearance, the greater the effect of entrainment between the TLF and the mainstream flow. Goto [15] numerically analyzed the interaction mechanism of secondary flow and the formation mechanism of jet wake structure in the end wall region of the mixed-flow pump for four different tip clearances using the three-dimensional RANS equations and found that the reverse flow caused by the TLF at larger clearances is the main reason for thickening of the boundary layer in the end wall region resulting in the deterioration of the entire flow field. Zhong et al. [16] studied the mixed-flow water-jet pump and analyzed the influence of different blade tip clearances on the performance and internal flow field of the water-jet pump and improved the blade profile in order to reduce the losses. Shi and Zhang [17, 18] studied the external flow characteristics as well as the internal flow field of the mixed-flow pump for different tip clearances through the combination of numerical simulation and experiment to analyze the evolution process of TLV and determine the influence of different tip clearances on formation and evolution of TLV and its effect on hydraulic performance and cavitation characteristics of the pump. Bing et al. [19] experimentally studied the efficiency drop of a mixed-flow pump ($\Delta\eta/\Delta\delta$) due to an increase in tip clearance $\Delta\delta$. It was found that the head, power, and efficiency decrease with an increase in tip clearance. A large number of studies have shown that TLF has a significant effect on the hydraulic performance of mixed-flow pumps. These studies provide a great deal of information toward the understanding of the complex three-dimensional flow field due to TLF in a mixed-flow pump.

### 2.2.2 Rotating Stall

The concept of the rotating stall in turbomachinery first appeared in 1955; it was first proposed by Emmons, [20] who provided a classical explanation as described in Fig. 2.1. Emmons considered an in-line cascade as an example to explain the rotating stall. The disturbances in the circumferential direction will easily occur due to the decrease of flow rate, and then resulting in an asymmetric flow. In addition, due to the uneven manufacturing or installation of blades, some flow channels in the impeller can generate separated flow, leading to stall and channel blockage. Assuming that blade C in Fig. 2.1 stalls first, flow separation will occur in the flow channel so that the flow in the flow channel between the adjacent blades D and C, opposite to the direction of rotation of the impeller, will be squeezed and the flow deformation shown in Fig. 2.1 will occur to avoid the extrusion area increasing the angle of attack of blade D and as a result the flow capacity in channel D will be weakened and the blockage in the flow channel will be exacerbated gradually entering into stall. The inlet angle of attack of adjacent blade B with the same direction of rotation as the impeller will decrease. If there is a stall in the flow channel of blade B, the blockage of the flow channel will be

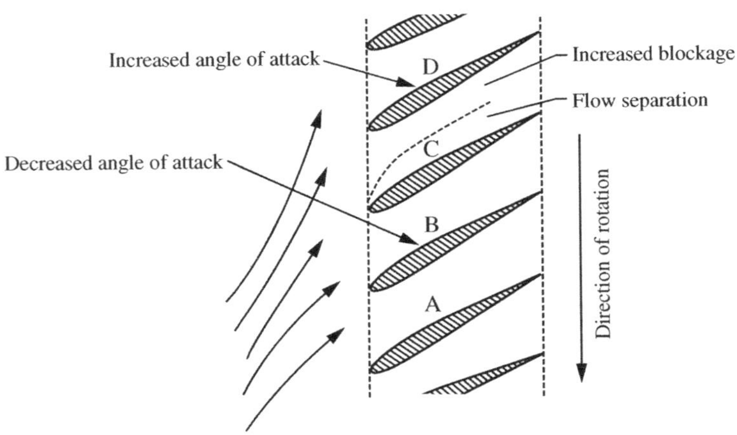


**Figure 2.1** Diagram showing circumferential development of rotating stall. Source: Emmons et al. [20]/The American Society of Mechanical Engineers.

alleviated to exit the stall state. This leads to the propagation and passage of stall nuclei, which will develop and propagate in the direction opposite to blade rotation.

From 1955 to the 1980s and 1990s, the research on the stall phenomenon was mainly focused on compressors and fans. With the rapid development of the water pump industry, users also had higher and higher requirements for the operational stability of pumps. Researchers began to pay attention to the stall phenomenon in water pumps. Hu et al. [21] studied the mixed-flow water-jet propulsion pump by numerical simulation and found that stall occurred at small flow rates (much smaller than the design flow rate), and the performance curve of the pump showed a positive slope. At the same time, the pumps with and without guide vanes were studied, and their performance curves were compared. It was found that the stall in the impeller channel led to a sudden drop in the flow head curve regardless of whether there were guide vanes or not; however, the causes for the sudden drop and the subsequent small rise in the head curve were not thoroughly analyzed. Li et al. [22] studied the causes of positive slope in the performance curve of a diagonal flow pump by numerical simulation and experimental measurement and found that the flow head curve obtained by both the numerical calculation and experiment showed a positive slope and provided a thorough analysis of the causes of the sudden drop and small rise in the flow head performance curve. Masahiro et al. [23] also found that when the flow in the guide vane of mixed-flow pump is reduced to 65% of its optimal efficiency point, the flow head curve of the pump has a positive slope characteristic of rotational stall. Li et al. [24] studied the dynamic characteristics of the rotating stall of a mixed-flow pump and found that with the decrease in flow rate, the relative velocity of the fluid in the impeller channel decreases, and the angle of attack of the blade increases. As the flow rate continues to decrease, the flow becomes unstable and flow separation occurs on the suction surface side of the blade, resulting in rotational stall. Zhang et al. [25] summarized the research status of rotating stalls in turbomachinery and arrived at a similar conclusion. They found that when the flow in turbomachinery is relatively small, the boundary layer separation occurs on the suction surface of the blade due to the increase in inlet angle of attack angle, resulting in the rotating stall. Byskov et al. [26] found a stall zone in the flow channel of the diagonal flow pump using LES as well as a reflux zone on the pressure surface of the impeller. Yamade et al. [27] also found using LES that when the rotational stall occurs in the diagonal flow pump, the internal flow state of the pump becomes unstable, and there is a local high-pressure area at the outlet tip on the back of the impeller blade and flow separation occurs. Li et al. [28] found that when the rotating stall occurs in the mixed-flow pump, unstable vortices appear at the impeller inlet and cause blockage in the flow channel. The vortices occupy almost the entire inlet flow channel resulting in the deterioration of the inlet flow. Pan et al. [29] studied the rotating stall characteristics of the guide vane mixed-flow pump and found the stall core propagating along the circumferential direction at the outlet of the impeller and in the inlet of the guide vane.

The change in flow pattern and the static and dynamic interference effects, and other factors result in pressure pulsations in the turbomachinery. When the mixed-flow pump rotates and stalls at smaller flow rates relative to the design flow rate, it generates unstable pressure pulsations which affect the stability of the pump. Du et al. [30] conducted LES on the full three-dimensional flow channel in the guide vane mixed-flow pump and analyzed the time-frequency characteristics of the pressure pulsation at eight monitoring points at the inlet and outlet sections of the impeller. It was found that due to the interference effect of rotating stall and separated flow with stationary guide vane, the amplitude of pressure pulsations at the monitoring points at the outlet is larger than that at the inlet monitoring points, and the attenuation speed of pulsations is also less. Zhang et al. [31] analyzed the internal flow field and pressure fluctuation characteristics of the mixed-flow pump under different operating conditions and found that when the rotating stall occurs, a large number of vortices appear near the suction surface of the guide vane and the number of vortices first increases and then decreases along the guide vane channel, and the pressure fluctuation in the vortex core is the smallest. The pressure fluctuation amplitude in the impeller from the hub to the tip gradually increases, and the pressure fluctuation frequency distribution at the inlet and outlet of the impeller is almost similar. Wu et al. [32] compared the pressure pulsation at each monitoring point at various operating conditions of the pump and found that when the rotating stall occurs, the amplitude of pressure pulsation at the monitoring point is significantly higher compared to that when the stall does not occur, and the amplitude of axial frequency of the monitoring point in the guide vane is much higher than that at the design operating condition. Anciger et al. [33] predicted the rotating stall of the pump-turbine and found that before the rotating stall, high-frequency pressure pulsation appeared on the guide vane surface, which caused a change in the torque.

### 2.2.3 Cavitation Flow

Cavitation is a unique phenomenon that occurs in liquids; it occurs when the liquid in a pump turns to vapor at low pressure. It occurs because there is not enough pressure at the suction end of the pump or there is insufficient *NPSH* due to a variety of hydrodynamic factors [34]. When cavitation takes place, air bubbles are created at low pressure. As the liquid passes from the suction side of the impeller to the delivery side, the bubbles implode. This creates a shock wave that hits the impeller and creates vibrations and mechanical damage to the pump, which may lead to complete failure of the pump at some stage.

In 1942, Thomas proposed the use of a dimensionless number $\sigma$ to describe the cavitation phenomenon; it is called the cavitation number which is an important parameter in hydrodynamics and hydraulics. The cavitation number $\sigma$ is defined as:

$$\sigma = \frac{p_\infty - p_\mathrm{v}}{\frac{1}{2}\rho_1 u_\infty^2} \tag{2.10}$$

where $p_\infty$ and $u_\infty$ denote the static pressure and velocity, respectively, in the reference section, and $\rho_1$ and $p_\mathrm{v}$ denote the density and saturated vapor pressure of the liquid, respectively.

The types of cavitation in hydraulic machinery are usually classified according to their physical characteristics; these are called the bubble cavitation, the sheet cavitation, the cloud cavitation, the supercavitation, and the vortex cavitation. Cavitation is a common phenomenon in hydraulic machinery including the mixed-flow pump [35–38], where in the process of accelerated startup of the pump it is the main cause for flow instability in the pump, resulting in its components' damage as well as reduction in efficiency and service life [39, 40]. At present, the research on cavitation of mixed-flow pumps employs both experiments and numerical calculations.

With improvements in the precision of experimental equipment and experimental methods, more research is now being conducted on cavitation in hydraulic machinery. Experiments conducted by Lu et al. [41] demonstrated that the occurrence of cavitation was strongly correlated with the vibration of the pump. It was found that when cavitation occurred in the pump, high amplitude vibrations began to appear in the frequency band above 1000 Hz, concentrated in the range 1200–1400 Hz. Zhang et al. [42] experimentally studied the spatiotemporal evolution of cavitation in a mixed-flow pump for different blade numbers during pump startup. They found that the cavitation flow structures in the blade tip region emerged at different times and locations at different scales with increase in blade numbers, as shown in Fig. 2.2. Yang et al. [43, 44] measured the vibrations generated by the mixed-flow pump for various operating conditions using an accelerometer and predicted the occurrence of cavitation by employing a specific vibration frequency. Horie et al. [45] compared the effect of different impeller blade numbers and inlet edge

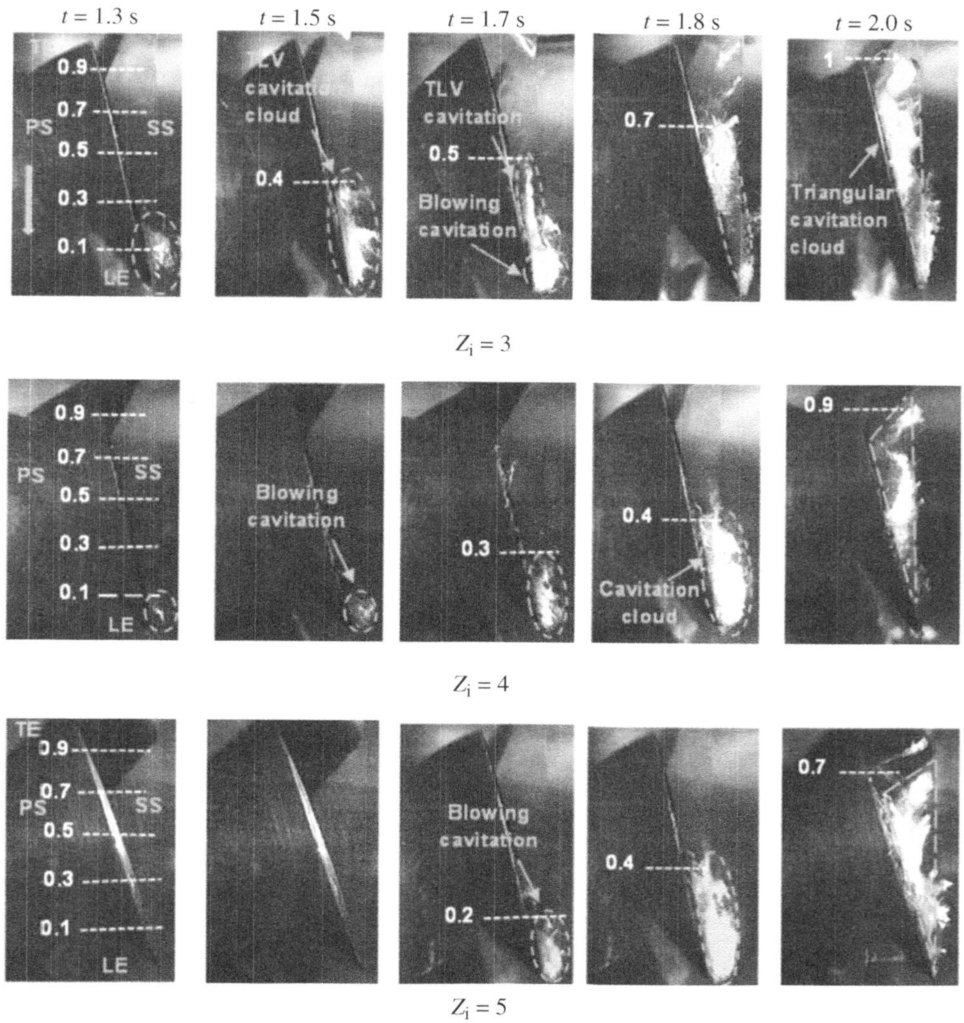


**Figure 2.2** Comparison of the snapshot images of cavitation in the tip region of the impeller at startup with different blade numbers. Source: Zhang et al. [42]/IOP Publishing/CC BY 3.0.

shapes on pump performance and cavitation characteristics. The results showed that the inlet edge shape has little effect on the overall performance of the pump but has significant effect on the cavitation characteristics. Chen et al. [46] tested and studied the hydrodynamic characteristics and cavitation characteristics of the high-speed mixed-flow pump during transient operation and found that the cavitation during startup has a significant impact on the pump head and power. Xu et al. [47] measured and analyzed the influence of *NPSH* and runaway speed at different angles on the performance of the mixed-flow pump. Sano et al. [48] studied the influence of leading-edge shape on cavitation employing both the numerical simulation and experiment. The results showed that the elliptical leading-edge shape can effectively reduce the hole length. Tosin et al. [49] studied the dependence of a reverse rotating mixed-flow pump on change in speed ratio and its impact on cavitation. The test results confirmed that different speed ratios can optimize the cavitation-free range of the pump. In addition, they also independently developed a variable-specific head drop efficiency curve, which highlighted the advantages of the new design compared to the conventional pump, especially under off-design operating conditions.

With the development of computer hardware and computational fluid dynamics (CFD) technology, the prediction of cavitation in hydraulic machinery by numerical simulation has become an important area of research in recent years [50]. Xu et al. [51] studied the transient characteristics under different conditions of cavitation and the influence of tip clearance on cavitation and found that a larger tip clearance significantly reduces the performance of the pump. Sedlar et al. [52] employed CFD to analyze the cavitation flow in the mixed-flow pump. The numerical results accurately captured the unsteady cavitation flow in the pump within the range of large flow coefficient and *NPSH*. In addition, the frequency obtained from the calculations was quite consistent with the measured noise, pressure pulsation, and vibration. Kobayashi et al. [53] used LES to calculate the torque acting on the blade of the mixed-flow pump and studied the relationship between the fluctuation in torque and cavitation. The results showed that the cavitation occurred near the leading edge of the suction side of the blade surface, and the peak value of the torque pulse acting on the blade was caused by the cavitation. Kim et al. [54] studied the influence of blade leading-edge shape on cavitation, and the results showed that with increase in ellipticity of the leading edge, the cavitation suction performance was better, and the blockage was less, as shown in Fig. 2.3.

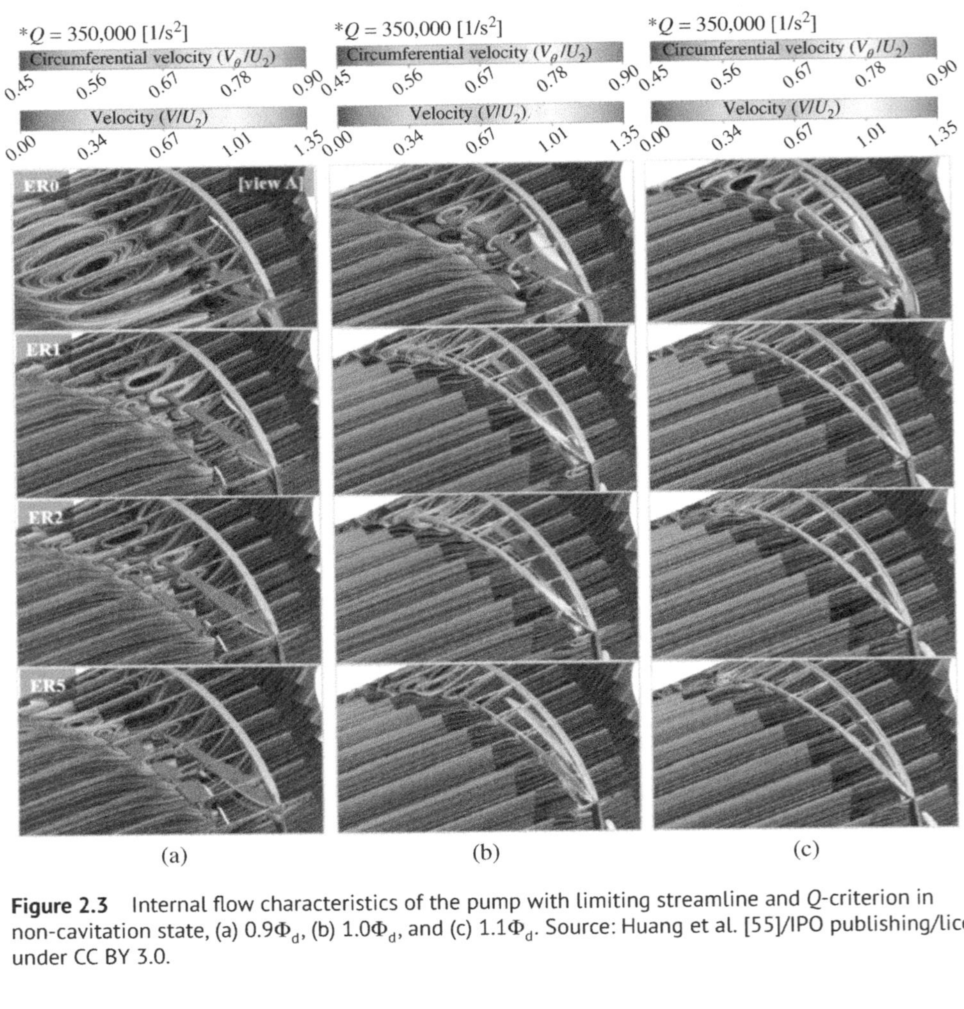


(a) (b) (c)

**Figure 2.3** Internal flow characteristics of the pump with limiting streamline and *Q*-criterion in non-cavitation state, (a) $0.9\Phi_d$, (b) $1.0\Phi_d$, and (c) $1.1\Phi_d$. Source: Huang et al. [55]/IPO publishing/licensed under CC BY 3.0.

Cavitation is an obstruction causing the flow separation and is the main reason leading to the drop in pump head.

Huang et al. [55] used the SST $k$–$\omega$ turbulence model with RANS equations for numerical simulation of cavitation flow in a mixed-flow pump with a single-blade channel. The predicted hydraulic characteristics were consistent with the experimental results. The numerical results showed that the blade thickness and inlet angle are closely related to the hydraulic performance of the pump and the severity of cavitation. Yang et al. [56] analyzed the distribution of cavitation position and gas phase volume fraction on the impeller blade and predicted the performance curve between $NPSH_a$ and efficiency of the pump. The results showed that the cavitation phenomenon directly affected the pressure distribution on the impeller blade surface and led to changes in the external characteristics of the pump. Zhao et al. [57] employed the $k$–$\varepsilon$ turbulence model for simulation of the cavitation. They found that the cavitation occurred on the blade suction surface near the leading edge of the blade tip. With a decrease in $NPSH$, the cavitation region expands. Gupta et al. [58] employed the CFX software to predict the performance of mixed-flow pump in the presence of cavitation; their numerical results accurately captured the process of cavitation formation and showed that its volume fraction increases with a decrease in $NPSH_r$.

## 2.3 Summary

In this chapter, the basic parameters of the mixed-flow pump are described including flow rate $Q$, head $h$, speed $n$, net positive suction head $NPSH$, power $P_e$, and efficiency $\eta$. In addition, some typical flow structures are introduced in detail, including the TLF, the rotating stall, and the cavitation. Furthermore, the literature review of these typical flow characteristics is presented to set the stage for the material presented in the book in the subsequent chapters.

## Nomenclature

| | |
|---|---|
| CFD | computational fluid dynamic |
| LES | large Eddy simulation |
| RANS | Reynolds-averaged Navier-stokes |
| SST | shear stress transport |
| TLF | tip leakage flow |
| TLV | tip leakage vortex |

## References


1 Rains, D.A. (1954). *Tip Clearance Flows in Axial Compressors and Pumps*. California: Division of Engineering and Applied Science.

2 Chen, G.T., Greitzer, E.M., Tan, C.S., and Marble, F.E. (1991). Similarity analysis of compressor tip clearance flow structure. *Journal of Turbomachinery* 113 (2): 260–271.

3 You, D., Wang, M., Moin, P., and Mittal, R. (2006). Effects of tip-gap size on the tip-leakage flow in a turbomachinery cascade. *Physics of Fluids* 18 (10): 105102.

4 Inoue, M. and Furukawa, M. (2002). Physics of tip clearance flow in turbomachinery. *ASME 2002 Joint U.S.-European Fluids Engineering Division Conference* (14–18 July 2002). Montreal, Quebec, Canada. 777–789.

5 Yi, B.L., Peng, L.H., Ren, N.L., and Zhen, B. (2014). Numerical analysis of the effect of different tip clearance on the performance of diagonal flow pump. *Transactions of the Chinese Society of Agricultural Engineering* 30 (23): 86–93.

6 Liu, Y.B. and Tan, L. (2018). Study on morphological characteristics and influencing factors of leakage vortex in blade tip clearance of mixed-flow pump. *Symposium on Development Strategy of Hydraulic Machinery and Proceedings of the 11th National Annual Conference on Hydraulic Machinery and its Systems.*

7 Wu, H.X., Miorini, R.L., and Katz, J. (2011). Measurements of the tip leakage vortex structures and turbulence in the meridional plane of an axial water jet pump. *Experiments in Fluids* 50 (4): 989–1003.

8 Wu, H.X., Tan, D., Miorini, R.L., and Katz, J. (2011). Three-dimensional flow structures and associated turbulence in the tip region of a water jet pump rotor blade. *Experiments in Fluids* 51 (6): 1721–1737.

9 Miorini, R.L., Wu, H.X., and Katz, J. (2012). The internal structure of the tip leakage vortex within the rotor of an axial water jet pump. *Journal of Turbomachinery* 134 (3): 031018.

10 Miyabe, M., Furukawa, A., Maeda, H., and Umeki, I. (2009). Investigation of internal flow and characteristic instability of a mixed-flow pump. *Proceedings of the ASME 2009 Fluids Engineering Division Summer Meeting*, (2–6 August 2009). Vail, Colorado USA: FEDSM 2009-78277. 315–321.

11 Hah, C. (2012). Investigation of turbulent tip leakage vortex in an axial water jet pump with large eddy simulation. *Proceedings of the ASME Fluids Engineering Division Summer Meeting* (8–12 July 2012). Rio Grande, PR, FEDSM 2012. 135–142.

12 Li, Y.J., Shen, J.F., Yan, H.J., and Zhang, Z.M. (2014). Large eddy simulation of the effect of tip clearance on flange leakage flow of axial flow pump. *Journal of Hydraulic Engineering* 45 (2): 235–242.

13 Li, W., Qiao, W.Y., Xu, K.F. et al. (2008). Numerical simulation of active control on tip leakage flow in axial turbine. *Journal of Aerospace Power* 7: 1260–1265.

14 Zhang, W.W., Yu, Z.Y., Zhu, B.S., and Yang, C. (2017). Numerical study on the effect of tip clearance on the performance and internal flow field of low specific speed mixed-flow pump. *Journal of Mechanical Engineering* 53 (22): 182–189.

15 Goto, A. (1992). Study of internal flows in a mixed-flow pump impeller at various tip clearances using three-dimensional viscous flow computations. *Journal of Turbomachinery* 114 (2): 373–382.

16 Zhong, J.J., Wang, H.R., and Han, J. (2015). Influence of blade tip clearance and swept modification on mixed-flow jet pump. *Journal of Dalian Maritime University* 41 (3): 77–81. https://doi.org/10.13243/j.cnki.slxb.2014.02.014.

17 Shi, W.D., Zhang, H., Chen, B. et al. (2010). Numerical calculation of internal flow field of axial flow pump with different tip clearance. *Journal of Drainage and Irrigation Machinery Engineering* 28 (5): 374–377.

18 Shi, W.D., Li, T.T., Zhang, D.S., and Li, H. (2013). Effects of different tip clearances on cavitation performance and flow field of axial flow pump. *Journal of Huazhong University of Science and Technology (Natural Science Edition)* 41 (4): 21–25.

19 Bing, H., Cao, S.L., He, C.L., and Li, L. (2013). Experimental study of the effect of blade tip clearance and blade angle error on the performance of mixed-flow pump. *Science China: Technological Sciences* 56 (2): 293–298.

20 Emmons, H.W., Pearson, C.E., and Grant, H.P. (1955). Compressor surge and stall propagation. *Transactions of the ASME* 77 (3): 455–469. https://doi.org/10.1115/1.4014389.

21 Hu, F.F., Wu, P., Wu, D.Z., and Wang, L.Q. (2014). Numerical study on the stall behavior of a water jet mixed-flow pump. *Journal of Marine Science and Technology* 19 (4): 438–449.

22 Li, J.J. (2012). Research on positive slope performance curve and its suppression of mixed flow pump with vanned diffuser. PhD thesis. Zhenjiang, Jiangsu University. Retrieved on 11 June 2012, Research Direct Database.

23 Masahiro, M., Hideaki, M., Isamu, U., and Yoshinori, J. (2006). Unstable head-flow characteristic generation mechanism of a low specific speed mixed-flow pump. *Journal of Thermal Science* 15 (2): 115–120.

24 Li, X.J., Yuan, S.Q., Pan, Z.Y. et al. (2013). Dynamic characteristics of rotating stall in mixed-flow pump. *Journal of Applied Mathematics* 7 (2): 234.

25 Zhang, L., Wang, S.L., and Hu, C.X. (2014). Research progress on rotating stall of turbomachinery. *Thermal Power Generation* 43 (1): 1–5.

26 Byskov, P.K., Jacobsen, C.B., and Pedersen, N. (2003). Flow in a centrifugal pump impeller at design and off-design conditions—Part II: large eddy simulations. *ASME Journal of Fluids Engineering* 125: 73–83. https://doi.org/10.1115/1.1524586.

27 Yamade, Y., Kato, C., Shimizu, H., and Nagahara, T. (2009). Large eddy simulation of internal flow of a mixed-flow pump. *ASME 2009 Fluids Engineering Division Summer Meeting. American Society of Mechanical Engineers*. 1: 407–416.

28 Li, Y.B., Li, R.N., Wang, X.Y. et al. (2013). Numerical simulation of flow instability in mixed-flow pump. *Journal of Drainage and Irrigation Machinery Engineering* 31 (5): 384–389.

29 Pan, Z.Y., Li, J.J., Li, X.J., and Yuan, S.Q. (2012). Study on instability characteristics and rotating stall of diagonal flow pump. *Transactions of the Chinese Society for Agricultural Machinery* 43 (5): 64–68.

30 Du, Y.Y., Gao, Q., Shang, C.C. et al. (2016). Analysis of pressure pulsation in impeller of guide vane mixed-flow pump based on large eddy simulation. *Journal of Xihua University (Natural Science Edition)* 35 (4): 75–78.

31 Zhang, L., Shi, W.D., Zhang, D.S., and Shi, L. (2017). Numerical study of unsteady flow in mixed-flow pump based on large eddy simulation. *Journal of Drainage and Irrigation Machinery Engineering* 35 (4): 303–308.

32 Wu, T.T. (2013). Study on rotating stall and fluid structure coupling of oblique flow water jet propulsion pump. PhD thesis. Zhenjiang, Jiangsu University, Retrieved on 6 July 2013, Research Direct database.

33 Anciger, W.H.M. and Hao, X.R. (2012). Prediction of rotating stall and incipient cavitation of pump turbine. *Large Electric Machine and Hydraulic Turbine* 1 (9): 63–66.

34 Pan, S.S. and Peng, X.X. (2013). *Cavitation Mechanism*. Frankfurter: National Defense Industry Press.

35 Chang, S.P. and Wang, Y.S. (2012). Study on cavitation characteristics of mixed-flow pump based on CFD. *Journal of Drainage and Irrigation Mechanical Engineering* 30 (2): 171–175.

36 Li, J.Y., Lai, X.D., Zhu, L. et al. (2016). Analysis of unsteady cavitation characteristics of mixed-flow pump impeller under multiple working conditions. *Journal of Hydropower* 35 (7): 106–111.

37 Pan, Z.Y., Ni, Y.Y., Yuan, S.Q., and Chao, Y.J. (2000). Research progress of diagonal flow pump. *Fluid Machinery* 37 (9): 37–41.

38 Li, W., Li, E.D., Ji, L.L. et al. (2020). Mechanism and propagation characteristics of rotating stall in a mixed-flow pump. *Renewable Energy* 153 (3): 74–92.

39 Li, W., Zhou, L., Shi, W.D. et al. (2017). PIV experiment of the unsteady flow field in mixed-flow pump under part-loading condition. *Experimental Thermal and Fluid Science* 83 (9): 191–199.

40 Ji, L.L. (2017). Study on unsteady flow mechanism in mixed-flow pump under Alford effect. PhD thesis. Zhenjiang, Jiangsu University, Retrieved on 6 March 2017, Research Direct database.

41 Lu, Y.P., Tan, L., Han, Y.D., and Liu, M. (2022). Cavitation-vibration correlation of a mixed-flow pump under steady state and fast start-up conditions by experiment. *Ocean Engineering* 251 (4): 111158.

42 Zhang, D.S., Zhou, Q., Gu, Q. et al. (2022). Experimental and numerical investigation on the transient cavitating flows in a mixed-flow pump with different number of blades at startup. *Journal of Fluids Engineering* 144 (5): 394.

43 Yang, H.M., Lee, K.Y., Kim, J.H., and Choi, Y.S. (2018). Analysis of performance variation with cavitation test for a mixed-flow pump. *IOP Conference Series: Earth and Environmental Science*. IOP Publishing. 163 (1).

44 Yang, H.M., Lee, K.Y., Kim, J.H., and Choi, Y.S. (2019). Vibration and flow characteristics near cavitation occurrence in mixed-flow pump. *IOP Conference Series: Earth and Environmental Science*. IOP Publishing. 240 (6).

45 Horie, C. and Oshims, M. (1964). Experimental study on cavitation in mixed-flow pump impellers. *Bulletin of JSME* 7 (25): 62–71.

46 Chen, T., Li, S.R., Li, W.Z. et al. (2013). Numerical simulation of the cavitation characteristics of a mixed-flow pump. *IOP Conference Series: Materials Science and Engineering*. IOP Publishing. 52 (6).

47 Xu, B.F. and Sun, S.S. (2013). Research on cavitation and runaway tests of mixed-flow pump with different blade angles. *Advanced Materials Research* Trans Tech Publications Ltd 805 (1): 1708–1711.

48 Sano, T., Iino, M., and Maeda, S. (2021). Effect of leading-edge profile on cavitation performance of mixed-flow impeller. *Journal of Physics: Conference Series* IOP Publishing 1909 (1): 012018.

49 Tosin, S., Friedrichs, J., Dreiss, A., et al. (2015). Experimental investigation of a counter-rotating mixed-flow single stage pump under cavitation conditions. *Fluids Engineering Division Summer Meeting*. American Society of Mechanical Engineers.

50 Gan, J.Y., Xue, Y.F., and Wu, K.Q. (2007). Numerical calculation of cavitation flow in mixed-flow pump impeller. *Journal of Engineering Thermoplastics* 28 (S1): 165.

51 Xu, Y., Tan, L., Liu, Y.B., and Cao, S.L. (2017). Pressure fluctuation and flow pattern of a mixed-flow pump with different blade tip clearances under cavitation condition. *Advances in Mechanical Engineering* 9 (4): 178.

52 Sedlar, M., Sputa, O., and Komarek, M. (2010). CFD analysis of cavitation phenomena in mixed-flow pump. *International Journal of Fluid Machinery and Systems* 5 (1): 18–29.

53 Kobayashi, K. and Chiba, Y. (2010). Computational fluid dynamics of cavitating flow in mixed-flow pump with closed type impeller. *International Journal of Fluid Machinery and Systems* 3 (2): 113–121.

54 Kim, Y.I., Yang, H.M., Lee, K.Y., and Choi, Y.S. (2022). Numerical investigation on blockage-related cavitation surge and pressure gain of a mixed-flow pump with influence of blade leading edge shape on suction performance. *Journal of Fluids Engineering* 144 (9): 291.

55 Huang, R.F., Zhai, Z.H., Zhou, J.J., and Lou, X.W. (2018). Cavitating flow analysis and design optimization for a waterjet mixed flow pump. *IOP Conference Series: Earth and Environmental Science*. IOP Publishing. 163 (1): 251.

56 Yang, F., Liu, C., and Tang, F.P. (2012). Cavitation performance prediction of mixed-flow pump based on CFD. *IOP Conference Series: Earth and Environmental Science*, IOP Publishing. 15 (3): 298.

57 Zhao, Y.X., Kim, C.G., and Lee, Y.H. (2013). CFD study on flow characteristics of pump sump and performance analysis of the mixed flow pump. *IOP Conference Series: Materials Science and Engineering*. IOP Publishing. 52 (7): 378.

58 Gupta, A. and Kshirsagar, J.T. (2005). Numerical and experimental investigation of cavitation in a pump. *ASME Fluids Engineering Division Summer Meeting*. 41987: 1535–1541.

# 3

# Brief Review of Computational Fluid Dynamics

## 3.1 CFD as a Flow Simulation Tool

Theoretical hydrodynamics, experimental hydrodynamics, and computational fluid dynamics (CFD) are the three approaches for the study of problems in hydrodynamics. CFD as a flow simulation tool requires physical models described by the governing partial differential equations (PDEs) of fluid dynamics, appropriate, accurate, and stable numerical algorithms for discretization of the governing equations on a mesh generated inside a chosen computational domain, and computer hardware for the solution of the discretized algebraic equations in the computational domain with appropriate boundary conditions. The discretized algebraic equations are coded to generate a computer code in a computer language (such as Fortran, Fortran 90, C, C++, Java, and Python). The computer code is debugged and validated by conducting benchmark simulations; this computer code is run on computer hardware to obtain the numerical solution of the flow field. In this context, CFD can be considered a multidisciplinary field involving fluid dynamics, numerical mathematics, and computer science, as shown in Fig. 3.1, with the ability to solve complex three-dimensional fluid flow problems [1]. In other words, it employs the computer as a tool to solve the discretized linear algebraic equations in an iterative manner (note that the original equations are non-linear, which are linearized and solved iteratively). Generally speaking, theoretical fluid dynamics provides the physical and mathematical models to describe the conservation laws governing fluid motion, experimental fluid mechanics provides measurements and visualization to discover and explore the physical consequences of fluid motion-related phenomena in an application, and CFD builds a bridge between the theoretical models and experiments since it is difficult to obtain the analytical solutions of the complex 3D fluid motions using the theoretical models and it is costly to build the experimental apparatus and conduct experiments for a wide variety of flows of interest. As a result, CFD has become one of the most important technologies for conducting hydrodynamics/fluid dynamics research. In some sense, CFD can be considered similar to the experimental research, since CFD includes the geometry and physical modeling, solution procedure/process, and debugging in writing a computer code to conduct the numerical simulation with the code; thus, numerical simulation is also sometimes called "numerical experiment." However, in contrast to experiments, CFD can often provide very detailed three-dimensional flow field data for analysis, design, and optimization of industrial products.

CFD solves the governing equations of fluid flow in an integral or differential form with appropriate boundary conditions in a suitably chosen computational domain to compute and

*Mixed-flow Pumps: Modeling, Simulation, and Measurements*, First Edition.
Wei Li, Leilei Ji, Ramesh Agarwal, Weidong Shi, and Ling Zhou.

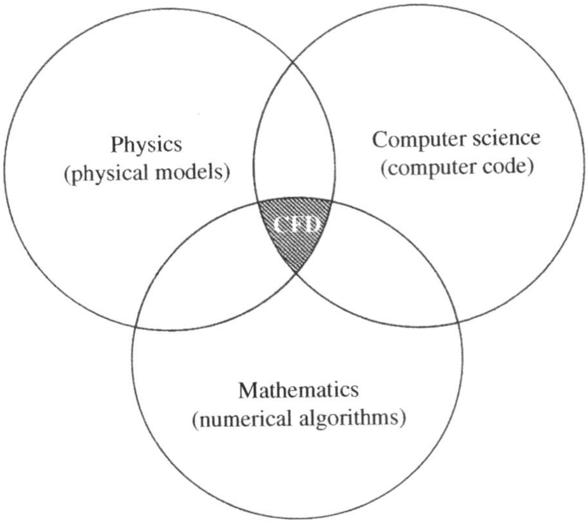


**Figure 3.1** CFD as a multidisciplinary field.

visualize the desired flow field data for the purpose of analysis, optimization, and improvements in the design of a product. The fundamental governing equations and boundary conditions of the flow in CFD are, of course, the same as in the theoretical hydrodynamics [2]. Employing the CFD, therefore, it is possible to bridge the gap between the experimental hydrodynamics and the theoretical hydrodynamics using the vast amount of "numerical data" and thus gain additional insight into the hitherto hidden flow features/phenomena. In particular, once the CFD can accurately reproduce some experimental phenomena (that is, the CFD results have been validated/verified), it can be employed to further study and analyze the theoretical models and physical laws describing these flow phenomena and explore them further by analyzing the conditions of their emergence, development, and evolution, and thereby deepen the understanding of the mechanisms behind the phenomenon from the theoretical models [3]. In the past three decades, many commercial and noncommercial codes have been developed for industrial applications. Among the commercial codes, the most well-known are ANSYS Fluent, CFX, COMSOL, etc. Among the noncommercial open-source codes are OpenFOAM, SU2, and others. There are also many industries and government-funded proprietary codes in many countries around the world. Figure 3.2 shows the entire computational solution process and its building blocks, which are briefly discussed below.

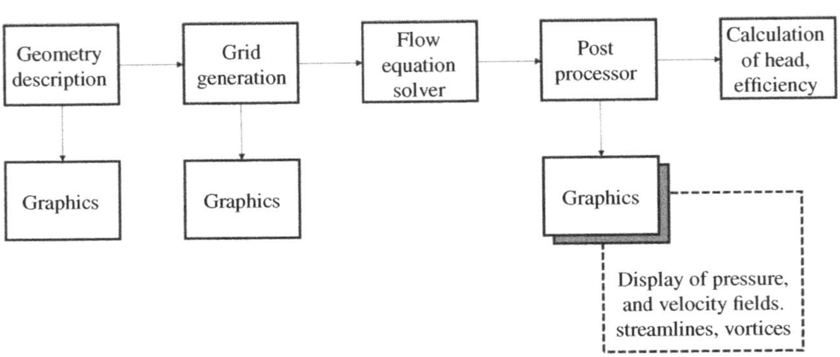


**Figure 3.2** Computational solution process.

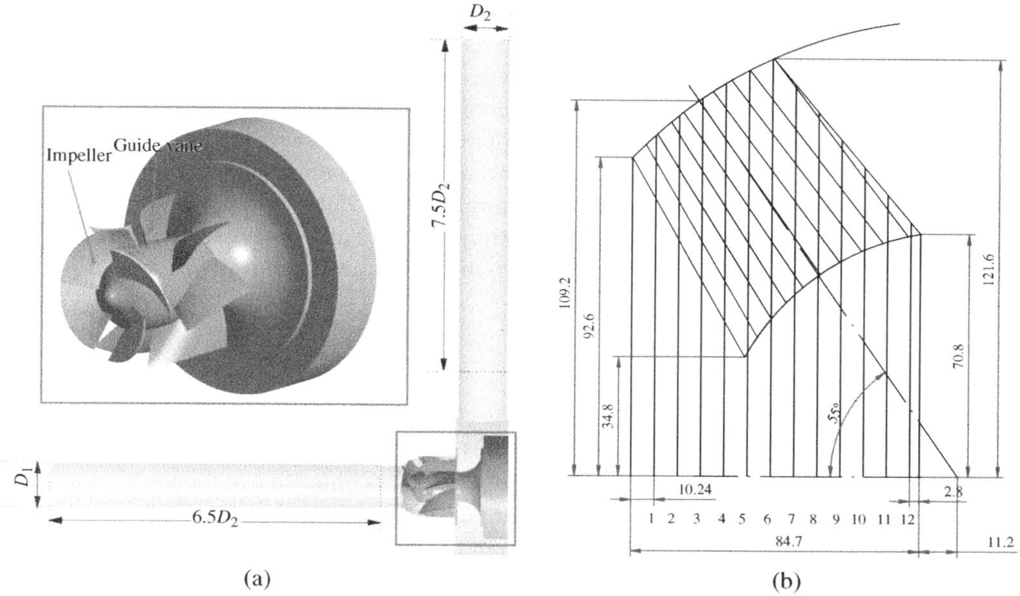


**Figure 3.3** Mixed-flow pump model: (a) 3D diagram of mixed-flow pump and (b) design of impeller.

**Table 3.1** Parameters of the mixed-flow pump model.

| Parameter | Value |
|---|---|
| *Design operating condition* | |
| Design flow rate $Q_{des}$ ($m^3$/h) | 380 |
| Rated design head $H$ (m) | 6 |
| Rated speed $n$ (r/min) | 1450 |
| Specific speed $n_s$ | 480 |
| *Impeller parameters* | |
| Number of impeller blades $Z$ | 4 |
| Width of the impeller $b_2$ (mm) | 67.2 |
| Inlet diameter of the impeller $D_1$ (mm) | 185.2 |
| Outlet diameter of the impeller $D_2$ (mm) | 243.2 |
| Average inlet blade angle $\beta_1$ (°) | 24.9 |
| Average outlet blade angle $\beta_2$ (°) | 31.8 |
| *Diffuser parameters* | |
| Number of guide vane blades $Z_d$ | 7 |
| Inlet diameter $D_3$ (mm) | 266.6 |
| Outlet diameter $D_4$ (mm) | 281.8 |
| Average inlet blade angle $\alpha_3$ (°) | 45.6 |
| Average outlet blade angle $\alpha_4$ (°) | 86.5 |

## 3.2 Geometry Modeling

The specific research model of the pump considered in this book is a low-specific-speed mixed-flow pump with guide vanes. The three-dimensional geometry of the pump is modeled by the Pro/E software, and then the simulation domains for computations are extracted from it, which include the inlet domain, tapered joint pipe domain (which connects the impeller domain and the inlet domain), impeller domain, guide vane domain, annular volute domain, and the outlet domain. Finally, all the solid domains of the mixed-flow pump for simulation are assembled. The pump model from inlet domain to outlet domain is shown in Fig. 3.3, and several key parameters of the mixed-flow pump are listed in Table 3.1.

## 3.3 Mesh Generation

The entire computational domain is shown in Fig. 3.3, which is divided into six parts to perform mesh generation, and these six parts are finally assembled. In order to avoid reverse flow from the inlet and outlet boundaries, the inlet and outlet pipe domains are extended for the purpose of numerical simulations beyond the actual geometry of the pump, as shown in Fig. 3.3. Since mesh generation is important for accurate simulation, the refined hexahedral meshes are employed, especially in the tip clearance region of the impeller. The O-block topology for mesh is used in the inlet pipe section and discharge pipe section, and the Y-block mesh topology is used in the coupling section. The J/O topological structure and the H/O topological structure are employed in meshing the impeller and guide vanes, respectively. The flow passages are arrayed periodically. The mesh is refined in the boundary layer and the tip region of the impeller, as shown in Fig. 3.4. The automatic near-wall treatment is employed to suit the requirements of a turbulence model. There are over 15 nodes in the blade tip clearance region to accurately resolve the flow field, and the $y+$ in the impeller section on the blade surface is less than 100 using the wall functions with a turbulence model. Most commercial packages come with the mesh generation software, e.g., ANSYS has Gambit, ICEM, DesignModeler, and SpaceClaim for geometry modeling and mesh generation and can automatically generate a 3D single or multi-block structured, unstructured, or hybrid grid with a desired topology for very complex external and internal flow configurations.

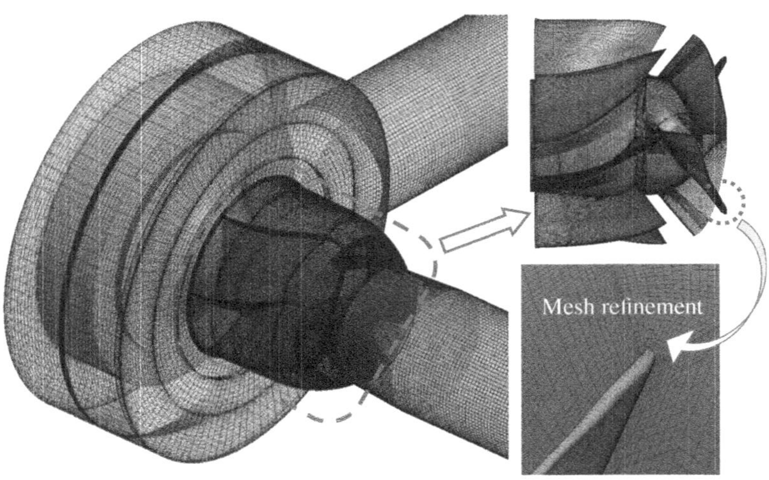


**Figure 3.4** Whole computational domain and domain discretization.

## 3.4 Governing Equations of Fluid Dynamics

The conservation equations of fluid dynamics, namely the continuity equation, momentum equations, and energy equation, are the basic equations governing fluid motion based on the law of conservation of mass, Newton's second law, and the law of conservation of energy. They are the fundamental equations governing the fluid motion, which are numerically solved by CFD [4].

In an inertial frame of reference in Cartesian coordinate system, the basic governing equations describing the internal instantaneous flow state in hydraulic machinery include the following two equations [5].

Continuity equation

$$\frac{\partial \rho}{\partial t} + \frac{\partial (\rho u_i)}{\partial x_i} = 0 \tag{3.1}$$

and the Navier–Stokes equations

$$\frac{\partial (\rho u_i u_j)}{\partial x_j} + \frac{\partial (\rho u_i)}{\partial t} = -\frac{\partial p}{\partial x_i} + \frac{\partial}{\partial x_j} \mu \left( \frac{\partial u_i}{\partial x_j} \right) + S_i \tag{3.2}$$

where $\rho$ denotes the fluid density, $u_i$ is the velocity component in the $i$th direction ($i = x$, $y$, or $z$), $p$ is the pressure, $t$ is the time, $x_i$ is the spatial coordinate ($x$, $y$, or $z$), $\mu$ is the dynamic viscosity, and $S_i$ is the source term in the $i$th direction ($i = x$, $y$, or $z$). In Eqs. (3.1) and (3.2), $i$ and $j$ represent the directional component ($i$ or $j = x$, $y$, or $z$) of a variable and follow the summation convention in tensor algebra.

## 3.5 Simulation of Turbulent Flows

The internal flow of hydraulic components such as an impeller and a guide vane in hydraulic machinery can be considered as the three-dimensional incompressible unsteady turbulent flow.

The numerical simulation of turbulent flow remains the pacing item and most difficult problem in CFD. The main turbulent flow simulation methods can be categorized as (a) the direct numerical simulation (DNS) and (b) the large eddy simulation (LES) of the Navier–Stokes equations, and (c) the simulation based on the Reynolds-averaged Navier–Stokes (RANS) equations in conjunction with a turbulence model needed to accomplish the closure of RANS equations by modeling the Reynolds or turbulent stresses. This classification of turbulent flow solution methods is shown schematically in Fig. 3.5 [1, 6–10].

DNS refers to the direct numerical solution of instantaneous or unsteady Navier–Stokes equations [11–14] without any approximation or modeling. All other (non-DNS) methods aim to analyze the turbulent flow by employing some degree of approximation and simplification to instantaneous or unsteady Navier–Stokes equations which in principle include all length and time scales of turbulence. Based on different approximation and simplification methods, they can be divided into LES, Reynolds time-averaging, and the statistical averaging methods. In order to use DNS to calculate some highly complex turbulent fluid motions at high Reynolds numbers (HRNs), very small time and spatial discretization steps must be employed in solving the Navier–Stokes equations in order to obtain the detailed local spatial structure and temporal characteristics of turbulent fluid motion. For the complete three-dimensional complex geometric model like those in hydraulic machinery, the existing supercomputing power is not sufficient to realize such calculations using DNS in the foreseeable future since the DNS of turbulence requires

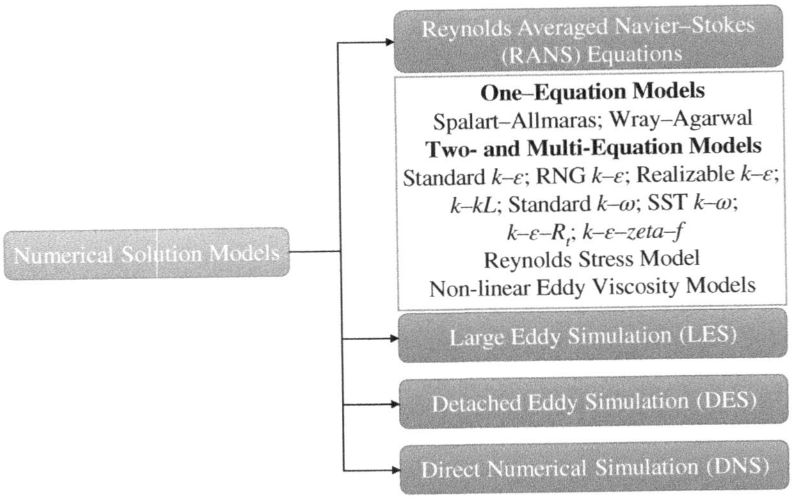


**Figure 3.5** Numerical solution models for turbulent flow simulations.

very large computer memory and operational speed. At present, DNS cannot be employed for practical engineering calculations; however, a great deal of exploratory work is underway as more and more computing power of exascale computers is becoming available. The LES method on the other hand employs the unsteady Navier–Stokes equation to directly simulate the large-scale eddies of turbulent flow, but the small-scale eddies are not directly simulated. Thus, in LES, conceptually, the turbulent fluid motion is considered to be composed of large- and small-scale eddies. The influence of small eddies on large eddies is considered by an approximate model. The standard method is to filter the small-scale eddies with smaller scales of turbulence compared to the specified filter function from the instantaneous Navier–Stokes equations so that the filtered equation then describes the flow field with large eddies. In this process, an additional stress term is added to the equations describing the motion of large eddies which reflects the influence of filtered small eddies on large eddy motion. The stress term is similar to the Reynolds stress term in the RANS equations; this stress term is called the subgrid scale stress in LES. Most subgrid Reynolds stress models in the literature since the original model of Smagorinsky are based on an eddy viscosity concept, which relates the subgrid scale stress to strain tensor using the Boussinesq type approximation. Although the requirements of LES for computer memory and operational speed are still relatively high for many routine 3D industrial applications, they are, however, much lower than those for DNS. As a result, LES is being increasingly used in 3D simulations at HRNs using the currently available high-end computing hardware.

Instead of solving the Navier–Stokes equations by DNS or LES, the Reynolds time-averaging method is employed to obtain the RANS equations for incompressible flows or the mass-weighted RNAS equations for compressible flows (also called the Favre equations) are solved in conjunction with a turbulence model in most commercial CFD codes for industrial applications; these methods give results of acceptable engineering accuracy on currently available computing hardware. Thus, the RANS method has become the most widely used approach for turbulent flow simulations in hydraulic machinery.

Finally, the statistical averaging method is based on the statistical theory of turbulent fluid motion extended from the turbulence correlation function. It mainly uses the correlation function and spectral analysis method to analyze the turbulence structure, which primarily involves the flow of small-scale eddies. It is hardly used in engineering practice.

Now, the most widely used RANS method, which requires modeling of Reynolds stresses in order to achieve closure of the governing equations, is described in detail. The problem of closure in RANS equations is referred to in the literature as "Turbulence Modeling." Over a century, a wide variety of turbulence models of different levels of complexity have been developed for use with RANS equations; they vary in accuracy and efficiency depending upon the application, but none of them can be considered to approach universality. Most of the turbulence models are based on the concept of "eddy viscosity" in Boussinesq approximation, which postulates that the turbulent stress tensor is proportional to the strain-rate tensor with a proportionality constant defined as eddy viscosity analogous to Stokes' hypothesis used in the derivation of Navier–Stokes equations from Newton's second law. Depending upon the complexity, most linear eddy-viscosity models are classified as the zero-equation, one-equation, and two-equation models depending upon the number of transport equations used for defining the turbulence variables such as turbulent kinetic energy $k$, turbulent dissipation rate $\varepsilon$, turbulent specific dissipation rate $\omega$ or turbulent length scale $L$ or a combination of these. At present, two-equation models are widely used in engineering applications. Among them, the oldest two-equation model is the standard $k$–$\varepsilon$ model, which includes the transport equations of turbulent kinetic energy $k$ and dissipation rate $\varepsilon$. In all the linear eddy-viscosity models, turbulence is assumed isotropic. On the other hand, the full Reynolds stress model (RSM) or the second-moment closure model, which has six transport equations for six Reynolds stresses, can calculate the anisotropic turbulence; however, it is relatively complex and requires significantly more computational effort. Therefore, at present, one- or two-equation turbulence models are widely used in engineering practice. For additional information on turbulence modeling, the reader is referred to the classic book by Wilcox [15].

In what follows, the Reynolds-averaging of the Navier–Stokes equations is briefly described, and the derivation of the RANS equations is presented. In Reynolds-averaging, the instantaneous flow variables in the Navier–Stokes equations are decomposed into two parts: the average value over a certain time period and the time-dependent fluctuation. For example, the velocity components can be decomposed as:

$$u_i = \overline{u}_i + u_i' \tag{3.3}$$

where $\overline{u}_i$ and $u_i'$ represent the average and fluctuating velocity components, respectively ($i = 1, 2, 3$).

Similarly, for scalar variables $\phi$ such as pressure and energy, one can write:

$$\phi = \overline{\phi} + \phi' \tag{3.4}$$

By introducing the decomposition of variables given in Eqs. (3.3) and (3.4) into the continuity and Navier–Stokes equations and performing the Reynolds-averaging on the equations over a time period $T$, the continuity equation and the momentum equations can be written in Cartesian coordinate system (by dropping the overbar on the mean velocity $\overline{u}$) in tensor notation as:

$$\frac{\partial \rho}{\partial t} + \frac{\partial}{\partial x_i}(\rho u_i) = 0 \tag{3.5}$$

$$\frac{\partial (\rho u_i)}{\partial t} + \frac{\partial}{\partial x_j}(\rho u_i u_j) = -\frac{\partial p}{\partial x_i} + \frac{\partial}{\partial x_j}\left[\mu\left(\frac{\partial u_i}{\partial x_j} + \frac{\partial u_j}{\partial x_i} - \frac{2}{3}\delta_{ij}\frac{\partial u_k}{\partial x_k}\right)\right] + \frac{\partial}{\partial x_j}\left(-\rho\overline{u_i' u_j'}\right) \tag{3.6}$$

Equations (3.5) and (3.6) are called the RANS equations. They have basically the same form as the unsteady Navier–Stokes equations except that the flow variables including the velocity components are replaced by their time averages, and there is an additional term $-\rho\overline{u_i' u_j'}$ in Eq. (3.5) which represents the Reynolds or turbulent stress due to the influence of turbulent fluctuations.

In Eq. (3.5), the Reynolds stress term is unknown and is not defined in terms of the other flow variables such as velocity and pressure. This presents the so-called "closure problem" for the solution of Eqs. (3.5) and (3.6). In order to close the equations, the Reynolds stress term $-\rho\overline{u_i'u_j'}$ needs to be modeled in terms of the flow variables. The modeling of Reynolds stress term in Eq. (3.6) is referred to in the literature as "Turbulence Modeling." The most used method to model the Reynolds stress is to employ the Boussinesq hypothesis, which relates the Reynolds stress tensor with the strain-rate tensor via an eddy viscosity $\mu_t$ as shown below:

$$-\rho\overline{u_i'u_j'} = \mu_t\left(\frac{\partial u_i}{\partial x_j} + \frac{\partial u_j}{\partial x_i}\right) - \frac{2}{3}\left(\rho k + \mu_t\frac{\partial u_k}{\partial x_k}\right)\delta_{ij} \tag{3.7}$$

Boussinesq hypothesis is applied in derivation of all one- and two-equation turbulence models described in the next section.

## 3.6 Turbulence Modeling

This section describes some of the most widely used one- and two-equation turbulence models. It should be noted that there are several other one-equation (e.g., Baldwin–Barth) and two-equation (e.g., k–kL) models that are not described here since they are infrequently used in simulation of flows in hydraulic machinery. Furthermore, there are more complex models, e.g., the three-equation models with three transport equations based on various turbulence quantities and full Reynolds stress (RSM) seven-equation models, which are not discussed because it has been found that they do not improve the accuracy but add to the computational cost for simulation of flows in hydraulic machinery. All categories of turbulence models are listed on the NASA Turbulence Modeling Resource (TMR) (https://turbmodels.larc.nasa.gov/) [16]. For additional information on turbulence modeling, the reader is referred to the classic book by Wilcox [15]. In the following sections, the mathematical details of some of the most commonly used one- and two-equation turbulence models are given.

### 3.6.1 Standard $k$–$\varepsilon$ Model

The standard two-equation $k$-$\varepsilon$ model employs the transport equation for turbulent kinetic energy $k$ and the transport equation for turbulent dissipation rate $\varepsilon$ to define the eddy viscosity. The model assumes that the flow is fully turbulent [17].

The equations of turbulent kinetic energy $k$ and dissipation rate $\varepsilon$ for the standard $k$–$\varepsilon$ model can be written as:

$$\frac{\partial(\rho k)}{\partial t} + \frac{\partial(\rho k u_i)}{\partial x_i} = \frac{\partial}{\partial x_j}\left[\left(\mu + \frac{\mu_t}{\sigma_k}\right)\frac{\partial k}{\partial x_j}\right] + G_k + G_b - \rho\varepsilon - Y_M + S_k \tag{3.8}$$

$$\frac{\partial(\rho\varepsilon)}{\partial t} + \frac{\partial(\rho\varepsilon u_i)}{\partial x_i} = \frac{\partial}{\partial x_j}\left[\left(\mu + \frac{\mu_t}{\sigma_\varepsilon}\right)\frac{\partial \varepsilon}{\partial x_j}\right] + C_{1\varepsilon}\frac{\varepsilon}{k}(G_k + C_{3\varepsilon}G_b) - C_{2\varepsilon}\rho\frac{\varepsilon^2}{k} + S_\varepsilon \tag{3.9}$$

where the turbulent (or Eddy) viscosity $\mu_t$ is computed by combining $k$ and $\varepsilon$ as follows:

$$\mu_t = \rho C_\mu \frac{k^2}{\varepsilon} \tag{3.10}$$

In Eqs. (3.8) and (3.9), $G_k$ represents the generation of turbulent kinetic energy due to the mean velocity gradients (it is described in the next section), $G_b$ is the generation of turbulence due to buoyancy, which can be defined as

$$G_{\mathrm{b}} = \beta g_i \frac{\mu_t}{Pr_t} \frac{\partial T}{\partial x_i} \tag{3.11}$$

where $Pr_t$ is the turbulent Prandtl number for the turbulent kinetic energy and $g_i$ is the component of the gravitational vector in the $i$th direction. For the standard $k$–$\varepsilon$ models, the default value of $Pr_t$ is 0.85.

The coefficient of thermal expansion $\beta$ is defined as:

$$\beta = -\frac{1}{\rho}\left(\frac{\partial \rho}{\partial T}\right)_{\mathrm{p}} \tag{3.12}$$

In Eq. (3.8), $Y_M$ represents the effect of the expansion of compressible turbulence on the total dissipation rate; this term is modeled based on a proposal by Sarkar [18] as:

$$Y_M = 2\rho\varepsilon M_t^2 \tag{3.13}$$

where $M_t$ is the turbulent Mach number defined as:

$$M_t = \sqrt{\frac{k}{a^2}} \tag{3.14}$$

where $a$ is the speed of sound.

In Eqs. (3.8) and (3.9), $C_{1\varepsilon}$, $C_{2\varepsilon}$, and $C_{3\varepsilon}$ are constants. $\sigma_k$ and $\sigma_\varepsilon$ are the turbulent Prandtl numbers for $k$ and $\varepsilon$, respectively. $S_k$ and $S_\varepsilon$ are user-defined source terms.

In ANSYS software, the default values of the model constants are $C_{1\varepsilon} = 1.44$, $C_{2\varepsilon} = 1.92$, and $C_\mu = 0.09$, and the turbulent Prandtl numbers associated with the turbulent kinetic energy $k$ and dissipation rate $\varepsilon$ are $\sigma_k = 1.0$ and $\sigma_\varepsilon = 1.3$, respectively. For $C_{3\varepsilon}$, it becomes 1 for the buoyant shear layers for which the main flow direction is aligned with the direction of gravity [19] and for buoyant shear layers that are perpendicular to the gravitational vector, $C_{3\varepsilon}$ becomes 0. For simulations in the fluid machinery, the buoyant term is not considered, and it is 0 in all the calculations reported in this book.

### 3.6.2 Shear Stress Transport (SST) $k$–$\omega$ Model

The SST $k$–$\omega$ is a widely used two-equation linear Eddy viscosity model developed by Menter. It combines the features of the standard $k$–$\varepsilon$ model and Wilcox's $k$–$\omega$ model [20]. It has been established by researchers over four decades that Wilcox's $k$–$\omega$ model is more accurate in the boundary layers near the solid boundaries, while the $k$–$\varepsilon$ model is more accurate in the free-stream and other shear flow regions in the flow field. The switching function $F_1$ in the model allows the model to switch between the $k$–$\omega$ model in the near-wall region and the $k$–$\varepsilon$ model in the free-stream region. This characteristic of the SST $k$–$\omega$ model allows the model to be accurate throughout the flow field. Since the first published version of the model in 1994 by Menter [21], various modifications to the model have been reported [22]. Here, the following version of the model is described, which has been recognized as relatively more accurate compared to the other versions for turbulent flow simulations in the field of rotating machinery. The relevant transport equations are as follows [23]:

$$\frac{\partial(\rho k)}{\partial t} + \frac{\partial(\rho k u_i)}{\partial x_i} = \frac{\partial}{\partial x_j}\left(\Gamma_k \frac{\partial k}{\partial x_j}\right) + G_k - Y_k + S_k \tag{3.15}$$

$$\frac{\partial(\rho \omega)}{\partial t} + \frac{\partial(\rho \omega u_i)}{\partial x_i} = \frac{\partial}{\partial x_j}\left(\Gamma_\omega \frac{\partial \omega}{\partial x_j}\right) + G_\omega - Y_\omega + S_\omega \tag{3.16}$$

where $G_k$ represents the production term of turbulent kinetic energy $k$; $G_\omega$ represents the production term of the turbulent dissipation rate $\omega$; $\Gamma_k$ and $\Gamma_\omega$ represent the coefficients of diffusion term for $k$ and $\omega$, respectively; $Y_k$ and $Y_\omega$ represent the dissipation terms of $k$ and $\omega$, respectively; and $D_\omega$ represents the cross-diffusion term calculated as described below. $S_k$ and $S_\omega$ are user-defined source terms for the $k$ and $\omega$ equations, respectively.

The expressions for the diffusion coefficients can be written as:

$$\Gamma_k = \mu + \frac{\mu_t}{\sigma_k} \tag{3.17}$$

$$\Gamma_\omega = \mu + \frac{\mu_t}{\sigma_\omega} \tag{3.18}$$

where, $\sigma_k$ and $\sigma_\varepsilon$ represent the turbulent Prandtl numbers in $k$ and $\omega$ equations, respectively. The turbulent viscosity coefficient $\mu_t$ is defined as:

$$\mu_t = \frac{\rho k}{\omega} \frac{1}{\max\left[\frac{1}{\alpha^*}, \frac{SF_2}{a_1 \omega}\right]} \tag{3.19}$$

where $S$ is the strain-rate magnitude and

$$\sigma_k = \frac{1}{\frac{F_1}{\sigma_{k,1}} + \frac{(1-F_1)}{\sigma_{k,2}}} \tag{3.20}$$

$$\sigma_\omega = \frac{1}{\frac{F_1}{\sigma_{\omega,1}} + \frac{(1-F_1)}{\sigma_{\omega,2}}} \tag{3.21}$$

In Eq. (3.19), $\alpha^*$ is the low Reynolds number correction coefficient (also known as the intermittent coefficient) given by:

$$\alpha^* = \alpha^*_\infty \left(\frac{\alpha^*_0 + \frac{Re_t}{R_k}}{1 + \frac{Re_t}{R_k}}\right) \tag{3.22}$$

where $Re_t = \frac{\rho k}{\mu \omega}$, $R_k = 6$, $\alpha^*_0 = \frac{\beta_i}{3}$, and $\beta_i = 0.072$. Note that, in HRN flows, $\alpha^* = \alpha^*_\infty = 1$.

In Eq. (3.20), $\sigma_{k,1} = 1.176$, $\sigma_{k,2} = 1.0$ and in Eq. (3.16), $\sigma_{\omega,1} = 2.0$ and $\sigma_{\omega,2} = 1.168$.

The functions $F_1$ and $F_2$ are defined as follows:

$$F_1 = \tan h\left(\phi_1^4\right) \tag{3.23}$$

$$\phi_1 = \min\left[\max\left(\frac{\sqrt{k}}{0.09\omega y}, \frac{500\mu}{\rho y^2 \omega}\right), \frac{4\rho k}{\sigma_{\omega,2} D^+_\omega y^2}\right] \tag{3.24}$$

$$\mathrm{F}_2 = \tan h\left(\phi_2^2\right) \tag{3.25}$$

$$\phi_2 = \max\left[2\frac{\sqrt{k}}{0.09\omega y}, \frac{500\mu}{\rho y^2\omega}\right] \tag{3.26}$$

$$D_\omega^+ = \max\left[2\rho\frac{1}{\sigma_{\omega,2}}\frac{1}{\omega}\frac{\partial k}{\partial x_j}\frac{\partial \omega}{\partial x_j}, 10^{-10}\right] \tag{3.27}$$

where $y$ is the distance from the wall and $D_\omega^+$ is the positive component of the cross-diffusion term.

In Eq. (3.15), the turbulence kinetic energy production term $G_k$ is defined as:

$$G_k = -\rho\overline{u_i'u_j'}\frac{\partial u_j}{\partial x_i} \tag{3.28}$$

$G_k$ is evaluated in a manner consistent with the Boussinesq hypothesis, i.e.,

$$G_k = \mu_t S^2 \tag{3.29}$$

In Eq. (3.16), the turbulence dissipation rate term $G_\omega$ is defined as:

$$G_k = \alpha\frac{\omega}{k}G_k \tag{3.30}$$

In Eq. (3.30), the coefficient $\alpha$ is given by:

$$\alpha = \frac{\alpha_\infty}{\alpha^*}\left(\frac{\alpha_0 + \frac{Re_t}{R_\omega}}{1 + \frac{Re_t}{R_\omega}}\right) \tag{3.31}$$

In Eq. (3.15), the dissipation term for $k$ is defined as:

$$Y_k = \rho\beta^* f_{\beta^*} k\omega \tag{3.32}$$

where

$$f_{\beta^*} = \begin{cases} 1 & \chi_k \le 0 \\ \dfrac{1 + 680\chi_k^2}{1 + 400\chi_k^2} & \chi_k > 0 \end{cases} \tag{3.33}$$

where

$$\chi_k \equiv \frac{1}{\omega^3}\frac{\partial k}{\partial x_i}\frac{\partial \omega}{\partial x_j} \tag{3.34}$$

and

$$\beta^* = \beta_i^*[1 + \xi^* F(M_t)] \tag{3.35}$$

$$\beta_i^* = \beta_\infty^*\left(\alpha_\infty^*\left(\frac{\frac{4}{15} + \left(\frac{Re_t}{R_\beta}\right)^4}{1 + \left(\frac{Re_t}{R_\beta}\right)^4}\right)\right) \tag{3.36}$$

In Eq. (3.16), the dissipation term for $\omega$ is defined as

$$Y_\omega = \rho\beta f_\beta \omega^2 \tag{3.37}$$

where

$$f_\beta = \frac{1 + 70\chi_\omega}{1 + 80\chi_\omega} \tag{3.38}$$

$$\chi_\omega = \left| \frac{\Omega_{ij}\,\Omega_{jk}\,S_{ki}}{(\beta^*_\infty \omega)^3} \right| \tag{3.39}$$

$$\Omega_{ij} = \frac{1}{2}\left(\frac{\partial u_i}{\partial x_j} - \frac{\partial u_j}{\partial x_i}\right) \tag{3.40}$$

$$\Omega_{ij} = \frac{1}{2}\left(\frac{\partial u_i}{\partial x_j} - \frac{\partial u_j}{\partial x_i}\right) \tag{3.41}$$

$$\beta = \beta_i \left[1 - \frac{\beta_i^*}{\beta_i}\xi^* F(M_t)\right] \tag{3.42}$$

In Eq. (3.35), the compressibility function, $F(M_t)$, is given by:

$$F(M_t) = \begin{cases} 0 & M_t \leq M_{t0} \\ M_t^2 - M_{t0}^2 & M_t > M_{t0} \end{cases} \tag{3.43}$$

where

$$M_t^2 \equiv \frac{2k}{a^2} \tag{3.44}$$

$$M_{t0} = 0.25 \tag{3.45}$$

$$a = \sqrt{\gamma RT} \tag{3.46}$$

Note that, in the high Reynolds number form of the SST $k$–$\omega$ model, $\beta_i^* = \beta^*_\infty$. In incompressible flow, $\beta^* = \beta^*_\infty$.

The other constants are $\alpha^*_\infty = 1$, $\alpha_\infty = 1$, $\alpha_0 = \frac{1}{9}$, $\beta^*_\infty = 0.09$, $R_\beta = 8$, $R_\omega = 2.95$, $\xi^* = 1.5$, $\sigma_k = 2.0$, and $\sigma_\omega = 2.0$.

### 3.6.3 Spalart–Allmaras (SA) Model

The SA turbulence model is an industry standard one-equation linear eddy-viscosity turbulence model developed for the prediction of aerodynamic flows. It is a low Reynolds number model and requires the boundary layer near the wall for accuracy. The SA model is given by the following equation:

$$\frac{\partial \tilde{\upsilon}}{\partial t} + u_j \frac{\partial \tilde{\upsilon}}{\partial x_j} = c_{b1}(1 - f_{t2})\tilde{S}\tilde{\upsilon} - \left[c_{\omega 1} f_\omega - \frac{c_{b1}}{\kappa^2} f_{t2}\right]\left(\frac{\tilde{\upsilon}}{d}\right)^2 + \frac{1}{\sigma}\left[\frac{\partial}{\partial x_j}\left\{(\upsilon + \tilde{\upsilon})\frac{\partial \tilde{\upsilon}}{\partial x_j}\right\} + c_{b2}\frac{\partial \tilde{\upsilon}}{\partial x_i}\frac{\partial \tilde{\upsilon}}{\partial x_i}\right] \tag{3.47}$$

The turbulent eddy-viscosity is given by the equation

$$\mu_t = \rho \tilde{\upsilon} f_{\upsilon 1} \tag{3.48}$$

Near-wall blocking is accounted for by the damping function $f_{\upsilon 1}$

$$f_{\upsilon 1} = \frac{\chi^3}{\chi^3 + c_{\upsilon 1}^3} \tag{3.49}$$

$$\chi = \frac{\tilde{\upsilon}}{\upsilon} \tag{3.50}$$

The remaining functions are defined as follows

$$\tilde{S} = \Omega + \frac{\tilde{\upsilon}}{\kappa^2 d^2} f_{\upsilon 2}, f_{\upsilon 2} = 1 - \frac{\chi}{1 + \chi f_{\upsilon 1}} \tag{3.51}$$

where $\Omega = \sqrt{2W_{ij}W_{ij}}$ is the magnitude of the vorticity, and $d$ is the distance of the flow field point to the nearest wall. $W_{ij}$ is defined by the following equation:

$$W_{ij} = \frac{1}{2}\left(\frac{\partial u_i}{\partial x_j} - \frac{\partial u_j}{\partial x_i}\right) \tag{3.52}$$

$$f_\omega = g\left[\frac{1 + c_{\omega 3}^6}{g^6 + c_{\omega 3}^6}\right]^{\frac{1}{6}}, \quad g = r + c_{\omega 2}(r^6 - r), \quad r = \min\left[\frac{\tilde{v}}{\tilde{S}\kappa^2 d^2}, 10\right] \tag{3.53}$$

$$f_{t2} = c_{t3}\exp(-c_{t4}\chi^2) \tag{3.54}$$

where $c_{b1} = 0.1355, \sigma = \frac{2}{3}, c_{b2} = 0.622, \kappa = 0.41, c_{\omega 2} = 0.3, c_{\omega 3} = 2, c_{v1} = 7.1, c_{t3} = 1.2, c_{t4} = 0.5$, and $c_{\omega 1} = \frac{c_{b1}}{\kappa^2} + \frac{1 + c_{b2}}{\sigma}$.

### 3.6.4 Wray–Agarwal (WA) Model

The Wray–Agarwal (WA) turbulence model is a one-equation linear eddy-viscosity model derived from the $k$–$\omega$ closure. It has been shown to have improved accuracy over the SA model and is competitive with the SST $k$–$\omega$ model in a large number of 2D and 3D complex flows. It behaves like the two-equation $k$–$\omega$ model in the near-wall region and like the standard two-equation $k$–$\varepsilon$ model in the free-stream and shear flow regions. The inclusion of the cross-diffusion term in the $R\,(= k/\omega)$ equation and a blending function that allows smooth switching between the two destruction terms are the most innovative features of this model. The transport equation for the variable $R$ in WA model is as follows:

$$\frac{\partial R}{\partial t} + u_j\frac{\partial R}{\partial x_j} = \frac{\partial}{\partial x_j}\left[(\sigma_R R + \nu)\frac{\partial R}{\partial x_j}\right] + C_1 RS + f_1 C_{2k\omega}\frac{R}{S}\frac{\partial R}{\partial x_j}\frac{\partial S}{\partial x_j} - (1 - f_1)C_{2k\varepsilon}R^2\left(\frac{\frac{\partial S}{\partial x_j}\frac{\partial S}{\partial x_j}}{S^2}\right) \tag{3.55}$$

In Eq. (3.55), $S$ is the mean strain given by:

$$S = \sqrt{2S_{ij}S_{ij}} \quad \text{where } S_{ij} = \frac{1}{2}\left(\frac{\partial u_i}{\partial x_j} + \frac{\partial u_j}{\partial x_i}\right) \tag{3.56}$$

The turbulent kinematic eddy viscosity is given by the equation:

$$\nu_T = f_\mu R \tag{3.57}$$

The wall-blocking effect is accounted for by the damping function $f_\mu$ as:

$$f_\mu = \frac{\chi^3}{\chi^3 + C_w^3}, \chi = \frac{R}{\nu} \tag{3.58}$$

The switching function $f_1$ allows the WA model to behave as a one-equation $k$–$\omega$ or one-equation $k$–$\varepsilon$ model and is defined as:

$$f_1 = \min\left(\tan h\left(arg_1^4\right), 0.9\right) \tag{3.59}$$

$$arg_1 = \frac{1 + \frac{d\sqrt{RS}}{\nu}}{1 + \left[\frac{\max(d\sqrt{RS}, 1.5R)}{20\nu}\right]^2} \tag{3.60}$$

where $d$ is the minimum distance to the nearest wall.

**Table 3.2** Model constants of WA 2017 model.

| Constant | Value |
|---|---|
| $C_{1k\omega}$ | 0.0829 |
| $C_{1k\varepsilon}$ | 0.1127 |
| $C_1$ | $f_1(C_{1k\omega} - C_{1k\varepsilon}) + C_{1k\varepsilon}$ |
| $\sigma_{k\omega}$ | 0.72 |
| $\sigma_{k\varepsilon}$ | 1.0 |
| $\sigma_R$ | $f_1(\sigma_{k\omega} - \sigma_{k\varepsilon}) + \sigma_{k\varepsilon}$ |
| $C_{2k\omega}$ | $C_{2k\omega} = \dfrac{C_{1k\omega}}{\kappa^2} + \sigma_{k\omega}$ |
| $C_{2k\varepsilon}$ | $C_{2k\varepsilon} = \dfrac{C_{1k\varepsilon}}{\kappa^2} + \sigma_{k\varepsilon}$ |
| $\kappa$ | 0.41 |
| $C_\omega$ | 8.54 |
| $C_m$ | 8 |

The various constants in the WA model are given in Table 3.2.

### 3.6.5 Detached Eddy Simulation Model

The DES is a modification of the original RANS turbulence model in which the model switches to a subgrid-scale formulation in regions fine enough for LES calculations. The regions near the solid boundaries and where the turbulent length scale is less than the maximum grid dimension are assigned the RANS mode of solution, but as the turbulent length scale exceeds the grid dimension, the regions are solved using the LES mode. Therefore, the grid resolution is not as demanding as pure LES, thereby considerably cutting down the cost of the computations with increased accuracy compared to RANS simulation. Although DES was initially formulated for the SA model, it can be implemented with other RANS models described above by appropriately modifying the length scale which is explicitly or implicitly involved in the RANS turbulence model. The construction of the DES model for the one-equation SA model is described below.

The transport equation for the kinematic eddy viscosity for the SA model can be rewritten as:

$$\frac{\partial}{\partial t}(\rho\hat{v}_T) + \frac{\partial}{\partial x_i}(\rho\hat{v}_T u_i) = \frac{1}{\sigma}\left\{\frac{\partial}{\partial x_i}\left[(\rho v_l + \rho\hat{v}_T)\frac{\partial\hat{v}_T}{\partial x_i}\right] + c_{b2}\rho\frac{\partial\hat{v}_T}{\partial x_i}\frac{\partial\hat{v}_T}{\partial x_i}\right\} + c_{b1}\rho\hat{S}\hat{v}_T - c_{\omega 1}f_\omega\rho\left(\frac{\hat{v}_T}{d}\right)^2 \tag{3.61}$$

where $x_i$, $t$ are space and time coordinates, respectively, $\hat{v}_T$ is the kinematic eddy viscosity coefficient, and $d$ is the distance to the nearest wall. The values of other coefficients in Eq. (3.61) are given in Reference [24].

The DES method modifies the dissipation term in the SA model, that is $d$ in the dissipation term (last term on RHS of Eq. (3.62)) is replaced by $\tilde{d}_{DES}$, so that the calculated flow field is related to the grid scale. $\tilde{d}_{DES}$ is defined as:

$$\tilde{d}_{DES} = \min(d, C_{DES}\Delta) \tag{3.62}$$

where $C_{\text{DES}}$ is a constant = 0.65 and $\Delta$ is the filtering scale of the subgrid-scale model with the maximum scaling values $\Delta_i = \max(\Delta x_i, \Delta y_i, \Delta z_i)$. This modification to the SA model, while very simple, changes the interpretation of the model substantially. This modified distance function causes the model to behave like a RANS model in regions close to the walls, and as a LES (e.g. Smagorinsky-like) model away from the walls.

Near the wall, $d_1 < C_{\text{DES}}\Delta_1$ and $d_2 < C_{\text{DES}}\Delta_2$, therefore, $\tilde{d}_{\text{DES},i} = d_i$, where the DES method matches the RANS method, while away from the wall, $d_3 > C_{\text{DES}}\Delta_3$ and $d_4 > C_{\text{DES}}\Delta_4$, therefore, $\tilde{d}_{\text{DES},i} = C_{\text{DES}}\Delta_i$ and the reduction in the turbulent eddy viscosity coefficient is determined by the local grid-scale $\Delta$. When the production term and dissipation term in the transport equation, Eq. (3.56) reach equilibrium, one obtains,

$$c_{b1}\rho\hat{S}\hat{v}_T = c_{\omega 1}f_\omega\rho\left(\frac{\hat{v}_T}{C_{\text{DES}}\Delta}\right)^2 \tag{3.63}$$

which gives $\hat{v}_T \propto \hat{S}(C_{\text{DES}}\Delta)^2$. Thus, when the local equilibrium occurs, the eddy viscosity coefficient is proportional to the amplitude of the deformation rate multiplied by the square of the filtering scale, which is the requirement of the Smagorinsky subgrid-scale eddy-viscosity model realizing the effect similar to LES implicit filtering.

It should be noted that while the SA model-based DES acts as LES with a wall model, DES based on other turbulence models (like the two equation models) behaves as a hybrid RANS-LES model. Grid generation is more complicated than for a simple RANS or LES case due to the switch between RANS and LES. Furthermore, DES is a non-zonal approach and provides a single smooth velocity field across the RANS and the LES regions of the solution.

### 3.6.6 Large Eddy Simulation (LES) Model

The LES is a well-established model for simulating turbulent flows. A theory developed by Kolmogorov in 1941 showed that the large eddies in a turbulent flow are dependent on the geometry while the smaller eddies are universal. This concept is used in the formulation of LES model from the Navier–Stokes equations [25].

In LES, the turbulent motion is assumed to be composed of vortex structures of different scales. Through a filter function, the instantaneous motion of the turbulent flow field is decomposed into two parts: large-scale vortex motion and subgrid-scale vortex motion. The large-scale vortex structures (large eddies) are calculated directly, while the subgrid-scale vortex structures (small eddies) are modeled. Thus, LES requires much less time and space resolution compared to the DNS method and can calculate the turbulent flow field characteristics at HRN on the currently available computer hardware, accurately reflecting the movement, development, and dissipation process in turbulent flow.

Filtering in LES is defined as the convolution of a function $u$ with a filtering Leonard term $G$ as [26]:

$$\overline{u}_i(\vec{x}) = \int G(\vec{x} - \vec{\xi})u(\vec{\xi})d\vec{\xi} \tag{3.64}$$

It results in

$$u_i = \overline{u}_i + u_i' \tag{3.65}$$

where $\overline{u}_i$ is the resolvable large-scale part and $u_i'$ is the subgrid-scale part of the variable $u$. However, most practical implementations of LES utilize the computational grid as the filter (the box filter) and do not employ explicit filtering.

The filtered equations can be obtained from the incompressible Navier–Stokes equations given below:

$$\frac{\partial u_i}{\partial t} + u_j\frac{\partial u_i}{\partial x_j} = -\frac{1}{\rho}\frac{\partial p}{\partial x_i} + \frac{\partial}{\partial x_i}\left(\upsilon\frac{\partial u_i}{\partial x_j}\right) \tag{3.66}$$

Substituting $u_i = \overline{u}_i + u'_i$ and $p = \overline{p} + p'$ in Eq. (3.61) and then filtering the resulting equation using Eq. (3.59) results in the equation for the resolved field:

$$\frac{\partial \overline{u}_i}{\partial t} + \overline{u}_j\frac{\partial \overline{u}_i}{\partial x_j} = -\frac{1}{\rho}\frac{\partial \overline{p}}{\partial x_i} + \frac{\partial}{\partial x_i}\left(\upsilon\frac{\partial \overline{u}_i}{\partial x_j}\right) + \frac{1}{\rho}\frac{\partial \tau_{ij}}{\partial x_j} \tag{3.67}$$

The new term $\frac{\partial \tau_{ij}}{\partial x_j}$ in Eq. (3.62) arises from the non-linear advection terms since

$$\overline{u_j\frac{\partial u_i}{\partial x_j}} \neq \overline{u}_j\frac{\partial \overline{u}_i}{\partial x_j} \tag{3.68}$$

Therefore,

$$\tau_{ij} = \overline{u}_i\overline{u}_j - \overline{u_i u_j} \tag{3.69}$$

Similar equations can be derived for the subgrid-scale field.

Boussinesq hypothesis is used in subgrid-scale (SGS) modeling of turbulence. The SGS turbulent stress can be modeled as:

$$\tau_{ij} - \frac{1}{3}\tau_{kk}\delta_{ij} = -2\mu_t\overline{S}_{ij} \tag{3.70}$$

where $\overline{S}$ is the strain-rate tensor for the resolved scale defined as:

$$\overline{S}_{ij} = \frac{1}{2}\left(\frac{\partial \overline{u}_i}{\partial x_j} - \frac{\partial \overline{u}_j}{\partial x_i}\right) \tag{3.71}$$

In Eq. (3.70), $\mu_t$ is the subgrid-scale turbulent eddy viscosity. Employing Eqs. (3.66)–(3.71), the Navier–Stokes equations become:

$$\frac{\partial \overline{u}_i}{\partial t} + \overline{u}_j\frac{\partial \overline{u}_i}{\partial x_j} = -\frac{1}{\rho}\frac{\partial \overline{p}}{\partial x_i} + \frac{\partial}{\partial x_i}\left([\upsilon + \upsilon_t]\frac{\partial \overline{u}_i}{\partial x_j}\right) \tag{3.72}$$

where the incompressibility constraint has been used to simplify the equation, and the pressure is modified to include the trace term $\frac{1}{3}\tau_{kk}\delta_{ij}$.

In order to close the equations and determine the filtered velocity field $\overline{u}(x,t)$ and the filtered pressure $\overline{p}(x,t)$, one needs to model the anisotropic residual-stress tensor $\tau^r_{ij}(x,t)$. The Smagorinsky model is the simplest model which has been proven to perform reasonably well. In this model, the residual subgrid-scale eddy viscosity $\upsilon_t$ is modeled to represent the motion of subgrid-scale eddies. $\upsilon_t$ is modeled as [27]:

$$\upsilon_t = l_s^2(2\overline{S_{lk}S_{lk}})^{\frac{1}{2}} = (C_S\Delta)^2(2\overline{S_{lk}S_{lk}})^{\frac{1}{2}} \tag{3.73}$$

where the Smagorinsky length scale is defined by $l_s = C_S\Delta$, and $C_S$ is the Smagorinsky coefficient and $\Delta$ is the local grid scale. In ANSYS FLUENT, $\Delta$ is computed according to the volume of the computational cell using

$$\Delta = V^{\frac{1}{3}} \tag{3.74}$$

Lilly derived a value of 0.17 for $C_S$ for homogeneous isotropic turbulence in the inertial subrange [28, 29]. However, this value was found to cause excessive damping of large-scale fluctuations in the presence of mean shear and in transitional flows near the solid boundary and needed to be reduced in such regions. In short, $C_S$ is not a universal constant, which is the most serious shortcoming of this simple model. Nonetheless, a $C_S$ value of around 0.1 has been found to yield the best results for a wide range of flows and is used as the default value in FLUENT.

In OpenFOAM, the Smagorinsky coefficient $C_S$ is calculated by the two coefficients $C_k$ and $C_\varepsilon$ as follows:

$$(C_S)^2 = \left(\frac{C_k}{C_\varepsilon}\right)^{\frac{1}{2}} \tag{3.75}$$

The two coefficients are determined based on the behavior of the model when applied to canonical free shear flows and a turbulent boundary layer. The lower bound of $C_k$ and $C_\varepsilon$ are 0.085 and 0.984, respectively, and the upper bound of $C_k$ and $C_\varepsilon$ are 0.103 and 1.152, respectively. The standard values of $C_k$ and $C_\varepsilon$ are 0.094 and 1.048 respectively.

## 3.7 Numerical Solution Algorithms

### 3.7.1 Numerical Discretization Method

Before conducting the simulations of turbulent flow in a specified pump geometry like the mixed-flow pump, the computational domain of the pump needs to be appropriately discretized into many discrete grid elements. Then, the governing RANS equations and the turbulence modeling PDEs are discretized on these grid elements, which turn from PDEs to the algebraic equations. At present, there are three commonly used discretization methods which include the finite-difference method, the finite element method, and the finite-volume method [30].

The finite-difference method is an early discretization method which was successfully used primarily for flow problems with simple geometries. Its basic feature is to replace the derivative terms in the governing equations with the corresponding difference expressions at each node of the grid created by the intersection of the three grid lines representing the three coordinate directions. The finite-difference equations result in algebraic equations at each node; algebraic equations at each node include the unknown flow variables at this node and from neighboring nodes. The required numerical solutions for the flow variables are then obtained by solving these algebraic equations with appropriate boundary conditions.

The main idea behind the finite element method is to divide the computational domain into a series of small elements and add several points on each element as nodes, and then obtain the discrete equations by integrating the governing equations on these finite elements.

The basic idea of the finite control volume method is to divide the computational domain into many control volumes, and each control volume is represented by a node. The discrete equations are derived by integrating the governing equations over the control volume. In this process, some assumptions should be made on the function representing the control volume – linear, quadratic, or cubic – and for its first-order derivatives, which are required on the interface of the two control volumes. Thus, different formulations for various numerical algorithms have been generated in the past 50 years. The finite-volume method is the most widely used method in CFD in both the commercial (ANSYS, COMSOL, CFX, etc.) and noncommercial open-source codes (OpenFOAM, SU2, etc.) as well as in the proprietary industrial and government codes. In this chapter, the focus is on the commercial code ANSYS Fluent/CFX.

### 3.7.2 Flow Field Solution Method

In incompressible flow, there is no pressure term in the continuity equation, and therefore pressure and velocity are not strongly coupled in the governing equations. This problem is addressed by employing the so-called "coupled" or "uncoupled" algorithm. Here, the "coupled" algorithm refers to the pressure–velocity coupling algorithm of Patankar and Spalding [31] (also known as the pressure correction approach) and the non-coupled algorithm refers to the solution method requiring the solution of the Poisson equation for pressure along with the momentum equations. The "Pressure Correction" based algorithms are known as the SIMPLE (Semi-Implicit Method for Pressure Linked Equations) family of algorithms in the literature [32]. Among SIMPLE algorithms, SIMPLEC is highly recommended since the influence of the velocity correction values of the adjacent grid points is omitted which makes the algorithm simpler; it has good convergence properties [33]. There are other variations of the algorithm; they are not discussed here and the reader is referred to Ref. [34].

After the discrete algebraic equations are established by the finite-volume method using the "coupled" or "uncoupled" solution approach, an "implicit," "explicit–implicit," or "explicit" solution technique can be used to solve the equations. In the implicit method, all equations are solved simultaneously at the same time for the flow variables $u$, $v$, $w$, and $p$. The solution can also be obtained by using an explicit–implicit method in which the solution algorithm has both the explicit and implicit parts; the values of the flow variables in the explicit parts are known, while in the implicit part, all flow variables are computed simultaneously. In the explicit solution algorithm, the algebraic equations are solved sequentially, one by one at a time. The choice of a particular algorithm is made based on the memory requirements and its accuracy and stability on a given mesh defined in terms of the CFL (Courant–Fredrick–Levy) condition or Courant number. CFL condition for the implicit or explicit–implicit algorithm may be relatively larger compared to that for the purely explicit algorithm; however, the implicit algorithm is more complex since it requires matrix inversion and substantially more memory requirement compared to the explicit algorithm.

## 3.8 Near-wall Flow Treatment

The turbulence modeling near the wall has a substantial effect on the accuracy of the numerical solutions since the walls are the primary source of vorticity and turbulence. Essentially, it is in the near-wall region where the flow variables have significant gradients, and the transport of momentum and other scalar quantities is most important. Consequently, a precise depiction of the flow in the near-wall region plays a crucial role in ensuring successful predictions of wall-bounded turbulent flows. For fully developed turbulent flow, numerous experiments have shown that the near-wall region can be subdivided into three layers: the viscous sublayer, the transition or buffer layer, and the logarithmic layer. The viscous sublayer is almost laminar, with the velocity component parallel to the wall having a linear distribution along the direction normal to the wall. The transition or buffer layer is located outside the viscous sublayer, where the influence of viscosity is comparable to the action of the turbulent shear stress. The flow conditions are more complex in this layer, and due to its small thickness, it is often not separately delineated in engineering calculations but is included in the logarithmic layer. The logarithmic layer is the outermost layer where turbulence is fully developed, the velocity profiles have logarithmic behavior, and the influence of molecular viscosity is not significant; turbulent shear stress dominates the flow field. Figure 3.6 shows these layers in the turbulent velocity profile in the near-wall region.

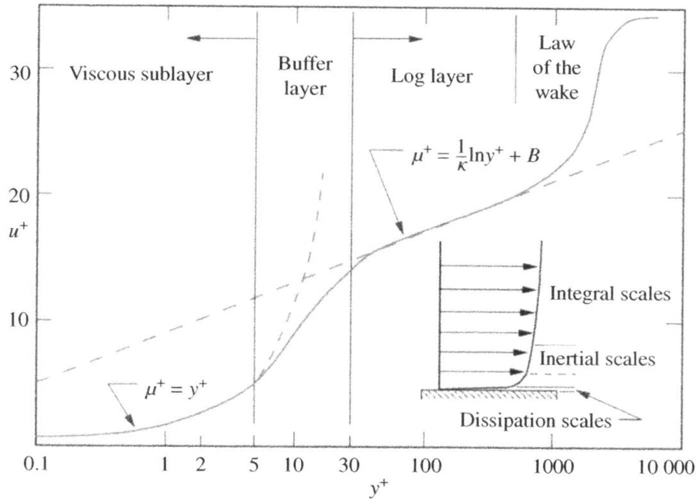


**Figure 3.6** Velocity profiles in the near-wall region of the turbulent boundary layer.

In order to describe the velocity profile in the viscous sublayer and the logarithmic layer, as shown in Fig. 3.6, two dimensionless parameters $y^+$ and $u^+$ are introduced to represent the velocity and the wall distance defined as:

$$u^+ = \frac{u}{u_\tau} \tag{3.76}$$

$$y^+ = y\frac{u_\tau}{\rho} \tag{3.77}$$

where $\rho$ is density, $u_\tau$ is the shear velocity (or friction velocity), which can be written as:

$$u_\tau = \sqrt{\frac{\tau}{\rho}} \tag{3.78}$$

The shear stress $\tau$ is proportional to the velocity gradient,

$$\tau = \mu\frac{du}{dy} \tag{3.79}$$

where $\mu$ is the molecular viscosity of the fluid.

In the viscous sublayer ($0 < y^+ < 5$), near the wall, the viscous effects dominate the flow and the velocity profile is linear given by:

$$u^+ = y^+ \tag{3.80}$$

In the buffer layer ($5 < y^+ < 30$), which serves as a transition layer between the viscous sublayer and the log-law layer, the viscous force due to molecular viscosity is close to the turbulent shear stress. It cannot be described by a simple formula. However, in most flow problems, its thickness is very small and it does not affect the flow field, and it is generally included in the log-law layer.

The log-law layer ($30 < y^+ < 300$), where the velocity profile follows a logarithmic behavior up to the end of the boundary layer, can be expressed as:

$$u^+ = \frac{1}{\kappa}\log Ey^+ + C^+ \tag{3.81}$$

where $\kappa \approx 0.41$ is the von Kármán constant and $E \approx 9.8$; these are empirical values that can slightly vary from case to case. $C^+$ is also an empirical parameter, and it changes according to the wall roughness (e.g., $C^+ \approx 5$ for smooth walls).

In the ANSYS software, the log-law is generally employed when $y^+ > 11.225$, and the laminar stress–strain relationship is applied when $y^+ < 11.225$.

There are two main approaches to deal with the near-wall region in the numerical simulation: the first approach is not to directly solve for the viscous sublayer and the buffer layer but to solve from the viscous sublayer to the outer layer of the turbulent boundary layer by using a semi-empirical formula for the entire boundary layer and employing the so-called wall function method (this method is employed with HRN turbulence models); and the second approach is to directly solve all the way to the wall including the laminar sublayer layer, buffer layer, and the log layer and analyze the influence of turbulent eddy viscosity all the way to the wall (this method is employed with low Reynolds number (LRN) turbulence models).

For many HRN flows, the wall function method can save a lot of computational resources. This is because in the near-wall region, the gradients in the flow variables are large, and therefore many more grid nodes are required for accurate resolution of the flow field, while only a few grid nodes are required when using the wall functions in the near-wall region. For many 3D industrial applications, such as the flows in the hydraulic machinery, the wall function method is often employed to obtain numerical solutions of acceptable engineering accuracy for quantities of interest, e.g., pump head and efficiency of a mixed-flow pump using substantially less computational resources. The commonly used wall functions include the so-called standard wall function and the enhanced wall function; the details are provided in the book by Wilcox [15].

When the LRN (HRN) turbulence model is employed, the flow field in the viscous sublayer is solved directly, which requires not only the dimensionless distance $y^+ < 1$ from the first layer of the grid to the wall but also at least 15–20 layers of grid in the near-wall region in order to accurately calculate the velocity and skin friction near the wall.

## 3.9 Boundary Conditions

The six gridded computational domains of the mixed-flow pump model in Fig. 3.4 are imported into ANSYS CFX to first conduct the steady-state flow simulations and then the transient flow simulations, which are executed by using the steady-state computational results as the initial data. The computational setup is shown in Table 3.3. The time step in the transient simulation is $3.448 \times 10^{-4}$ s, which corresponds to 3 degrees impeller rotation. The total time of transient simulation is 0.7448 s, which corresponds to the time required for the impeller to rotate 18 revolutions.

Based on the operating conditions of the mixed-flow pump, the boundary conditions are determined. The velocity inlet boundary condition is used at the inlet pipe; the velocity distribution in the inlet pipe is assumed to be uniform and is determined by the flow rate. The outlet boundary condition is set as an opening (pressure outlet) with a relative pressure of 30 000 Pa. The dynamic interface of the impeller cross-section is described by the multi-rotating reference frame method. The no-slip boundary condition is employed on all solid walls of the pump. The enhanced wall functions [35] are used near the wall region to calculate the turbulent kinetic energy and the turbulence resolution frequency near the wall. When the root mean square residuals drop below $10^{-5}$, the convergence of each physical time step is obtained in 4–10 iterations. The specific iterative process used in the numerical simulations in various pump working conditions is shown in Fig. 3.7. The high-resolution scheme is used for the convection terms, and the central difference scheme is used for the diffusion terms. For the time-dependent term in the transient flow simulations, the unsteady second-order implicit time integration scheme is used.

**Table 3.3** Computational setup.

| Location | Boundary type |
|---|---|
| Inlet of the model pump | Velocity inlet |
| Outlet of the model pump | Outlet pressure (30 000 Pa) |
| Wall surfaces of model pump | No-slip wall |
| *Impeller interfaces* | |
| Steady state | Frozen rotor |
| Transient state | Transient rotor-stator |
| *Solver control for transient simulation* | |
| Time step | $3.448 \times 10^{-4}$ s |
| Total time | $7.448 \times 10^{-1}$ s |
| Convergence criterion | $10^{-5}$ |

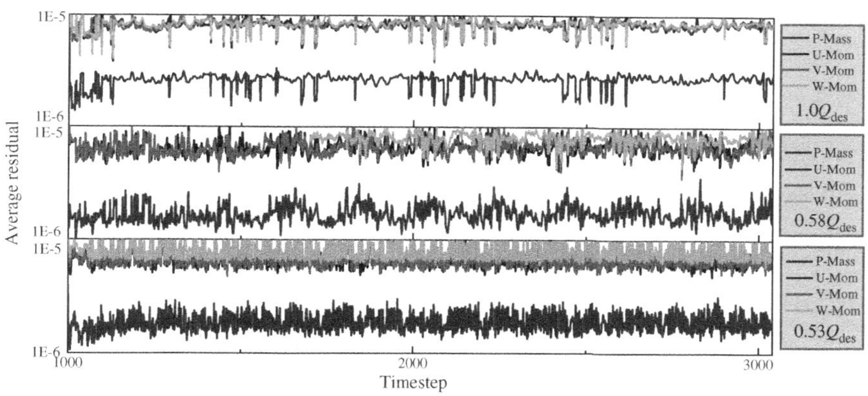


**Figure 3.7** Typical convergence characteristics of the residuals of flow variables at various flow rates.

## 3.10 Uncertainty Analysis

The solution independence of the mesh for steady flow in the mixed-flow pump in the design condition is tested and ensured by conducted simulations on a sequence of meshes (Table 3.4). The Grid Convergence Index (GCI) methodology is selected to conduct the uncertainty analysis [36, 37]. In the analysis of the solution independence of mesh, a set of meshes is generated using the same mesh topology, and the number of mesh nodes on the topological lines as well as the corresponding nodes, are adjusted to create meshes with consistent quality but different cell numbers. Table 3.5 shows the value of the pump head for mesh elements with different cell numbers. In general, the errors caused by the iterative process and other numerical factors are relatively small compared to the errors caused by the grid; thus, the numerical solution errors $\delta_{SN}$ and uncertainties $U_{SN}$ in the simulation are mainly determined by the grid errors $\delta_G$ and grid uncertainties $U_G$. Therefore, $\delta_{SN} \approx \delta_G$ and $U_{SN} \approx U_G$.

**Table 3.4** Head of the mixed-flow pump with different numbers of mesh elements.

| | Number of mesh elements | Head (m) |
|---|---|---|
| Case A | 797 860 | 6.208 |
| Case B | 564 173 | 6.112 |
| Case C | 398 930 | 5.951 |

**Table 3.5** Uncertainty parameters related to mesh convergence.

| Research target | $R_G$ | $p_G$ | $C_G$ | $U_G$ | $\delta_G^*$ | $U_{GC}$ | $S_c$ |
|---|---|---|---|---|---|---|---|
| | 0.596 | 1.482 | 0.677 | 0.2334 | 0.096 | 0.0458 | 6.112 |

Here, the errors in the total head $\varepsilon_{BA}$ and $\varepsilon_{CB}$ calculated by the various grids can be described by the following equations:

$$\varepsilon_{BA} = H_B - H_A = 0.096 \tag{3.82}$$

$$\varepsilon_{CB} = H_C - H_B = 0.161 \tag{3.83}$$

The convergence rate of the grid is calculated as:

$$R_G = \frac{\varepsilon_{BA}}{\varepsilon_{CB}} = 0.596 \tag{3.84}$$

The order of accuracy estimate $p_G$ is given by:

$$p_G = \frac{ln(\varepsilon_{CB}/\varepsilon_{BA})}{ln(r_G)} = 1.492 \tag{3.85}$$

The grid error $\delta_{RE_G}^*$ can be obtained as:

$$\delta_{RE_G}^* = \frac{\varepsilon_{BA}}{r_G^{p_G} - 1} = 0.1418 \tag{3.86}$$

The correction factor $C_G$ is given by,

$$C_G = \frac{r_G^{p_G} - 1}{r_G^{p_{Gest}} - 1} = 0.677 \tag{3.87}$$

The improved grid error $\delta_G^*$ is,

$$\delta_G^* = CG\delta_{RE_G}^* = 0.096 \tag{3.88}$$

The modified numerical simulation result $S_c$ is given by,

$$S_C = S_1 - \delta_G^* = 6.112 \tag{3.89}$$

The correction factor $C_G$ is 0.677, and it satisfies the following relationship, that is $|1 - C_G| \geq 0.125$ and $|1 - C_G| \geq 0.25$.

$$U_G = [2|1 - C_G| + 1] \left|\delta_{RE_G}^*\right| = 0.2334 \tag{3.90}$$

$$U_{CG} = [|1 - C_G| + 1] \left|\delta_{RE_G}^*\right| = 0.0458 \tag{3.91}$$

From the above calculations, it can be inferred that $\delta_G^*$, $U_{CG}$ are less than $U_G$.

**Table 3.6** Validation of grid independence of the solution.

| Research target | $E$ | $U_V$ | $E_C$ | $U_{VC}$ |
|---|---|---|---|---|
| | 0.273 | 0.2766 | 0.148 | 0.155 |

**Table 3.7** Number of mesh elements for each component of the pump.

| Component of the pump | Mesh elements |
|---|---|
| Inlet | 55 120 |
| Impeller | 399 792 |
| Guide vane | 232 596 |
| Annular volute chamber | 110 352 |

The simulation results of the whole pump head can be verified and analyzed. The numerical simulation value in the comparison error depends on the simulation value $S_1$ which is calculated on the finest mesh.

Comparing the error first, get

$$E = D - S_1 = 0.273 \tag{3.92}$$

where $D$ is the experimental value.

The uncertainty in error is given by,

$$U_V = \sqrt{U_{SN}^2 + U_D^2} = 0.2766 \tag{3.93}$$

where $U_D = 2.5\,\%\,D$.

The corrected comparison error is given by,

$$E_C = D - S_C = 0.148 \tag{3.94}$$

The uncertainty in the comparison error is given by,

$$U_{VC} = \sqrt{U_{ScN}^2 + U_D^2} = 0.155 \tag{3.95}$$

where $U_{ScN} = U_{GC}$.

In Table 3.5, $R_G = 0.848 < 1$, which indicates that the simulation results satisfy the monotone convergence condition. In Table 3.6, $|E_C| < U_{VC}$, thus the simulation results of the mixed-flow pump also meet the requirements of the CFD inspection, verification, and convergence. The final selected mesh number is 797 860, and the mesh numbers in different parts of the pump are shown in Table 3.7.

## 3.11 Summary

The primary focus of this chapter has been on the description of the basic methodology employed in CFD for numerical simulation of the flow field in a mixed-flow pump. An overview of the main structure of a model mixed-flow pump is provided along with the key parameters for each of its six major components including the impeller and the guide vane. A three-dimensional geometric model of the mixed-flow pump is established using the 3D design software, and a computational flow domain inside the pump is generated. The multi-block structured grid generation in various computational domains of the mixed-flow pump is described including the treatment of interfaces using the ANSYS ICEM mesh generation software. Subsequently, the chapter delves into some of the fundamental physical fluid flow models for numerical simulation of the turbulent flow fields in the fluid machinery, which include the RANS equations, the LES, the DES, and the DNS. For simulations with RANS equations, the chapter describes the four commonly used turbulence models in hydraulic machinery, namely the tow-equation standard $k$–$\varepsilon$ model, the two-equation SST $k$–$\omega$ model, and one-equation SA and WA models. The near-wall treatment of the boundary layer using these models – both LRN and HRN versions of the models – are discussed including the use of wall functions with HRN. The finite-volume method, numerical solution algorithms for incompressible flows, and the boundary conditions are discussed. Finally, the chapter introduces the concepts of error and uncertainty analysis in the numerical solution.

## Nomenclature

| | |
|---|---|
| CFD | computational fluid dynamics |
| DNS | direct numerical simulation |
| GCI | grid convergence index |
| HRN | high Reynolds number |
| LES | large Eddy simulation |
| LRN | low Reynolds number |

## References


1 Anderson, J.D. and Wendt, J. (1995). *Computational Fluid Dynamics*. New York: McGraw-Hill.
2 Spurk, J. and Aksel, N. (2007). *Fluid Mechanics*. Springer Science & Business Media.
3 Lomax, H., Pulliam, T.H., and Zingg, D.W. (2001). *Fundamentals of Computational Fluid Dynamics*. Berlin: Springer.
4 Hirsch, C. (2007). *Numerical Computation of Internal and External Flows: The Fundamentals of Computational Fluid Dynamics*. Elsevier.
5 Pulliam, T.H. and Zingg, D.W. (2014). *Fundamental Algorithms in Computational Fluid Dynamics*. New York: Springer.
6 Glatzel, T., Litterst, C., Cupelli, C. et al. (2008). Computational fluid dynamics (CFD) software tools for microfluidic applications—a case study. *Computers & Fluids* 37 (3): 218–235.
7 Versteeg, H.K. and Malalasekera, W. (2007). *An Introduction to Computational Fluid Dynamics: The Finite Volume Method*. Pearson Education.
8 Moser, R.D., Haering, S.W., and Yalla, G.R. (2021). Statistical properties of subgrid-scale turbulence models. *Annual Review of Fluid Mechanics* 53: 255–286.
9 Dick, E. (2009). Introduction to finite element methods in computational fluid dynamics. In: *Computational Fluid Dynamics*, 235–274. Berlin Heidelberg: Springer.
10 Wilcox, D.C. (1988). Multiscale model for turbulent flows. *AIAA Journal* 26 (11): 1311–1320.
11 Moin, P. and Mahesh, K. (1998). Direct numerical simulation: a tool in turbulence research. *Annual Review of Fluid Mechanics* 30 (1): 539–578.

12 Ishihara, T., Gotoh, T., and Kaneda, Y. (2009). Study of high-Reynolds number isotropic turbulence by direct numerical simulation. *Annual Review of Fluid Mechanics* 41: 165–180.

13 Lee, M. and Moser, R.D. (2015). Direct numerical simulation of turbulent channel flow up to. *Journal of Fluid Mechanics* 774: 395–415.

14 Eggels, J.G.M., Unger, F., Weiss, M.H. et al. (1994). Fully developed turbulent pipe flow: a comparison between direct numerical simulation and experiment. *Journal of Fluid Mechanics* 268: 175–210.

15 Wilcox, D.C. (1998). *Turbulence Modeling for CFD*. La Canada, CA: DCW Industries.

16 https://turbmodels.larc.nasa.gov/

17 Launder, B.E. and Spalding, D.B. (1972). *Lectures in Mathematical Models of Turbulence*. London, England: Academic Press.

18 Sarkar, S. and Lakshmanan, B. (1991). Application of a Reynolds stress turbulence model to the compressible shear layer. *AIAA Journal* 29 (5): 743–749.

19 Henkes, R., Van Der Vlugt, F.F., and Hoogendoorn, C.J. (1991). Natural-convection flow in a square cavity calculated with low-Reynolds-number turbulence models. *International Journal of Heat and Mass Transfer* 34 (2): 377–388.

20 Menter, F.R. (1994). Two-equation Eddy-viscosity turbulence models for engineering applications. *AIAA Journal* 32 (8): 1598–1605.

21 Menter, F.R., Kuntz, M., and Langtry, R. (2003). Ten years of experience with the SST turbulence model. In: *Turbulence, Heat and Mass Transfer* (ed. K. Hanjalic, Y. Nagano, and M. Tummers), 625–632. Begell House Inc.

22 https://turbmodels.larc.nasa.gov/sst.html

23 Fluent, A. (2011). *Fluent 14.0 User's Guide*. Ansys Fluent Inc.

24 Vatsa, V.N., Lockard, D.P., and Spalart, P.R. (2017). Grid sensitivity of SA-based delayed-detached-Eddy-simulation model for blunt-body flows. *AIAA Journal* 55 (8): 2842–2847.

25 Bailly, C. and Comte-Bellot, G. (2015). *Turbulence*. Springer.

26 Hou, Z. (2017). Uncertainty Quantification of Wall-Modeled Large Eddy Simulation (WMLES) Model in OpenFOAM.

27 Smagorinsky, J. (1963). General circulation experiments with the primitive equations: I. The basic experiment. *Monthly Weather Review* 91 (3): 99–164.

28 Scotti, A., Meneveau, C., and Lilly, D.K. (1993). Generalized Smagorinsky model for anisotropic grids. *Physics of Fluids A: Fluid Dynamics* 5 (9): 2306–2308.

29 Lilly, D.K. (1992). A proposed modification of the Germano subgrid-scale closure method. *Physics of Fluids A: Fluid Dynamics* 4 (3): 633–635.

30 Ashgriz, N. and Mostaghimi, J. (2002). An introduction to computational fluid dynamics. In: *Fluid Flow handbook*, Vol. 1, 1–49.

31 Patankar, S.V. and Spalding, D.B. (1983). A calculation procedure for heat, mass and momentum transfer in three-dimensional parabolic flows. In: *Numerical Prediction of Flow, Heat Transfer, Turbulence and Combustion*, 54–73. Pergamon.

32 Pantankar, S.V. and Spalding, D.B. (1980). Semi implicit method for pressure linked equations. *Numerical Heat Transfer and Fluid Flow* .

33 Yong, L., Zhiyou, L., and Yiran, A. (2001). A brief introduction to fluent—a general purpose CFD code. *Journal of Hydrodynamics* 16 (2): 254–258.

34 Petrila, T. and Trif, D. (2004). *Basics of Fluid Mechanics and Introduction to Computational Fluid Dynamics*. Springer Science & Business Media.

35 Kader, B.A. (1981). Temperature and concentration profiles in fully turbulent boundary layers. *International Journal of Heat and Mass Transfer* 24 (9): 1541–1544.

36 AIAA, Computational Fluid Dynamics Committee (1998). Guide: Guide for the Verification and Validation of Computational Fluid Dynamics Simulations, AIAA G-077-1998 (2002).

37 Celik, I.B., Ghia, U., Roache, P.J. et al. (2008). Procedure for estimation and reporting of uncertainty due to discretization in CFD applications. *Journal of Fluids Engineering-Transactions of the ASME* 130 (7).

# 4

# Pump Performance Analysis Methods

## 4.1 Entropy Production Analysis

In a mixed-flow pump, due to the viscosity of the working fluid medium and the existence of turbulent shear stresses during operation, the mechanical energy is converted into internal energy irreversibly which causes irreversible energy losses. Therefore, the local entropy production method is a proper analysis tool to measure the mechanical energy loss and to determine the energy dissipation in the blade tip region of the pump.

Entropy production can be considered as a kind of state parameter. For the single-phase incompressible flow in Cartesian coordinate system, without regard to the heat transfer in the system, the transport equation for the entropy production rate per unit mass can be expressed as follows [1]:

$$\rho \frac{\partial s}{\partial t} + \rho u_i \frac{\partial s}{\partial x_j} = \operatorname{div}\left(\frac{q}{T}\right) + \dot{S}_D''' \tag{4.1}$$

where indices $i, j$ denote the three coordinate directions ($i, j = 1, 2, 3$), $\rho$ the density, $t$ the time, $T$ the temperature, $u_j$ the velocity component in coordinate direction $j$, $x_j$ the coordinate direction $j$, $s$ the specific entropy, $q$ the heat flux, and $\dot{S}_D'''$ the specific entropy production rate.

The last term $\dot{S}_D'''$ on the right-hand side of Eq. (1) represents the entropy production generated by the viscous dissipation, and it impacts the mechanism of $\Phi_{\text{entropy}}$ generation in the process of its transport. It appears as a source term and is always positive. As the specific entropy is only a function of temperature and pressure which are constrained by the basic equations of fluid flow, namely the continuity equation, the momentum equations, and the energy equation, it can be obtained by the post-processing of the numerical simulation.

The specific entropy production rate can be written as given n Eq. (4.2) [2]:

$$\dot{S}_D''' = \frac{\dot{Q}}{T} = \frac{\Phi}{T} = m_{ji} \frac{\partial u_i}{\partial x_j} \tag{4.2}$$

where $\dot{Q}$ represents the energy transfer rate, $\Phi$ the energy dissipation rate, and $m_{ji}$ the viscous stress tensor. Since the flow in the pump is at a high Reynolds number, the specific expression for the local entropy production term can be obtained by time-averaging, which is similar to Reynolds-averaging used in the derivation of RANS equations. Different from the steady flow, the entropy production in transient flow involves two parts: one part is generated by a time-averaged process, and the other part is the dissipation produced by the velocity fluctuations. The local entropy production

*Mixed-flow Pumps: Modeling, Simulation, and Measurements*, First Edition.
Wei Li, Leilei Ji, Ramesh Agarwal, Weidong Shi, and Ling Zhou.

rate $\dot{S}_D'''$ is thus defined by Eq. (4.3) in which the entropy production rate $\dot{S}_{\overline{D}}'''$ can be calculated by Eq. (4.4)

$$\dot{S}_D''' = \dot{S}_{\overline{D}}''' + \dot{S}_{D'}''' \tag{4.3}$$

$$\dot{S}_{\overline{D}}''' = m_{ji}\frac{\partial \overline{u}_i}{\partial x_j} \tag{4.4}$$

where $m_{ji}$ is the viscous stress tensor obtained after the time-averaging, which can be obtained from Eq. (4.5).

$$m_{ji} = \mu\left(\frac{\partial \overline{u}_i}{\partial x_j} + \frac{\partial \overline{u}_j}{\partial x_i}\right) - \frac{2}{3}\mu\delta_{ij}\frac{\partial \overline{u}_i}{\partial x_i} \tag{4.5}$$

where $\delta_{ij}$ is the Kronecker delta symbol and $\mu$ is the dynamic viscosity.

For the incompressible fluid in fluid machinery, the continuity equation is given by

$$\frac{\partial \overline{u}_i}{\partial x_j} = 0 \tag{4.6}$$

Thus, Eq. (4.2) can be expanded into Eq. (4.7) as given below:

$$\dot{S}_{\overline{D}}''' = \frac{\mu_{\text{eff}}}{T}\left\{\begin{aligned} &2\left[\left(\frac{\partial \overline{u}_1}{\partial x_1}\right)^2 + \left(\frac{\partial \overline{u}_2}{\partial x_2}\right)^2 + \left(\frac{\partial \overline{u}_3}{\partial x_3}\right)^2\right] \\ &+\left[\left(\frac{\partial \overline{u}_2}{\partial x_1} + \frac{\partial \overline{u}_1}{\partial x_2}\right)^2 + \left(\frac{\partial \overline{u}_3}{\partial x_1} + \frac{\partial \overline{u}_1}{\partial x_3}\right)^2 + \left(\frac{\partial \overline{u}_2}{\partial x_3} + \frac{\partial \overline{u}_3}{\partial x_2}\right)^2\right] \end{aligned}\right\} \tag{4.7}$$

Likewise, the entropy production rate $\dot{S}_{D'}'''$ generated by the velocity fluctuations can be obtained by following the same approach, if the time-averaging is not employed in the calculation. However, the components of the velocity fluctuations cannot be obtained from a turbulence model such as the two-equation SST $k$–$\omega$ turbulence model used in many mixed-flow pump simulations using the RANS equations. According to Kock [3] and Mathieu [4], the local entropy production rate resulting from the velocity fluctuations has a close correlation with $\omega$ in the SST $k$–$\omega$ turbulence model. Thus, with the SST $k$–$\omega$ model, the local entropy production rate induced by the velocity fluctuations can be approximately calculated by Eq. (4.8):

$$\dot{S}_{D'}''' = \beta\frac{\rho\omega k}{T} \tag{4.8}$$

where $\beta = 0.09$, $\omega$ stands for the turbulent eddy frequency ($s^{-1}$), and $k$ stands for the turbulent kinetic energy ($m^2/s^2$).

Thus, the total entropy production in each computational domain can be obtained by integrating the local entropy production rate in the computational domain as given in Eqs. (4.9) and (4.10):

$$\Delta S_{\text{pro},\overline{D}} = \int_V \dot{S}_{\overline{D}}''' dV \tag{4.9}$$

$$\Delta S_{\text{pro},D'} = \int_V \dot{S}_{D'}''' dV \tag{4.10}$$

Furthermore, due to the high velocity and pressure gradient between the blade tip and pump wall region of the pump, there exists a strong wall effect on the entropy production which cannot be ignored. Hence, the entropy production rate in the tip region from the pump wall to the first grid point near it could be calculated by using Eq. (4.11):

$$\Delta S_{\text{pro},W} = \int_A \frac{\vec{\tau} \cdot \vec{v}}{T} dA \tag{4.11}$$

where $\vec{\tau}$ denotes the shear stress at the wall (Pa), $A$ the area ($m^2$) and $\vec{v}$ the velocity at the first grid point near the wall (m/s).

To sum up, the total entropy production $\Delta S_{\text{pro}}$ and the irreversible energy loss $I_{\text{pro}}$ of the whole system can be obtained using Eqs. (4.12) and (4.13):

$$\Delta S_{\text{pro}} = \Delta S_{\text{pro},\overline{D}} + \Delta S_{\text{pro},D'} + \Delta S_{\text{pro},W} \tag{4.12}$$

$$I_{\text{pro}} = \sum_i I_i = \sum_i (T\Delta S_{\text{pro},i}), i = \overline{D}, D', W \tag{4.13}$$

## 4.2 Vortex Identification and Vorticity Transport

### 4.2.1 Vortex Identification

At present, there is no general mathematical method to accurately describe the vortex structures in fluid flow in different flow regimes. Therefore, for different flow situations, researchers have proposed a variety of vortex identification methods [5–14]. In the post-processing of numerical calculation of fluid flow in the pump, the more commonly used vortex identification methods are the $Q$-criterion and the regularized helicity method.

#### 4.2.1.1 $Q$-Criterion

To represent the vortex structure, the $Q$-criterion for vortex visualization is generally employed to identify the vortex in the impeller [15, 16]. The formula for $Q$-criterion is defined as the second invariant of velocity gradient tensor expressed as:

$$Q = \frac{1}{2}[(\text{tr}(\nabla \mathbf{u}))^2 - \text{tr}((\nabla \mathbf{u})^2)] \tag{4.14}$$

where $\text{tr}(\nabla \mathbf{u})$ represents the trace of velocity gradient of $\mathbf{u}$. It can be expressed as

$$\text{tr}(\nabla \mathbf{u}) = \frac{\partial u}{\partial x} + \frac{\partial v}{\partial y} + \frac{\partial w}{\partial z} = \nabla \cdot \mathbf{u} \tag{4.15}$$

For incompressible fluid, the above formula can be further simplified. Using the continuity equation,

$$\text{tr}(\nabla \mathbf{u}) = \nabla \cdot \mathbf{u} = \mathbf{0} \tag{4.16}$$

The formula for $Q$-criterion becomes

$$Q = -\frac{1}{2}\text{tr}((\nabla \mathbf{u})^2) \tag{4.17}$$

The velocity gradient can decomposed into two parts as

$$\nabla \mathbf{u} = \underbrace{\frac{1}{2}(\nabla \mathbf{u} + (\nabla \mathbf{u})^T)}_{\mathbf{S}} + \underbrace{\frac{1}{2}(\nabla \mathbf{u} - (\nabla \mathbf{u})^T)}_{\mathbf{\Omega}} \tag{4.18}$$

In Eq. (4.18), $\mathbf{S}$ is a symmetric tensor (also known as strain rate tensor), and $\mathbf{\Omega}$ is an antisymmetric tensor (also known as the spin tensor).

Therefore, the $Q$-criterion formula can be expressed as:

$$Q = -\frac{1}{2}\text{tr}((\nabla \mathbf{u})^2) = -\frac{1}{2}\text{tr}(\mathbf{S}^2 + 2\mathbf{S\Omega} + \mathbf{\Omega}^2) = -\frac{1}{2}[\text{tr}(\mathbf{S}^2) + \text{tr}(\mathbf{S\Omega}) + \text{tr}(\mathbf{\Omega}^2)] \quad (4.19)$$

Using the symmetric and antisymmetric properties of symmetric and antisymmetric tensors,

$$\text{tr}(\mathbf{S}^2) = \text{tr}(\mathbf{S}^T\mathbf{S}), \text{tr}(\Omega^2) = -\text{tr}(\Omega^T\Omega) \quad (4.20)$$

and

$$\text{tr}(\mathbf{S\Omega}) = \text{tr}\left[\frac{1}{2}(\nabla\mathbf{u} + (\nabla\mathbf{u})^T)\frac{1}{2}(\nabla\mathbf{u}\text{-}(\nabla\mathbf{u})^T)\right] = \text{tr}\left[\frac{1}{4}((\nabla\mathbf{u})^2 - ((\nabla\mathbf{u})^T)^2)\right] = 0 \quad (4.21)$$

Therefore, the formula for $Q$-criterion can be written as:

$$Q = -\frac{1}{2}[\text{tr}(\mathbf{S}^T\mathbf{S}) - \text{tr}(\mathbf{\Omega}^T\mathbf{\Omega})] = -\frac{1}{2}[||\mathbf{\Omega}||^2 - ||\mathbf{S}||^2] \quad (4.22)$$

where $||\mathbf{\Omega}||$ and $||\mathbf{S}||$ can be expressed as Frobenius norm of $\mathbf{S}$ and $\mathbf{\Omega}$, respectively, which are defined as

$$||\mathbf{S}||_F = \sqrt{\text{tr}(\mathbf{S}^T\mathbf{S})}, ||\mathbf{\Omega}||_F = \sqrt{\text{tr}(\mathbf{\Omega}^T\mathbf{\Omega})} \quad (4.23)$$

When $Q > 0$, it indicates that there is vortex structure in this region of the flow.

#### 4.2.1.2 Regularized Helicity Method

The application of the regularized helicity method does not require strict control of the numerical calculation errors; the algorithm is very robust. It is widely used in vortex core identification in the field of vane pumps. It uses the regularized helicity $H_n$ to extract the vortex core:

$$H_n = \frac{\mathbf{v} \cdot \boldsymbol{\omega}}{|\mathbf{v}||\boldsymbol{\omega}|} \quad (4.24)$$

It can be seen from Eq. (4.24) that the normalized helicity $H_n$ is the cosine of the included angle between the velocity vector $\boldsymbol{v}$ and the vorticity vector $\boldsymbol{\omega}$. In the flow field region, $H_n$ is a scalar field defined everywhere except at the special point where the velocity vector $\boldsymbol{v}$ and vorticity vector $\boldsymbol{\omega}$ are zero, and its value is in the interval $[-1, 1]$. In the vortex core region, the direction of velocity vector and vorticity vector are nearly parallel, and the regularized helicity $H_n$ tends to $\pm 1$. The sign of the regularized helicity $H_n$ indicates the direction of the vortex rotation with the flow direction being the positive direction. If $H_n$ is positive, the vortex rotation direction is counterclockwise; if $H_n$ is negative, the vortex rotation direction is clockwise. The helicity method can accurately identify the flow of primary and secondary vortices and describe the vortex rupture phenomenon.

### 4.2.2 Vorticity Transport

In order to understand the growth and spatial distribution of the vortex structure in the stalled region of the flow field with changes in blade numbers of the impeller, the vortex transport equation can be used to analyze the stalled flow field in the impeller. The vorticity transport equation can be expressed as [17–19]

$$\frac{D\vec{\Omega}}{Dt} = (\vec{\Omega} \cdot \nabla)\vec{W} - \vec{\Omega}(\nabla \cdot \vec{W}) - 2\nabla \times (\omega \times \vec{W}) + \frac{\nabla \rho_m \times \nabla \rho}{\rho_m^2} + \nu\nabla^2\vec{\Omega} \quad (4.25)$$

where $\vec{\Omega}$ is the vortex vector, $\vec{W}$ the relative velocity vector, $\omega$ the rotational angular velocity, $\rho_m$ the fluid density, $\rho$ the air density, and $\nu$ the kinematic viscosity of the fluid.

The term on the left side of the equal sign in Eq. (4.25) represents the rate of change in vorticity caused by the fluid motion. There are five terms on the right side of the equal sign in Eq. (4.25) which respectively represent the vortex stretching (VS) caused by the velocity gradient, vorticity change caused by the fluid compressibility, vortex dilatation (VD) caused by Coriolis force under rotary motion, baroclinic torque (BT) caused by the pressure and density changes, and viscosity change rate (VCR) caused by the viscous diffusion of vorticity. Since generally the working medium used in the pump is clear water, the fluid medium can be approximately regarded as an incompressible fluid. Therefore, the second and fourth terms in Eq. (4.25) can be neglected.

## 4.3 Transient Flow Analysis Using the Wavelet Method

The pressure pulsation signals from a mixed-flow pump contain a variety of frequency signals caused by various flow fluctuations. Therefore, the pressure pulsation signal needs to be processed by an appropriate mathematical technique, since the traditional Fourier transform cannot determine the specific time of each frequency component of the pressure pulsation signal. Figure 4.1 shows the frequency spectrum of a stationary signal and a non-stationary signal after the Fourier transform. It can be seen from this figure that the frequency results of the two signals using the Fourier transform are basically similar, and the high-frequency components appear at the same frequency. Therefore, in order to extract the multi-time scale characteristics of pressure pulsation signals, wavelet transform is used to process the signals.

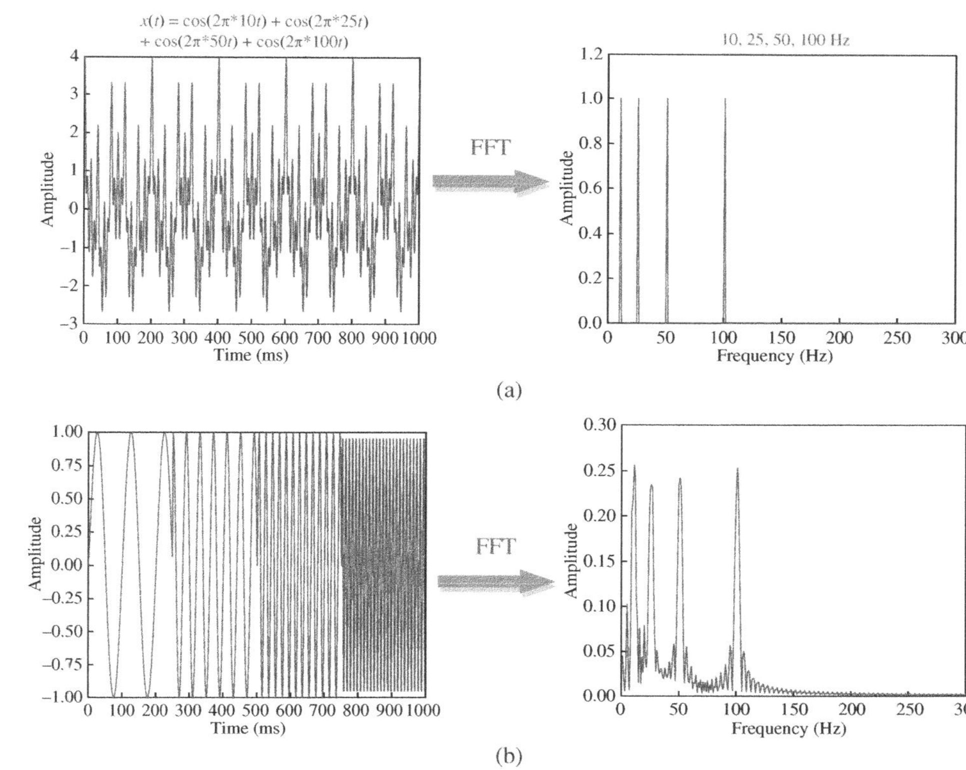


**Figure 4.1** Fourier analysis results of the two signals. (a) Stationary signal. (b) Non-stationary signal.

The wavelet transform was discovered and named by Haar in the early part of the 20th century. It was developed from the Window Fourier transform. After many years of development, it is now widely used in the fields of meteorology, early earthquake detection systems, electric power, military industry, image processing, etc. After many improvements by mathematicians and industrial application engineers, several types of wavelet transforms have been developed, namely the continuous wavelet transform (CWT), discrete wavelet transform (DWT), and wavelet reconstruction, which are widely used in variety of industrial applications.

Different from Fourier transform analysis, which only provides frequency resolution, wavelet transform also provides time resolution. A wavelet transform is the representation of a function by wavelets. The wavelets are scaled and translated copies (known as "daughter wavelets") of a finite-length or fast-decaying oscillating waveform (known as the "mother wavelet"). Thus, the time and frequency characteristics of the signal are obtained by a series of translations and scaling transformations of the mother wavelet. The basic idea behind wavelet analysis is to use a series of function systems to represent or approximate a signal function. Therefore, the wavelet function is the key to wavelet analysis. It is a function defined in a limited interval, and the average value of the wavelet amplitude is zero. It also has the characteristics of oscillation and rapid attenuation. A mother wavelet function can be defined as $\psi(t) \in L^2(\mathbf{R})$, where $L^2(\mathbf{R})$ is the signal space with limited energy and satisfies the following admissibility condition:

$$C_\psi = \int_{-\infty}^{+\infty} |\omega|^{-1} |\hat{\psi}(\omega)|^2 d\omega < +\infty \tag{4.26}$$

such that $C_\psi$ is bounded and $\hat{\psi}$ is the Fourier transform of $\psi$. If $\hat{\psi}$ is continuous, it is easy to deduce that,

$$\int_{-\infty}^{+\infty} \psi(t)dt = 0 \tag{4.27}$$

Therefore, $\psi(t)$ can be employed to obtain a cluster of wavelet functions through translation, stretching, and transformation as:

$$\psi_{a,b}(t) = \frac{1}{\sqrt{|a|}} \psi\left(\frac{t-b}{a}\right) \text{ with } a, b \in \mathbf{R}, a \neq 0 \tag{4.28}$$

where $\psi_{a,b}(t)$ is a sub-wavelet, $a$ is the scale factor representing wavelet period, $b$ is a translation factor, and $t$ is the translation response time.

In practical engineering applications, different mother wavelet functions need to be used for different situations; it is often the signal of the same time series. It is important to note that choosing different mother wavelet functions will lead to different results with large differences. At present, a mother wavelet function is mainly judged by comparing the errors between the results obtained by different wavelet analysis and signal processing methods and the theoretical results, and based on these comparisons, a mother wavelet function required for a particular application analysis is selected.

In the mixed-flow pump applications described in this book, CWT is used for signal processing since it collects high-frequency continuous signals. The basic principles and formulas of the CWT are introduced below, but the DWT and wavelet reconstruction are not described in detail. CWT was first proposed by Grossman and Morlet in 1984. It adopts a process similar to the Fourier transform, summing the product of the processed signal with the stretched and shifted wavelet generating function during the duration of the signal as follows:

$$W_f(a, b) = \int_{-\infty}^{+\infty} f(t)\overline{\psi_{a,b}(t)}dt \tag{4.29}$$

where $f(t)$ is the signal function or square-integrable function, $W_f(a, b)$ is the wavelet transform coefficient, $\overline{\psi_{a,b}(t)}$ is the conjugate function of the wavelet function $\psi(t)$ (also expressed as $\psi^*(t)$). From Eq. (4.29), one can obtain,

$$W_f(a,b) = \frac{1}{\sqrt{|a|}} \int_{-\infty}^{+\infty} f(t)\overline{\psi\left(\frac{t-b}{a}\right)}dt \tag{4.30}$$

In engineering applications, for continuous signal processing, the wavelet is expected to decay rapidly in the time domain, that is the wavelet has a small support set. If the wavelet has a vanishing moment of $K$-order, it is orthogonal to the polynomial up to degree $K-1$. Therefore, according to the above formula $\int_{-\infty}^{+\infty} \psi(t)dt = 0$ and

$$\int_{-\infty}^{+\infty} a^K \psi(t)dt = 0 \tag{4.31}$$

Therefore, if and only if there is a rapidly decaying function $\theta$ such that,

$$\psi(t) = (-1)^K \frac{\mathrm{d}^K \theta(t)}{\mathrm{d}t^K} \tag{4.32}$$

It can be shown that,

$$W_f(a,b) = a^K \frac{\mathrm{d}^K (f * \overline{\theta}_a)(t)}{\mathrm{d}t^K} \tag{4.33}$$

where $\overline{\theta}_a(t) = a^{-1/2}\theta(-t/a)$ and $\psi(t)$ has no more than the vanishing moment of $K$-order if and only if it satisfies the following condition:

$$\int_{-\infty}^{+\infty} \theta(t)dt \neq 0 \tag{4.34}$$

The above formula indicates that the essence of a wavelet transform is to calculate the $K$-order derivative after the smoothing signal; thus, it is equivalent to a multi-scale differential operator, namely

$$W_f(a,b) = \frac{1}{\sqrt{a}} \int_{-\infty}^{+\infty} f(t)\psi\left(\frac{t-b}{a}\right) dt = f * \overline{\psi}_a(t) \tag{4.35}$$

where $\overline{\psi}_a(t)$ can be solved by the following formula:

$$\overline{\psi}_a(t) = \psi(-t/a)/\sqrt{a} \tag{4.36}$$

In this manner, the formula for the wavelet after translation and stretching transformation can be reorganized to obtain:

$$W_f(a,b) = \frac{\mathrm{d}^P}{\mathrm{d}t^P}(f * a^P\theta)(t) = a^P\left(\frac{\mathrm{d}^P f}{\mathrm{d}t^P} * \theta\right)(t) \tag{4.37}$$

In Eq. (4.37), the superscript $P$ represents the derivative of the $P$-th order.

In this book mainly some disturbance factors occurring in the process of the rotating stall of the mixed-flow pump are studied. Tin this context, the proper selection of the wavelet function becomes very important in analyzing the results of the signal function. In the research on rotating stall signal diagnosis described in this book in Chapter 8, MORLET wavelet function and DOG wavelet function are primarily used. Their characteristics are given in Table 4.1.

**Table 4.1** Two types of generating functions and their properties.

| Name | $\psi(t)$ | $\hat{\psi}(a\omega)$ | Fourier wavelength $\lambda$ |
|---|---|---|---|
| MORLET | $\pi^{-1/4}e^{i\omega_0 t}e^{-t^2/2}$ | $\pi^{-1/4}H(\omega)e^{-(a\omega-\omega_0)^2/2}$ | $\dfrac{4\pi a}{\omega_0+\sqrt{2+\omega_0^2}}$ |
| DOG $m$th derivative | $\dfrac{(-1)^{m+1}}{\sqrt{\Gamma(m+1/2)}}\dfrac{d^m}{dt^m}(e^{-t^2/2})$ | $\dfrac{(-1)^{m+1}}{\sqrt{\Gamma(m+1/2)}}(a\omega)^m e^{-(a\omega)^2/2}$ | $\dfrac{2\pi a}{\sqrt{m+1/2}}$ |

Note: in the formula, $\omega_0$ is the dimensionless frequency, $m$ is the order, $H(\omega)$ is the Heaviside step function, if $\omega > 0$, otherwise $H(\omega) = 1$.

## 4.4 Summary

This chapter provides an overview of the research methods employed to elucidate the internal flow characteristics of the mixed-flow pumps. The methodologies include the entropy production analysis, vortex identification and its transport, and transient flow analysis based on the wavelet transforms. The entropy production method is utilized to quantify the internal flow losses and the energy losses induced by the wheel rim leakage flow. Vortex structures are identified using the $Q$-criterion and the regularized vorticity method. The vortex transport equation is applied to describe the evolution and spatial distribution of the vortex structures. The wavelet transform method, known for its excellent signal decomposition capability, is employed to analyze the pressure fluctuation signals with multiple time scale characteristics.

## References


1. Li, W., Agarwal, R.K., Zhou, L., et al. (2019). Effect of tip clearance on rotating stall in a mixed-flow pump. *Proceedings of ASME Turbo Expo 2019* (17–21 June 2019). Arizona, USA: Phoenix.
2. Bejan, A. (1994). *Entropy Production Through Heat and Fluid Flow*. New York: *John Wiley & Sons*.
3. Kock, F. and Herwig, H. (2004). Local entropy production in turbulent shear flows: a high-Reynolds number model with wall functions. *International Journal of Heat and Mass Transfer* 47 (10-11): 2205–2215.
4. Mathieu, J. and Scott, J. (2000). An introduction to turbulent flow. *Physics Today* 54 (9): 56–57.
5. Ming, J., Raghu, M., and David, T. (2005). Detection and virtualization of vortices. In: *The Visualization Handbook* (ed. C.D. Hansen and C.R. Johnson), 295–309. Amsterdam: Elsevier.
6. Levy, Y., Degani, D., and Seginer, A. (1990). Graphical visualization of vortical flows by means of helicity. *AIAA Journal* 28 (8): 1347–1352.
7. Berdahl, C.H. and Thompson, D.S. (1993). Education of swirling structure using the velocity gradient tensor. *AIAA Journal* 31 (1): 97–103.
8. Jeong, J. and Hussain, F. (1995). On the identification of a vortex. *Journal of Fluid Mechanics* (285): 69–94.
9. Banks, D.C. and Singer, B.A. (1995). A predictor-corrector technique for visualizing unsteady flow. *IEEE Transactions on Visualization and Computer Graphics* 1 (2): 151–163.
10. Sujudi, D. and Haimes, R. (1995). Identification of swirling flow in 3D vector fields. *AIAA 12th Computational Fluid Dynamics Conference, 1995*. 95-1715.

**11** Roth, M. and Peikert, R. (1998). A higher-order method for finding vortex core lines. *IEEE Visualization* 98: 143–150.

**12** Strawn, R.C., Kenwright, D.N., and Ahmad, J. (1999). Computer visualization of vortex wake systems. *AIAA Journal* 37 (4): 511–512. https://doi.org/10.2514/2.744.

**13** Sadarjoen, I.A., Post, F.H., Ma, B. et al. (2002). Selective visualization of vortices in hydrodynamic flows. *IEEE Visualization* 98: 419–422.

**14** Jiang, M., Machiraju, R. and Thompson, D.S. (2002). A novel approach to vortex core region detection. *Joint Euro graphics-IEEE TCVG Symposium on Visualization*. 217-225.

**15** Hunt, J.C., Wray, R.A., and Moin, P. (1988). *Eddies, Stream, and Convergence Zones in Turbulent Flows*. Center for Turbulence Research Report CTR-S88.

**16** Liu, M., Tan, L., and Cao, S.L. (2019). Cavitation–vortex–turbulence interaction and one-dimensional model prediction of pressure for hydrofoil ALE15 by large Eddy simulation. *Journal of Fluid Engineering* 141 (2).

**17** Ji, B. and Luo, X.W. (2013). Numerical analysis of unsteady cavitating turbulent flow and shedding horse-shoe vortex structure around a twisted hydrofoil. *International Journal of Multiphase Flow* 51 (5): 33–43.

**18** Huang, R., Ji, B., and Luo, X. (2015). Numerical investigation of cavitation-vortex interaction in a mixed-flow waterjet pump. *Journal of Mechanical Science and Technology* 29 (9): 3707–3716.

**19** Ji, B., Luo, X.W., and Arndt, R.E.A. (2014). Numerical simulation of three-dimensional cavitation shedding dynamics with special emphasis on cavitation-vortex interaction. *Ocean Engineering* 87: 64–77.

# 5

# Experimental Methods, Data, and Analysis

## 5.1 External Characteristics Experiment

### 5.1.1 Test Equipment and Instruments

The experimental measurements reported in this chapter were conducted on the closed mixed-flow pump test platform in the Research Center of Fluid Machinery Engineering and Technology in Jiangsu University. The closed test platform is made of the Φ250 mm stainless steel pipe. The experiments were conducted following the GB/T 18149-2000 mixed-flow pump and axial flow pump hydraulic performance test specification [1]. The test platform meets the accuracy requirement of the first class. The experimental setup and equipment used in the experiment are shown in Fig. 5.1.

(1) The model test pump
The physical diagram of the impeller, the guide vane, and the complete pump is shown in Fig. 5.2. The specific speed of the pump in the experimental model is $n_s = 480$. The blade tip clearance can be adjusted by replacing the mixed-flow pump impeller with blades of different heights. Here, impellers with four different blade tip clearances of $t = 0.2$ mm, $t = 0.5$ mm, $t = 0.8$ mm, and $t = 1.1$ mm are considered. For the particle image velocimetry (PIV) measurement, diffuse reflection is created on the surface of the impeller and guide vane, which affects the accuracy of shooting the laser beam. In order to avoid the influence of diffuse reflection as much as possible and enable the light source to enter the internal flow field smoothly, black paint is sprayed on the surface of the impeller and the guide vane to absorb the laser light. The shell used in the test is made of fully transparent Plexiglas, whose refractive index is close to that of the test medium, clean water. The shape of the shell is square outside and round inside in order to reduce the influence of light refraction from the outer surface and the inner wall of the shell.

(2) Test instruments
In the test, the LWGY-250 turbine flowmeter (Fig. 5.3) is used for flow measurement with a precision of 0.5 and a nominal pressure of 1.6 Mpa. The current signal is amplified by an amplifier (Fig. 5.4). The WT-1151 capacitive pressure transmitter (Fig. 5.5) installed at the inlet and outlet of the pump is used for head measurement. The measurement range of the inlet pressure transmitter is ±100 kPa, and the accuracy is 0.2. The measurement range of the outlet pressure transmitter is 0–600 kPa, and the accuracy is 0.2. The speed, torque, and shaft power are measured by a torque speed measuring instrument (Fig. 5.6) with a measurement error of ±0.2%. The whole system meets the accuracy requirements of level 1. TPAM-V (torque type) pump product parameter measuring instrument (Fig. 5.7) is used to receive the signal of the sensor, then transmit the data to the computer, and analyze and process

*Mixed-flow Pumps: Modeling, Simulation, and Measurements*, First Edition.
Wei Li, Leilei Ji, Ramesh Agarwal, Weidong Shi, and Ling Zhou.

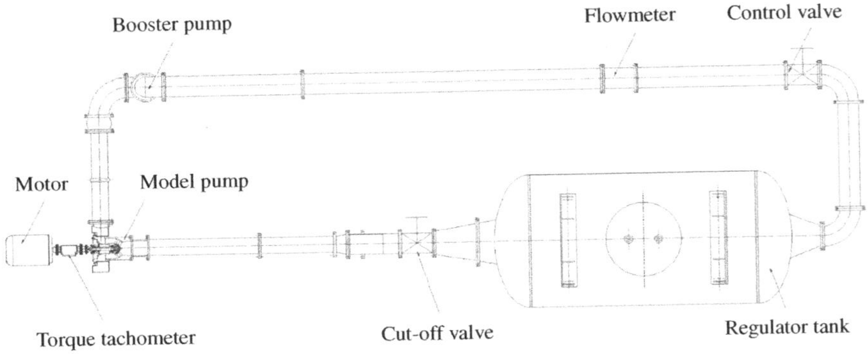


**Figure 5.1** Experimental test system.

**Figure 5.2** (a) Mixed-flow pump impeller, (b) Mixed-flow pump guide vane, and (c) The whole mixed-flow pump. Source: (a) Wei et al. (2019)/Sage Publications, Inc.

**Figure 5.3** Turbine flowmeter.

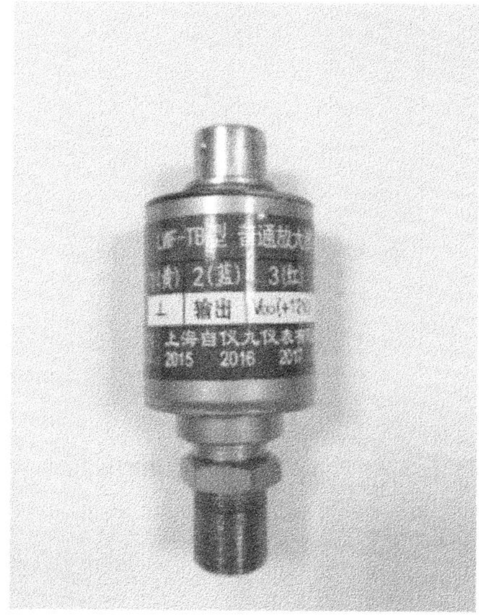


**Figure 5.4** Flowmeter.

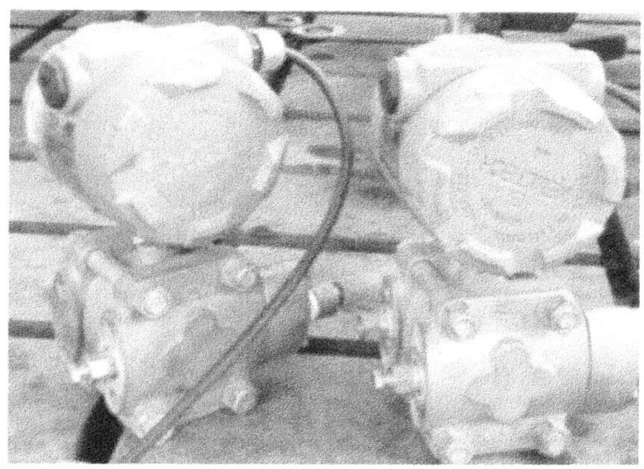

**Figure 5.5** Pressure transmitter.

**Figure 5.6** Torque measuring instrument.

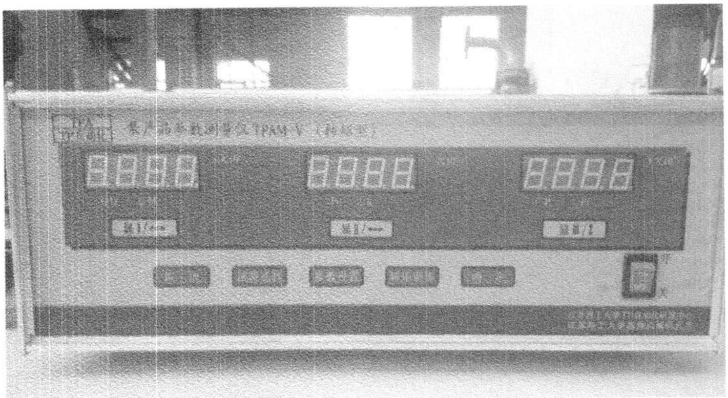

**Figure 5.7** Pump measuring instrument.

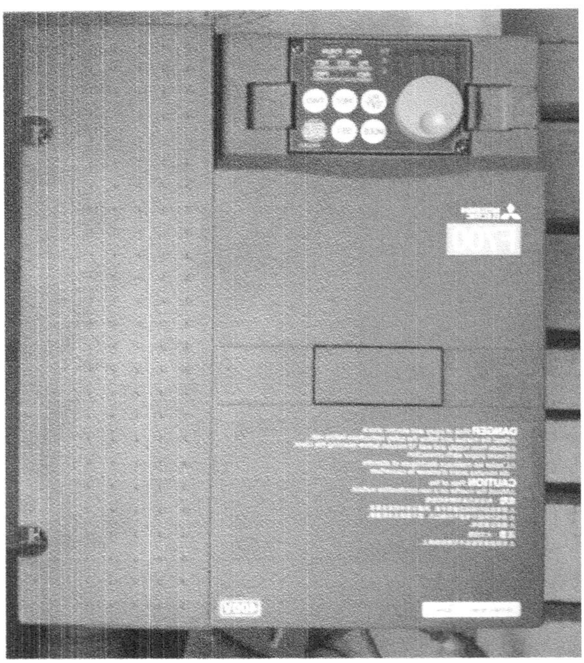

**Figure 5.8** Frequency converter.

the data through the supporting pump product analysis and processing software installed on the computer. In addition, the input voltage of the variable frequency motor selected for the power driving device is 380 V, the speed is 1500 r/min, the power is 15 kW and the frequency is 50 Hz. The speed of the inverter motor is adjusted by the Mitsubishi F700 frequency converter (Fig. 5.8).

### 5.1.2 Test Method for External Flow Characteristics of the Pump

Since the mixed-flow pumps have a certain stall range, therefore in the external characteristics test, from $1.2Q_{des}$ to $0.2Q_{des}$ flow rates where $Q_{des}$ is the design flow rate, test measurements are conducted in multiples of $0.2Q_{des}$ when approaching the stall interval, and after stall at every additional $0.1Q_{des}$. The test steps are as follows:

(1) After the test pipeline is filled with water, various instruments are installed as required and checked to see whether they can work properly.
(2) After the motor is installed, its rated speed is adjusted to 1450 r/min through the frequency converter with zero torque, and then it is connected to the main shaft of the pump.
(3) The inlet and outlet valves are adjusted to the maximum opening, and the pump speed is adjusted through the frequency converter until it reaches and stabilizes at 1450 r/min.
(4) The outlet valve is slowly adjusted, and the flow rate is adjusted from 1.2 times the rated flow rate to 0.2 times the rated flow rate and is adjusted every 0.2 times the rated flow rate. When the indicators on the pump's parameter measuring instruments and the system analysis software are stable, the test data is recorded.
(5) The external characteristics test is repeated more than three times to reduce the test error as much as possible, which may be caused by a variety of reasons. Finally, the arithmetic average of the recorded test data is taken as the final result from measurements.

The pump product analysis and processing software matched with the pump product parameter measuring instrument can directly calculate the pump head using the measured inlet and outlet pressure employing Eq. (5.1). Since the inlet and outlet installation heights of the test pump are the same, therefore $(Z_2 - Z_1) = 0$. In addition, the average flow velocities $v_1$ and $v_2$ of the fluid at the inlet and outlet of the pump can be obtained from the flow rate using the known inner diameter of the inlet and outlet pipeline, respectively. From the Bernoulli equation, the pump head $H$ is given by:

$$H = \left(\frac{p_2 - p_1}{\rho g}\right) + (Z_2 - Z_1) + \left(\frac{v_2^2 - v_1^2}{2g}\right) \tag{5.1}$$

where $\rho$ is the density of water, $\rho = 1000\ \text{kg/m}^3$, $g$ is the gravitational acceleration, $g = 9.8\ \text{m/s}^2$, $p_1$, $p_2$ are the pump inlet and outlet pressure respectively in Pa, $Z_1$, $Z_2$ are the installation height of pump at inlet and outlet respectively in m, and $v_1$, $v_2$ are the average velocity at pump inlet and outlet respectively in m/s.

Using the measured data of speed and torque, the pump product analysis software can directly calculate the shaft power $P$ from Eq. (5.2):

$$P = \frac{n \cdot (M - M_0)}{9552} \tag{5.2}$$

where $M$ is the measured axial torque in N m, $M_0$ is the no-load torque in N m and $n$ is the speed in revolutions r/min.

Using the obtained head (Eq. (5.1)) and shaft power (Eq. (5.2)), the efficiency of the pump can be calculated by Eq. (5.3):

$$\eta = \frac{\rho g(Q/3600)H}{1000P} \times 100\% \tag{5.3}$$

where $Q$ is the flow rate of the mixed-flow pump in $\text{m}^3/\text{h}$, $H$ is the head of the mixed-flow pump in m, and $P$ is the shaft power of the mixed-flow pump in kW.

### 5.1.3 Uncertainty Analysis of the Experiment

The test rig facility and test equipment used in the experiment meet the requirements and approval of the technology department of Jiangsu province, China. The test rig abides by the

**Table 5.1** Computational process of the experimental uncertainties.

| Notation/Note | | Q (m³/h) | H (m) | Ps | |
|---|---|---|---|---|---|
| | | | | M (N m) | n (r/min) |
| Test Number | 1 | 381.5 | 5.522 | 56.05 | 1440 |
| | 2 | 380.2 | 5.516 | 56.14 | 1439 |
| | 3 | 379.6 | 5.492 | 56.02 | 1439 |
| Random uncertainty | Average value | 380.433 | 5.510 | 56.07 | 1439.667 |
| | Standard deviation (%) | 0.9713 | 0.0159 | 0.0624 | 0.5774 |
| | $U_{ran}$ (%) | 0.0063 | 0.0072 | 0.0028 | 0.0010 |
| System uncertainty | Measurement accuracy (%) | 0.500 | 0.500 | 0.200 | 0.200 |
| | $U_{sys}$ (%) | 0.2887 | 0.2887 | 0.1155 | 0.1155 |
| Overall uncertainty | Components $U$ (%) | 0.2888 | 0.2888 | 0.1155 | 0.1155 |
| | Overall uncertainty $U$ (%) | 0.4399 | | | |
| Extended overall uncertainty $U_{ext}$ (%) | | 0.8798 | | | |

measurement requirements described in the International Standard ISO9906 and corresponds to Chinese national standards. According to ISO9906 hydraulic performance acceptance tests, the uncertainty in the pump efficiency can be calculated by the square root of the quadratic sum of the uncertainties in the measurement of each key variable of the pump, as given in Eq. (5.4):

$$U = \sqrt{U(Q)^2 + U(H)^2 + U(n) + U(M)^2} \tag{5.4}$$

where $U(Q)$, $U(H)$, $U(n)$, and $U(M)$ are the uncertainty in flow rate, head, rotating speed, and torque, respectively.

The measurement uncertainty of each component is calculated by the square root of the sum of the squares of the systematic and random uncertainties given by:

$$U = \sqrt{{U_{sys}}^2 + {U_{ran}}^2} \tag{5.5}$$

The Bessel method has been adopted to calculate the random uncertainty $U_{ran}$ and the method of general coefficient has been applied to compute the systematic uncertainty $U_{sys}$. Table 5.1 shows the detailed calculation of uncertainties at the design flow rate of the pump. The overall uncertainty in the experiment is calculated as 0.4399%. In order to include other not easily quantifiable uncertainties in the measurement, the overall uncertainty is increased by a factor of two and is reported here as 0.8798%. In conclusion, the experimental data described in this chapter meets the requirement of accuracy and can be trusted by the reader.

### 5.1.4 Test Results for External Flow Characteristics of the Pump

Figure 5.9 shows the test results of the energy performance of a mixed-flow pump under different tip clearances, which are compared with those obtained from CFD simulations. The simulations shown are based on the numerical solution of the Reynolds-Averaged Navier–Stokes (RANS) equations with SST $k$–$\omega$ turbulence model. It can be seen from Fig. 5.9 that under different tip clearances, the experimental test head of the mixed-flow pump gradually increases with a decrease

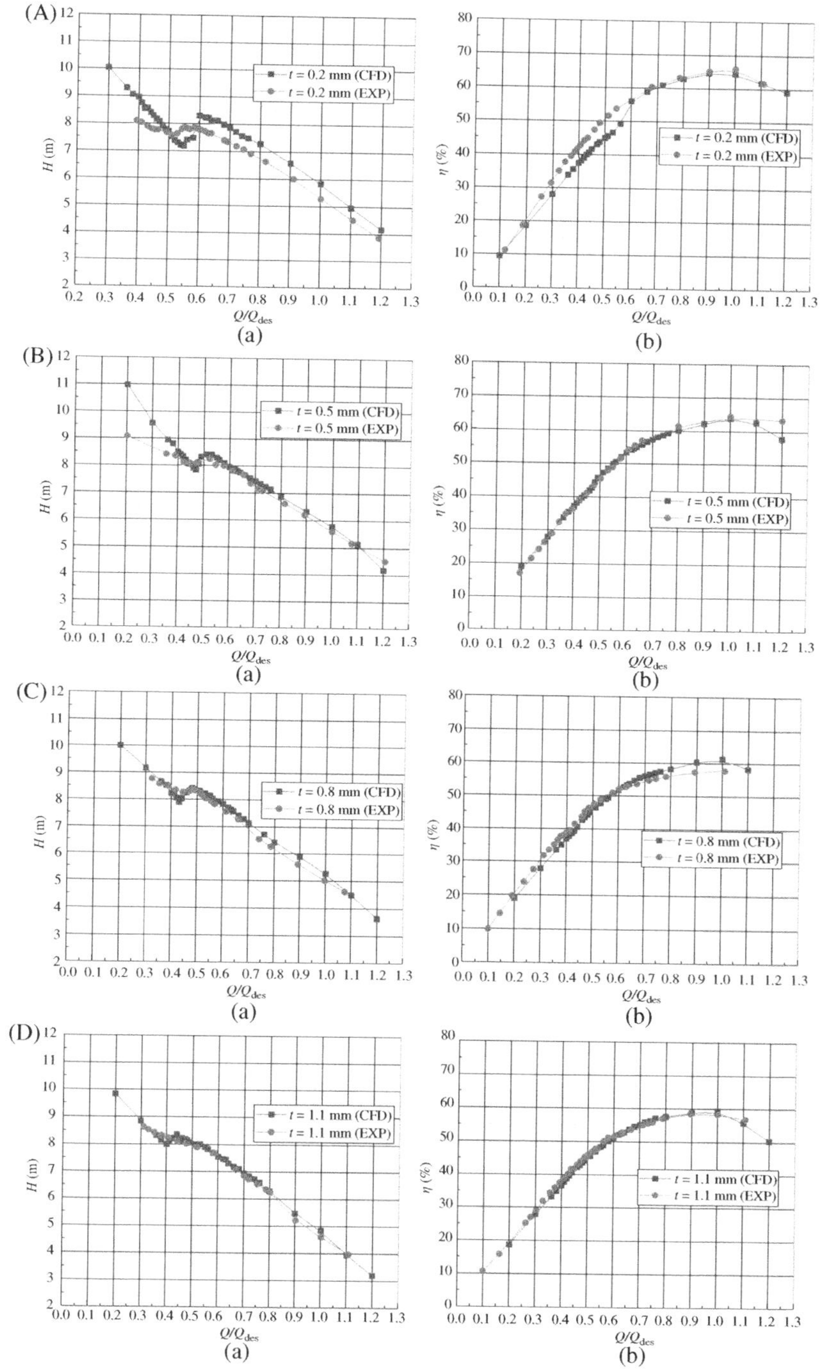


**Figure 5.9** Experimental energy performance of the mixed-flow pump with different tip clearances. (A) $t = 0.2$ mm, (B) $t = 0.5$ mm, (C) $t = 0.8$ mm, and (D) $t = 1.1$ mm.

in the flow rate, and when the flow rate is in the range 0.4–0.6$Q_{des}$, the "positive slope" interval appears in the head curve (called the "saddle area" of the mixed-flow pump) indicating that the rotational stall has occurred in the inner part of the mixed-flow pump. With increase in the tip clearance, the range of flow rate for positive slope of the head gradually decreases, indicating that the stall characteristics of the mixed-flow pump have gradually weakened. In the efficiency curve of the mixed-flow pump, there is the highest value at the design flow rate for each tip clearance. With decrease in flow rate, the efficiency of the mixed-flow pump decreases gradually. Compared to the CFD simulation results, the head and efficiency curves of the mixed-flow pump for each tip clearance are in good agreement with the experimental results indicating the reliability of the CFD simulations.

## 5.2 Experiment for Measuring Pressure Fluctuations

In the test and measurement of pressure pulsations in the mixed-flow pump, the MPM480 piezoresistive pressure sensor produced by the Mike Sensor Co., Ltd. is used as shown in Fig. 5.10. The piezoresistive OEM pressure sensor with 316L stainless steel isolation film is the main signal measuring element whose width range and temperature performance are compensated by the laser resistance regulation process. The sensor's protection grade is IP65, the power supply voltage is 15–28 V DC, the output signal is 4–20 mA, and the accuracy grade is ±0.5% FS (maximum value).

The high-frequency pressure sensors are installed on the impeller end wall at different positions to measure the local pressure pulsation. The positions of the sensors are shown in Fig. 5.11. The pressure pulsation of the mixed-flow pump with different tip clearances at design flow conditions is collected. In order to ensure the implementation of different tip clearances, four impellers with different tip clearances are manufactured. During the implementation of impeller tip clearance, the feeler gauge is utilized to measure the size of different positions along the blade chord to keep the tip clearance uniform. Three positions are selected for measurement during this test including the impeller inlet (A), middle part (B), and outlet (C). Four pressure sensors are utilized

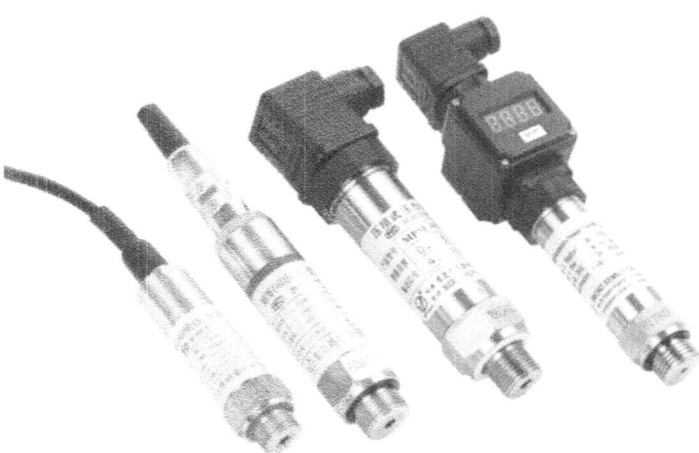

**Figure 5.10** MPM480 piezoresistive pressure sensor. Source: Micro Sensor/https://www.microsensor.cn/upload/article/191026/1_MPM480_E4_B8_AD_E6_96_87_E8_AF_B4_E[...]_8E_E4_B9_A6_EF_BC_882019_2e10_2e21_EF_BC_89_1572056932.pdf/Last accessed under 08 December 2023.

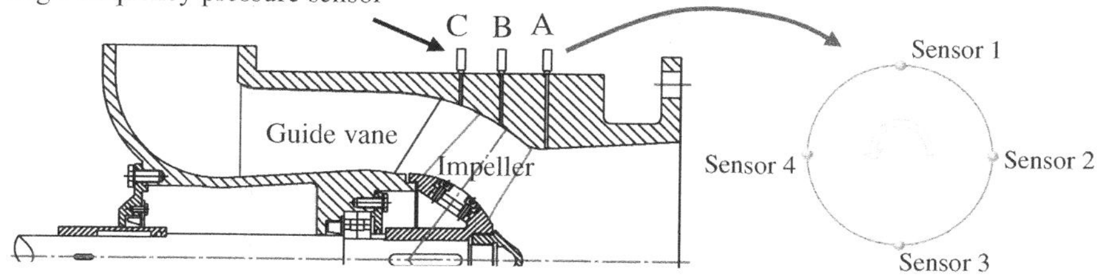


**Figure 5.11** Sensor installation positions.

and are equally spaced 90° apart in the circumferential direction at each position. Since all the sensors are installed on the pipe, the pressure signal obtained is the pressure pulsation near the wall. The output signal of the pressure sensor ranges from 4 to 20 mA, and the electric signal is converted to pressure signal by the hydraulic machinery tester and computer.

### 5.2.1 Experimental Procedure

The external flow characteristics test and the pressure pulsation experiment are conducted simultaneously. The following experimental procedures are applied to the mixed-flow pumps with different sizes of tip clearances. During the first procedure, all the experimental facilities are tested, and the supporting software of the hydraulic machinery tester is debugged to record the pressure data successfully. The opening of the inlet valve and outlet valve are adjusted to the maximum in the beginning stage of the experiment. The converter power source is turned on, and the rotating speed of the pump is kept at 1450 r/min. The flow rate conditions are changed by adjusting the opening of the outlet valve. All the parameters related to pump energy characteristics are recorded during the experiment. At the same time, the data from the pressure pulsation experiment is also recorded. After recording all the experimental data, the power source is turned off. The above-mentioned procedure is repeated three times after the fluid in the test system becomes stable. The experimental site (a) and the installation location of the pressure sensor (b) are shown in Fig. 5.12.

In this section, the effect of tip clearance size ($t = 0.2$ mm, $t = 0.5$ mm, $t = 0.8$ mm, and $t = 1.1$ mm) on the pressure pulsation of the mixed-flow pump is presented. The pressure pulsation distribution

**Figure 5.12** Arrangement of external characteristic and pressure pulsation measurement. Source: Leilei et al. (2020)/Sage Publications, Inc.

at the impeller inlet, impeller middle, and impeller outlet are acquired at the design flow rate. Since the transient relative pressure values at each monitoring point acquired by the high-frequency pressure sensors are different depending on the local position, the pressure pulsation coefficient is employed to normalize the relative pressure value. Since the pressure pulsation coefficient is dimensionless and therefore it eliminates the effect of flow rate conditions on pressure pulsation. Thus, the pressure pulsations at various positions could be compared with each other at the same level. The pressure pulsation coefficient $C_\mathrm{p}$ can be expressed as:

$$C_\mathrm{p} = \frac{\frac{1}{N}\sum_{i=1}^{N}[P_i(t) - \overline{P}_i]}{0.5\rho u_2^2} \tag{5.6}$$

where $P$ is the relative static pressure measured by the pressure sensor in Pa, $u_2$ is the circumferential velocity at the impeller outlet in m/s, $\rho$ is the density of water at room temperature (=1000 kg/m$^3$), and $N$ is the number of pressure pulsation signals.

Figure 5.13 shows the original time-domain characteristics of pressure pulsations of the pump at the inlet, middle section, and outlet of the impeller at design flow conditions with four tip clearances. For convenience in analysis, a short segment of pressure signals is selected. It can be seen from this figure that the time-domain response of pressure pulsation at each monitoring point is rather large. At the impeller inlet, the average amplitude of pressure pulsation in the time-domain curve is smaller than that at other positions, and its periodicity is also weaker. In the middle part of the impeller, the periodicity of the pressure coefficient pulsation curve is obvious; it has four peaks and troughs within one impeller rotation cycle (time period $T = 0.0414$ seconds), which coincides with the blade passing frequency (BPF) ($f \approx 96.68$ Hz). Furthermore, the time-domain amplitude of the pressure pulsation curve also increases significantly in the middle of the impeller. At the impeller outlet, the average pressure pulsation amplitude reduces compared to that in the middle portion of the impeller, but the influence of BPF is still obvious. In particular, the amplitude of the pressure pulsation curve is still obvious, and the frequency of peaks and troughs is still large. In addition, the rotor–stator interaction (RSI) of the impeller and guide vane is obvious at the outlet of the impeller. Four peak-to-peak differences in the pressure curve within a period are large, indicating that the unsteady effect of the RSI zone is strong.

Considering the effect of different tip clearances, the characteristics of the pressure pulsation curve for each tip clearance are quite different at the impeller inlet. When the tip clearance is smallest, the fluctuation intensity of time-domain curves varies greatly with time, and the maximum peak-to-peak difference of pressure pulsation curves in adjacent periods is about 0.1. With the increase in tip clearance, the peak-to-peak value of the pressure coefficient curve decreases first and then increases. When the tip clearance is 0.5 mm, not only the fluctuation amplitude of the pressure curve is low, but also the amplitude difference between the peak and trough values is not large. When the tip clearance is increased to 1.1 mm, the pressure coefficient curve begins to show good periodicity, and the fluctuation in the pressure coefficient is relatively stable. At this time, the time elapsed in one cycle in the pressure curve is about 0.06 seconds (as shown by the red dashed line in Fig. 5.13), which is slightly larger than the rotation period of the impeller. In the middle part of the impeller, the difference in pressure pulsation curves for different tip clearances is not obvious, and the period of the pressure pulsation curve is about one-fourth of the impeller rotation period due to the influence of BPF. However, the periodicity in the pressure pulsation curve for 0.5 mm tip clearance is better than that for the other three tip clearances. There is difference between the peak and trough values of the pressure amplitude in different impeller rotation cycles when the tip clearance is small. On the other hand, the harmonic components of the pressure curve increase

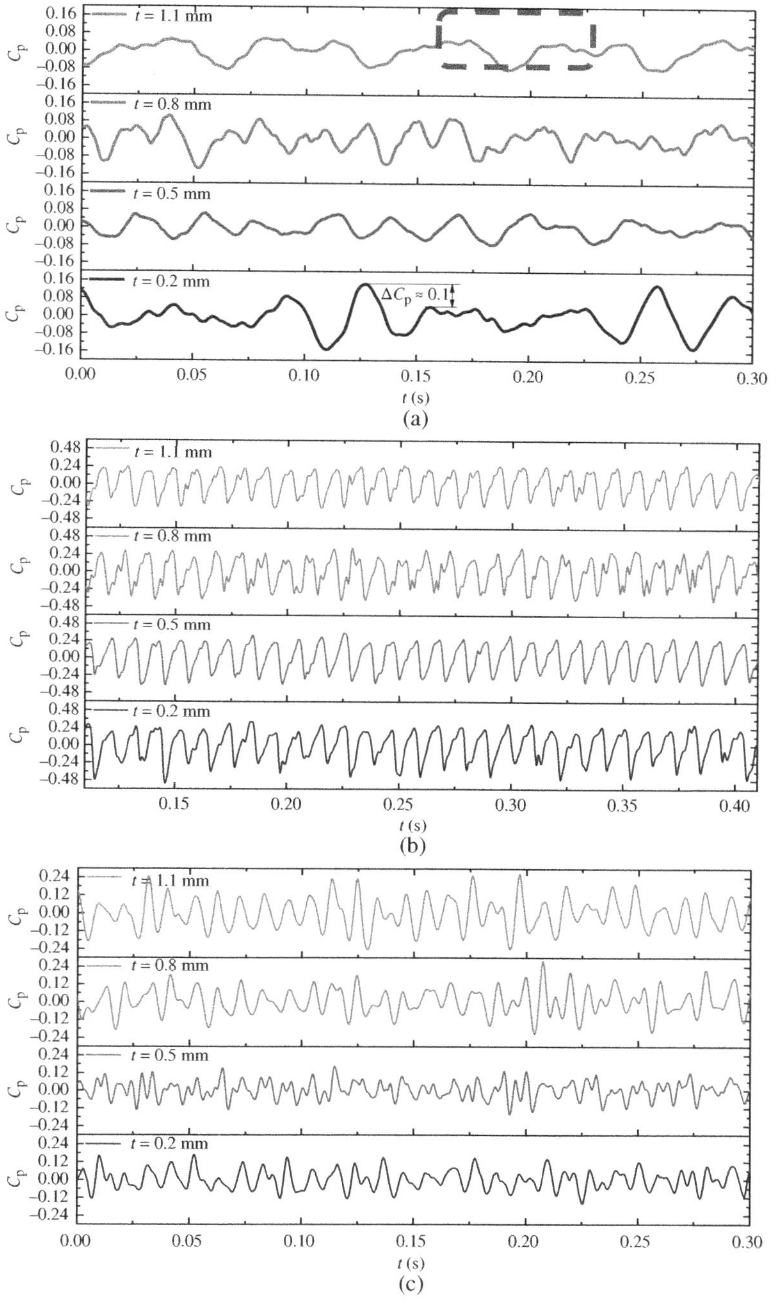


**Figure 5.13** Time-domain characteristic of the impeller inlet, middle, and outlet for different tip clearances of the impeller at (a) impeller inlet, (b) impeller middle, and (c) impeller outlet.

significantly when the tip clearance is large. At the impeller outlet, under the dual influence of BPF and RSI between the impeller and guide vane, there is more information available for the pressure coefficient curve for each tip clearance, which is difficult to distinguish among various tip clearances. In general, with the increase in tip clearance, the amplitude of pressure pulsation shows a decreasing trend first and then an increasing trend. In summary, the size of tip clearance does affect the pressure pulsation characteristics of the impeller inlet, middle part, and outlet part of the mixed-flow pump, and the impeller inlet and the impeller outlet are significantly affected compared to the middle portion of the impeller.

### 5.2.2 Wavelet Transform of Pressure Pulsation for Different Tip Clearances

It is known that the time-domain signals of pressure pulsation could only provide the characteristics in time domain but could not describe the characteristics in the frequency domain. In recent years, the emergence of continuous wavelet transforms (CWT) can not only describe the change in signals with time but also obtain the characteristics of signals in the frequency domain. Therefore, continuous wavelet transform is used here to analyze the pressure pulsation signals.

Figure 5.14 shows the wavelet transform results with wavelets of different frequency-space resolutions at the inlet, middle, and outlet of the impeller for different tip clearances. The time of pressure signals has been changed to the number of revolutions of the impeller, which is abbreviated here as Rev. The method of relating the wavelet scale $s$ to the Fourier frequency can be found in Ref. [2]. In order to weaken the edge effects and error due to the finite-length of the experimental signals [3], a number of pressure pulsation signals, which correspond to the integer power of 2 are selected for the analysis. The time-averaged wavelet spectrum is shown on the right side of each wavelet transform in Fig. 5.14. It can be seen from this figure that most of the energy in the wavelet spectrum is concentrated from 8 to 64 Hz for different tip clearances, and the modulus value of the first-order peak formed in the time-averaged wavelet spectrum curve is higher. However, compared to the other gaps, the frequency range of energy concentration is narrower when the tip clearance is 0.2 mm, nearly from 16 to 64 Hz, which indicates that the pressure signal at the impeller inlet for small tip clearance is more stable and periodic. With increase in tip clearance, the high-energy value region of the wavelet spectrum expands toward the low-frequency, and the second-order peak is formed when the tip clearance is 0.5 or 0.8 mm, and the second-order peak value at 0.8 mm is significantly higher than that at 0.5 mm. However, the time-averaged wavelet spectrum in the low-frequency band is significantly higher for the tip clearance of 0.5 mm, indicating that the low-frequency disturbance at the impeller inlet is present. With the increase in tip clearances, the energy of the wavelet spectrum changes significantly. For instance, when the tip clearance is 1.1 mm, the frequency corresponding to the first-order peak of the wavelet spectrum significantly reduces while the frequency corresponding to the second-order peak is higher, indicating that the main frequency pulsation of the impeller inlet has changed at large tip clearances. At the same time, comparing the time-averaged wavelet spectrum for the four tip clearances, it can be seen that the extreme value of the wavelet spectrum decreases rapidly and then increases slightly as the tip clearance increases, and the extreme value tends to be stable at the two large tip clearances. This indicates that, although the increase in TLF reduces the disturbance frequency at the impeller inlet for large tip clearances, the intensity of the disturbance does not increase much.

Similarly, Fig. 5.15 shows the wavelet modulus spectrum and the time-averaged wavelet modulus spectrum curve at the impeller middle part for different tip clearances. For different tip clearances, the first-order peaks of the wavelet spectrum range from 64 to 128 Hz, which is consistent with the

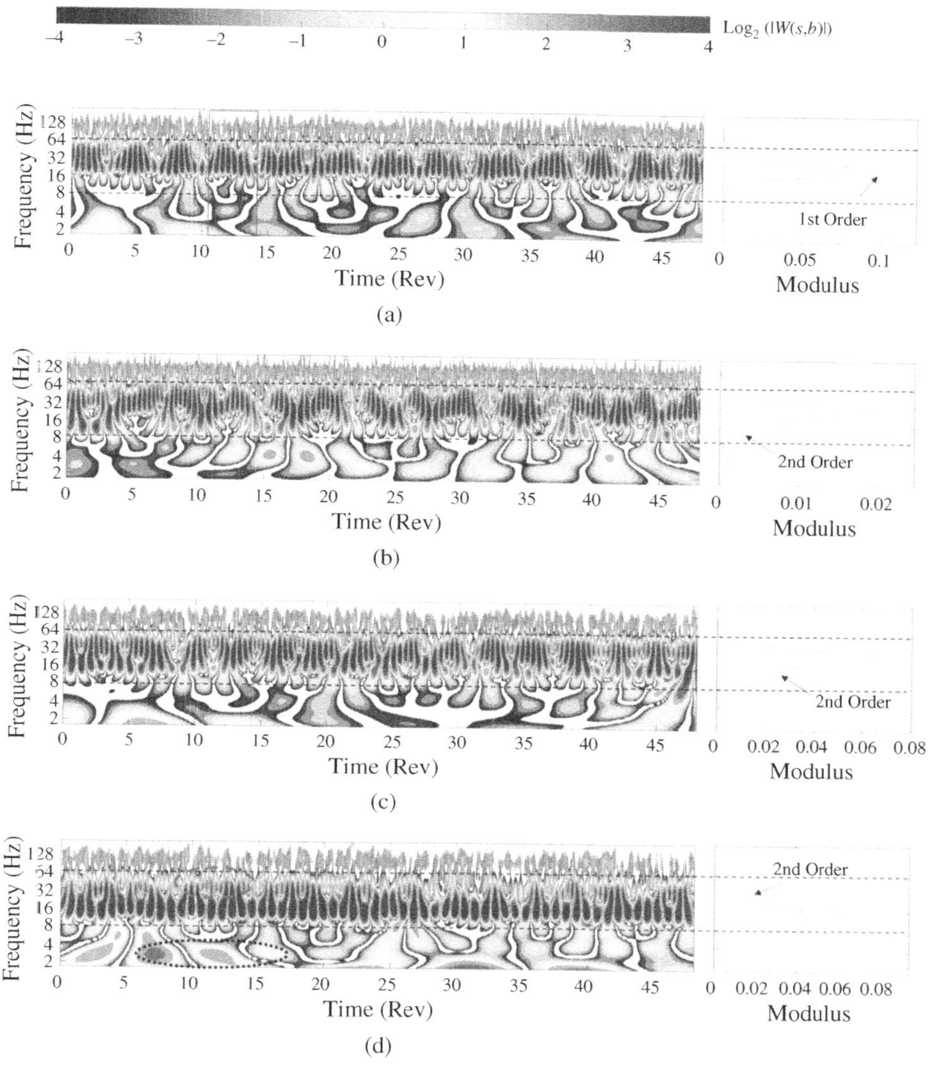


**Figure 5.14** Wavelet modulus spectrum of the pressure pulsation at the impeller inlet with different tip clearances. (a) $t = 0.2$ mm, (b) $t = 0.5$ mm, (c) $t = 0.8$ mm, and (d) $t = 1.1$ mm.

frequency band that includes BPF, indicating that the impeller middle part for four tip clearances is significantly affected by the BPF. However, it can be seen from the time-averaged wavelet spectrum that, as the tip clearance increases, the time-averaged value of the wavelet modulus spectrum gradually decreases, indicating that the intensity of pressure pulsation caused by the TLF from the impeller blade middle rim decreases, further explaining that the pressure difference on both sides of the blades at the impeller middle part reduces.

Figure 5.16 shows the wavelet modulus spectrum and the time-averaged wavelet modulus spectrum curve at the impeller outlet for different tip clearances. As can be seen from this figure, the energy distribution of the wavelet spectrum at the impeller outlet is quite different for different tip clearances; therefore, the wavelet spectrum is divided into high-frequency bands and

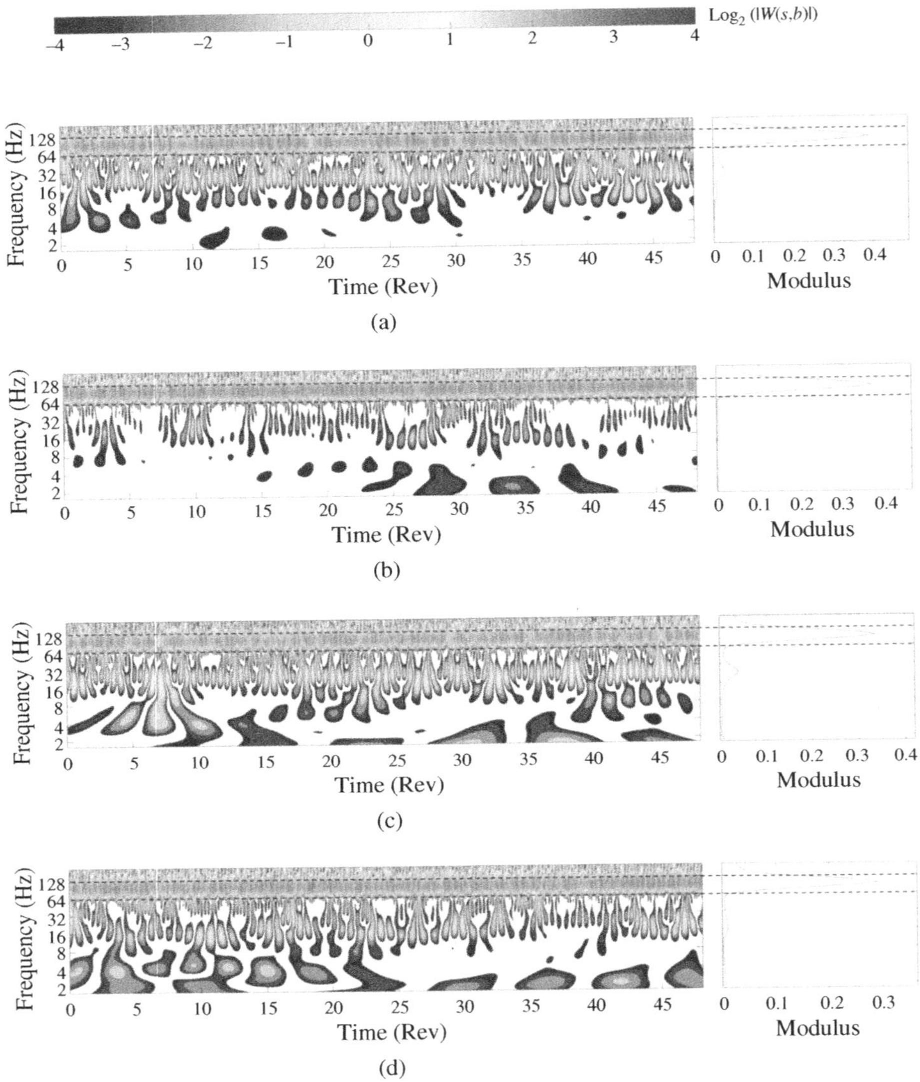


**Figure 5.15** Wavelet modulus spectrum of the pressure pulsation at the impeller middle with different tip clearances. (a) $t = 0.2$ mm, (b) $t = 0.5$ mm, (c) $t = 0.8$ mm, and (d) $t = 1.1$ mm.

low-frequency bands for analysis; it is marked with dashed lines in the figure. In general, for all four tip clearances, the wavelet energy is still concentrated in the high-frequency band, indicating that the influence of BPF is prominent. Among four tip clearances, the wavelet energy concentration in the high-frequency region is the highest for 0.5 mm tip clearance, and most of the high-frequency energy is in the interval 64–128 Hz. However, when the tip clearance decreases or increases, part of the energy in the high-frequency region flows toward the high-frequency direction. That is, when the frequency is greater than 128 Hz, the wavelet energy spectrum value is still high. At the same time, it can be seen from the first-order peak value of the time-averaged wavelet spectrum that the extreme value of the time-averaged wavelet energy also decreases first and then increases with the increase in tip clearance. Within the low-frequency band, the value of the wavelet energy spectrum

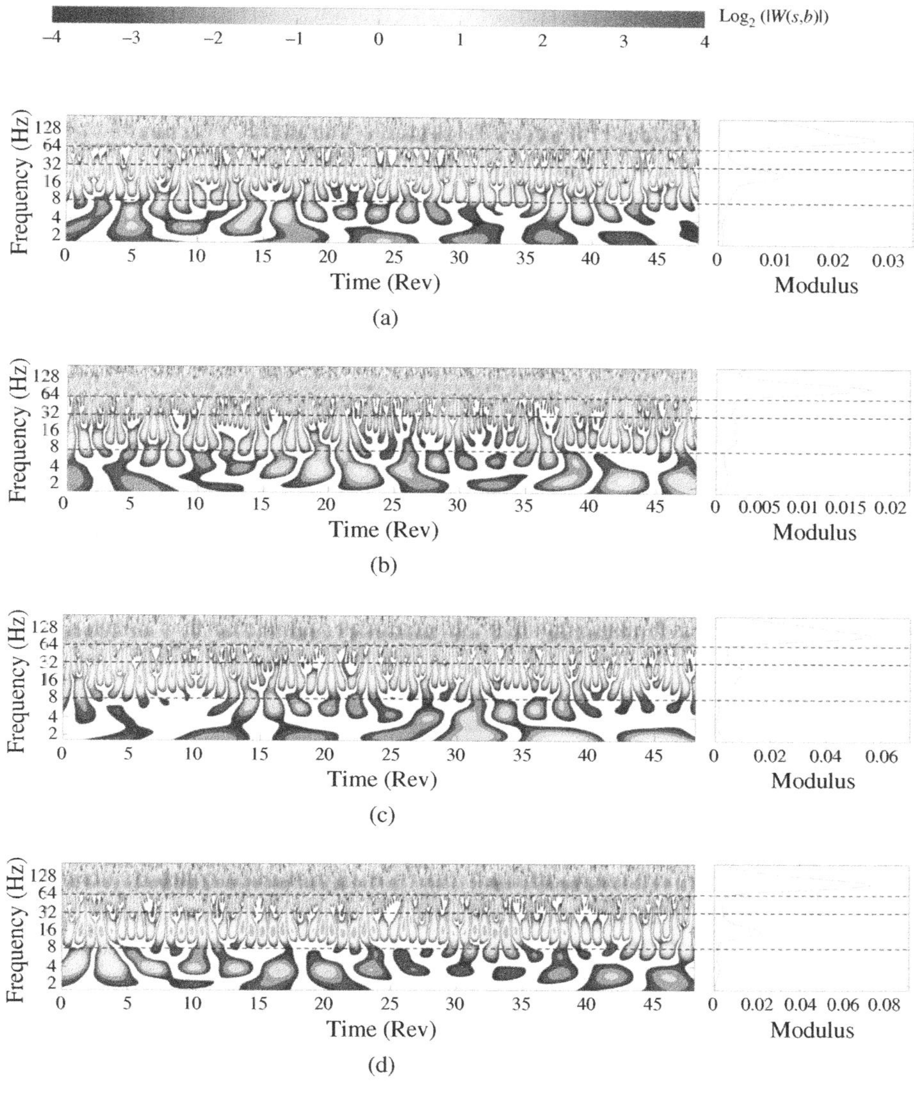


**Figure 5.16** Wavelet modulus spectrum of the pressure pulsation at the impeller outlet with different tip clearances. (a) $t = 0.2$ mm, (b) $t = 0.5$ mm, (c) $t = 0.8$ mm, and (d) $t = 1.1$ mm.

is also higher for the four tip clearances as a result of the RSI effect. At 0.5 mm tip clearance, the wavelet energy in the low-frequency band does not fluctuate much in the frequency space, and the time-averaged wavelet modulus value is small. However, for other tip clearances, the time-averaged wavelet energy fluctuates greatly, and the second-order peaks are evident. Similar to the wavelet energy distribution within the high-frequency band, the extreme value of wavelet energy decreases first and then increases within the low-frequency band as the tip clearance increases. These phenomena indicate that the size of tip clearance can indirectly affect the flow fields in the RSI zone. As the tip clearance increases, the high-frequency disturbance in the RSI zone decreases first and then increases; thus, there is an optimal tip clearance for the impeller for which there is the smallest disturbance.

### 5.2.3 Coherence of Pressure Pulsation for Different Tip Clearances

In the statistical literature, coherence refers to the degree of correlation between the two variables in the frequency domain. The wavelet coherence analysis has the ability to analyze the unsteady state models of the pressure pulsation signals. For the two time-series pressure pulsation signals $x(t)$ and $y(t)$, the wavelet coherence spectrum between them can be written as

$$R^2(a,b) = \frac{|S(a^{-1}W_{XY}(a,b))|^2}{S(a^{-1}W_X(a,b))S(a^{-1}W_Y(a,b))} \tag{5.7}$$

where $R^2(a,b)$ is the coherence coefficient, and $S$ is the smoothing operator.

In order to analyze the coherence of pressure signals at different locations, the wavelet transforms coherence (WTC) analysis is performed on the dynamic pressure pulsation signals collected for the tip clearance of 0.5 mm. In the analysis, adjacent pressure sensors (sensors 1 and 2) and the axis symmetrical pressure sensors (sensors 1 and 3) are selected for coherence analysis. Figures 5.17 and 5.18 provide the wavelet coherence spectrum obtained from adjacent pressure sensors and axis symmetrical pressure sensors in the same axial section, respectively, when the tip clearance is 0.5 mm. On the right side of the wavelet coherence spectrum, the time-averaged results of the wavelet coherence spectrum are also shown as described previously. In the analysis, the signals from different positions are divided into three regions. The first is from 2 to 16 Hz, which is named as the L-region. The second is from 16 to 64 Hz, which is named as M-region. The third is from 64 Hz to maximum value, which is named as the H-region. Because of the edge effect of the signal, the error is higher in the cone of influence (COI) region; therefore, only the wavelet coherence in the confidence interval is studied. It can be seen from Figs. 5.17 and 5.18 that, no matter whether the distance between the two sensors is far or close, the wavelet coherence is strong near the BPF and is in the 2–16 Hz range at the impeller inlet. It indicates that the similarity in the two time-series pressure signals is high in the frequency space. Furthermore, the wavelet coherence of the pressure pulsation becomes lower when the two pressure sensors are located asymmetrically; however, there is still a high coherence region, which indicates that the disturbance is stable at the impeller inlet. In addition, it can be noticed that the flow field at the impeller inlet is mainly affected by the disturbance of low-frequency and BPF, which is consistent with the analysis of the wavelet spectrum. However, the wavelet coherence spectrum shows that the low-frequency disturbance is relatively stable in different positions, and the influence of BPF on the impeller inlet is unstable.

At the impeller middle section, there is not significant difference between the two sets of wavelet coherence spectrum, both of which have high coherence between 64 and 128 Hz frequency range, indicating that the two sets of signals are affected by the stable BPF and the stability of the impeller rotational speed. In the intermediate frequency and low-frequency region, the wavelet coherence spectrum of the two groups shows weaker wavelet coherence and a strong coherence value appears only at some moments in time, such as in the interval 10–15 and 35–45 Rev. In fact, these two regions correspond to the high coherence in the low-frequency band of the impeller inlet from the coherence spectrum of axis symmetrical sensors, which indicates that the low-frequency disturbance of the impeller inlet will also affect the flow fields of the impeller middle section to some extent, but the impact will not be great. At the impeller outlet, the wavelet coherence value near BPF is high, and the intensity of the wavelet coherence shows a decreasing trend as the frequency drops in the intermediate frequency and low-frequency space. Comparing the wavelet coherence spectrum of two pairs of pressure sensors, the time-averaged coherence difference at the intermediate frequency band is significantly different, followed by the low-frequency band. Within the intermediate frequency band, when the sensors are adjacent, the interaction between the adjacent flow channels makes the correlation signal between the pressure pulsation signals such that the

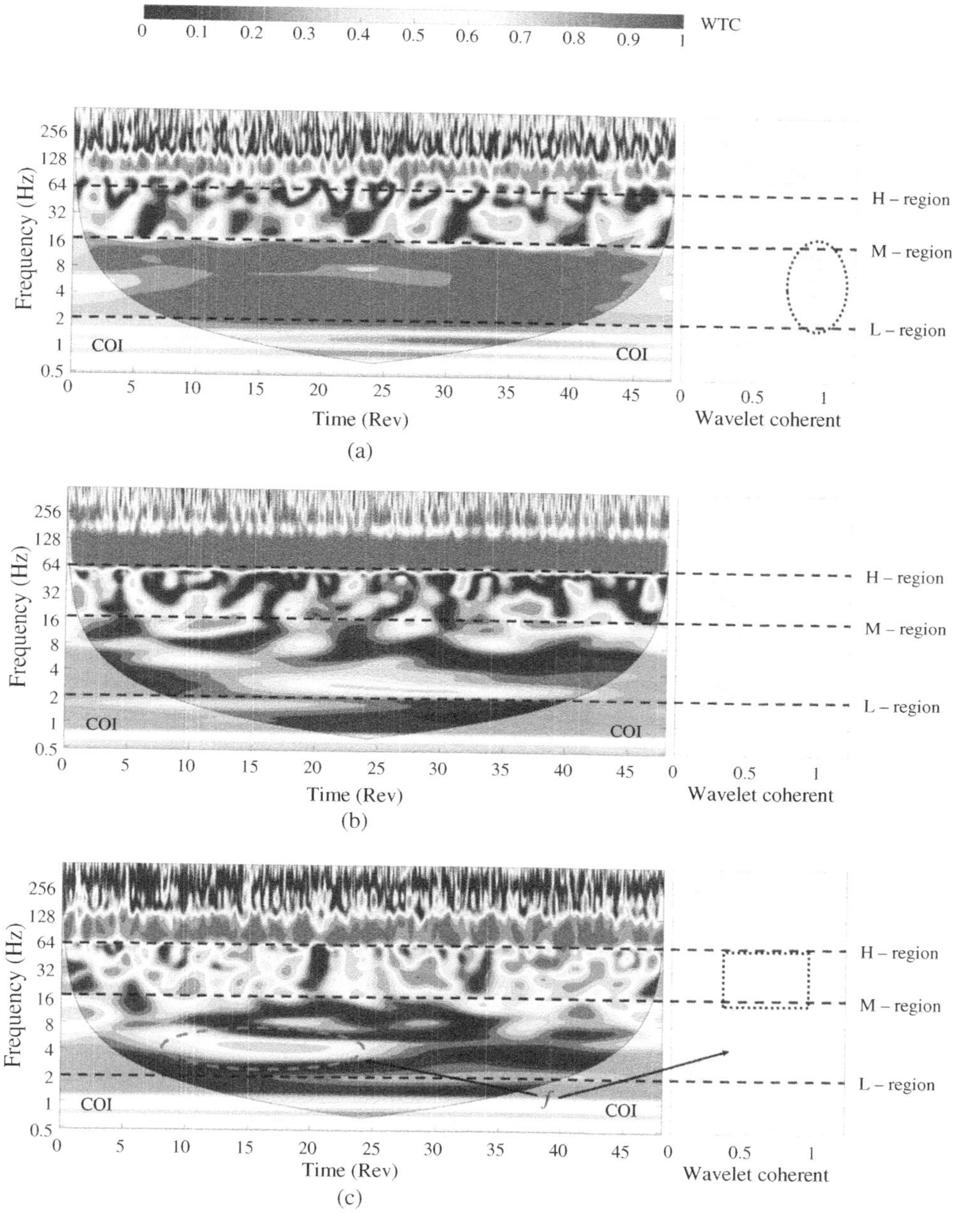


**Figure 5.17** Wavelet coherence spectrum of the pressure pulsation of adjacent pressure sensors with different tip clearances: (a) impeller inlet, (b) impeller middle, and (c) impeller outlet.

coherence of the entire frequency band becomes strong. When the two sensors are far away, there is an isolated flow channel between the corresponding flow channels, thus, the interaction effects reduce, and the coherence strength of the low-frequency region of the intermediate frequency band is weakened. However, the coherence in the high-frequency part of the intermediate frequency band is still strong, and it is located near the axial frequency of the impeller, indicating that the axial frequency disturbance is strong locally. Therefore, the distance between the sensors is also

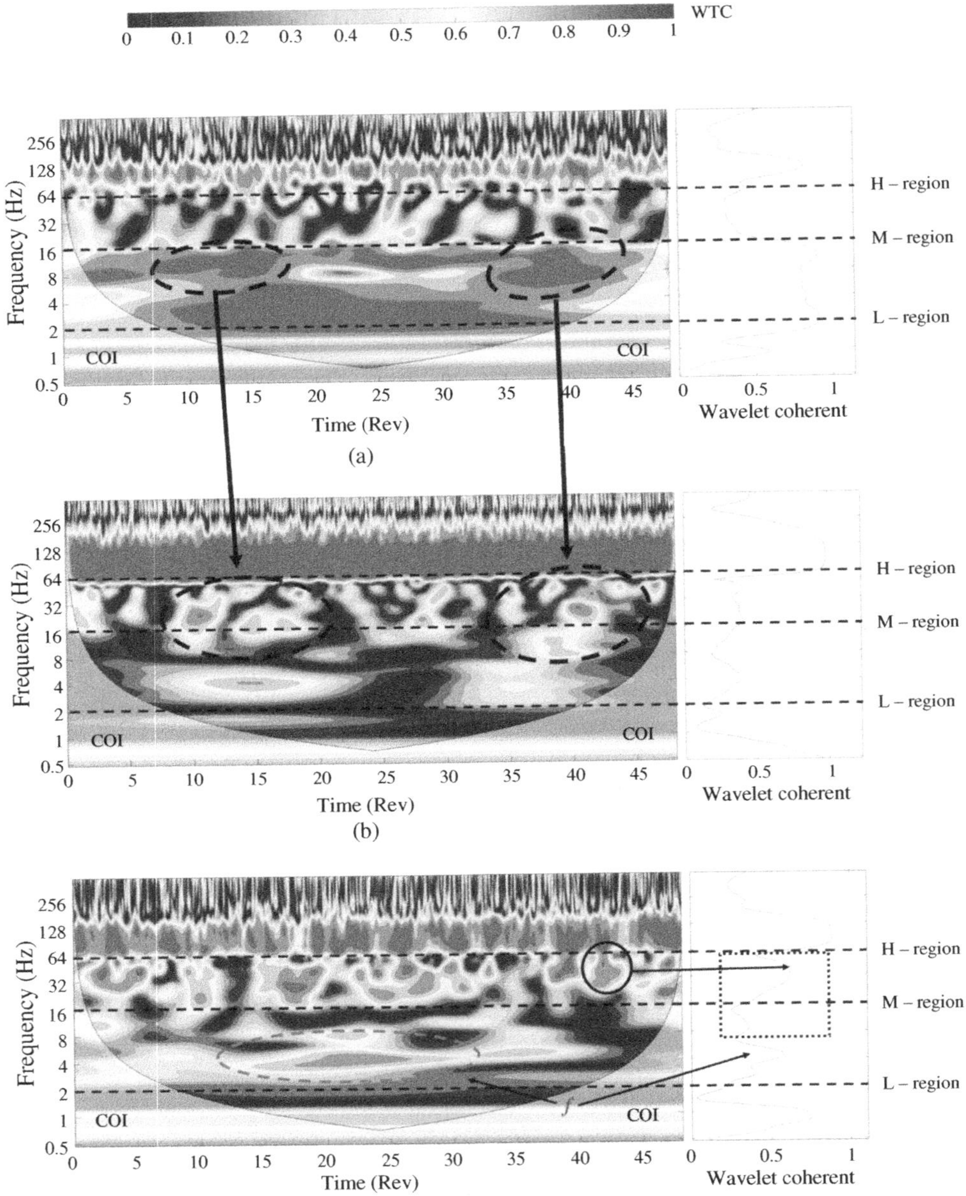


**Figure 5.18** Wavelet coherence spectrum of the pressure pulsation of axis symmetrical pressure sensors with different tip clearances: (a) impeller inlet, (b) impeller middle, and (c) impeller outlet.

one of the factors affecting the coherence. In the low-frequency band, the coherence between the two sets of pressure pulsation signals is further weakened, but the coherence is also greater at some moments in time. When two sensors are close, the maximum coherence appears near 18 Rev, and when the two sensors are far away, the maximum coherence appears near 22 Rev, indicating that the signal has a random pulsation.

## 5.3 PIV Measurement

### 5.3.1 Basic Principles of PIV Measurement

The basic principle of PIV measurement is shown in Fig. 5.19 [4]. First, the tracer particles with appropriate concentration are added to the measured flow field, and then the pulsed laser sheet beam is passed through the cylindrical mirror and the spherical mirror to form a very thin sheet light source (about 1 mm thick), which is then incident onto the test area of the flow field. The CCD camera perpendicular to the light source of the sheet is exposed continuously twice in time $\Delta t$, and the particles' picture is recorded on the film negative or CCD camera. The displacement information of particles is then identified through two different images, and then the velocity calculation methods such as the optical Young's fringe method, autocorrelation method, or cross-correlation analysis method are used to process the images recorded by the PIV film or CCD camera point by point to obtain the velocity distribution in the internal flow field of the pump [5].

As shown in Fig. 5.20, two consecutive photos are captured from the flow area illuminated by the laser sheet light source, and the position of the particles in the photos at time $t_1$ and $t_2$ is obtained through processing. Based on the displacement ratio, the following formula is used to obtain the speed of the particles [6, 7]:

$$U = \lim_{t_2 \to t_1} \frac{X_2 - X_1}{t_2 - t_1}, \quad V = \lim_{t_2 \to t_1} \frac{Y_2 - Y_1}{t_2 - t_1} \tag{5.8}$$

where $U$ and $V$ are the two velocity components. If the selected tracer particles are followed properly, the particle trajectory can accurately reflect the fluid motion. If the time interval between the two sets of images and the displacement of particles are small enough [8], the magnitude and direction of the particle velocity can be calculated using Eq. (5.8), and then the velocity field of the whole measured flow field can be obtained by processing the particles one by one.

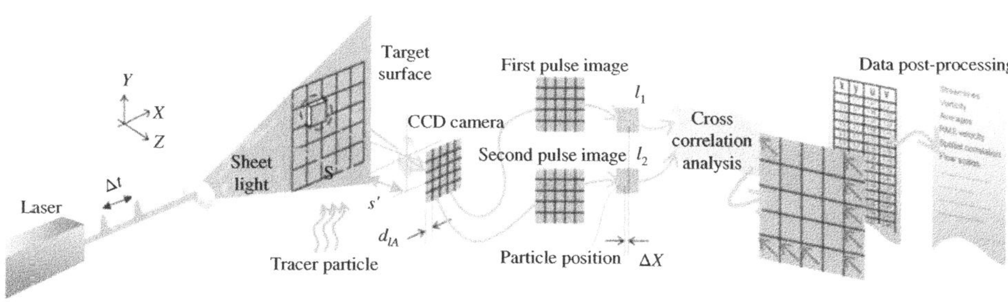


Figure 5.19 PIV measurement setup and principle.

Figure 5.20 Displacement of tracer particles.

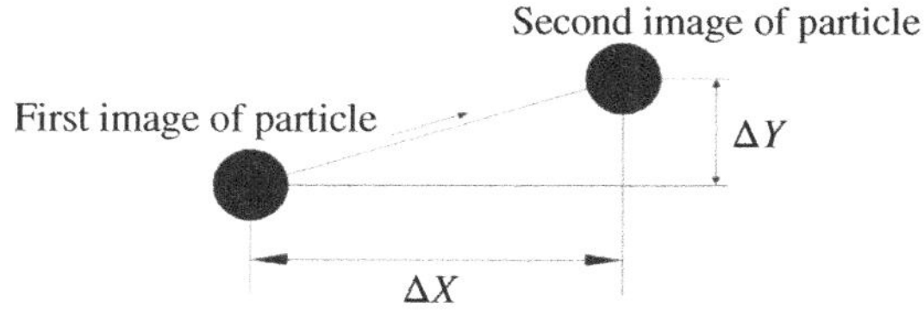

### 5.3.2 The PIV Image Processing System

PIV is a high-density image processing technique [9]; it generates images of many particles flowing in the fluid; therefore, the method of tracking a single particle trajectory cannot be used to obtain the velocity information, and only the statistical method can be employed. Optical methods and digital image technology are usually employed for the analysis. The optical method is Young's interference fringe method, and the digital image technology includes the Fourier transform method, direct spatial correlation method, and particle image spacing probability statistics method.

At present, the cross-correlation analysis method is the most widely used. The cross-correlation analysis needs three two-dimensional Fast Fourier transforms (FFTs). The image $F(x, y)$ in the inquiry area is the result of the superposition of the image $f_1(x, y)$ formed by the first pulse laser and the image $f_2(x, y)$ formed by the second pulse laser [10]. When the area of the inquiry is small enough, it can be considered that the particle velocity remains the same in the area of inquiry, and the image formed by the second pulse laser can be considered as the image formed by the first pulse laser after translation, i.e.

$$\int_2 (\mathrm{x}, y) = \int_1 (\mathrm{x} + \Delta\mathrm{x}, y + \Delta\mathrm{y}) \tag{5.9}$$

Therefore

$$F(x, y) = f_1(x, y) + f_1(x + \Delta x, y + \Delta y) \tag{5.10}$$

The first frame image is obtained by the FFT transforms,

$$\overline{f_1}(\omega_x + \omega_y) = \frac{1}{2\pi} \iint f_1(x, y) \mathrm{e}^{i(\omega_x x + \omega_y y)} \mathrm{d}x\mathrm{d}y \tag{5.11}$$

$$\overline{f_2}(\omega_x + \omega_y) = \frac{1}{2\pi} \iint f_2(x, y) \mathrm{e}^{i(\omega_x x + \omega_y y)} \mathrm{d}x\mathrm{d}y \tag{5.12}$$

Using the translational characteristics of the Fourier transforms, the following equation can be obtained:

$$\overline{f_2}(\omega_x + \omega_y) = \overline{f_1}(\omega_x + \omega_y) \mathrm{e}^{-i(\omega_x \Delta x, \omega_y \Delta y)} \tag{5.13}$$

Now, by applying the inverse Fourier transforms, the following equation is obtained:

$$F(x, y) = \frac{1}{2\pi} \iint \overline{f_1}(\omega_x + \omega_y) \overline{f_2}(\omega_x + \omega_y) \mathrm{e}^{-i(\omega_x x + \omega_y y)} \mathrm{d}\omega_x \mathrm{d}\omega_y \tag{5.14}$$

Substituting Eq. (5.13) into Eq. (5.14), obtain,

$$F(x, y) = \overline{f}(x + \Delta x, y + \Delta y) \tag{5.15}$$

where $\overline{f}$ is the inverse Fourier transform of $f_1(\omega_x, \omega_y)^2$, and $F$ has a maximum value only at $(x + \Delta x, y + \Delta y)$. If it is represented by the gray scale, due to the existence of background noise and other correlation quantities, there is a function,

$$R(S) = R_C(S) + R_D(S) + R_F(S) \tag{5.16}$$

where $R_D$ is the maximum gray value representing the displacement information, and $R_C$ and $R_F$ are random correlations representing background noise.

### 5.3.3 PIV Measurement Criteria

The most critical step in PIV experiment is to obtain a pair of correlated particle images; therefore, the following criteria must be followed in the PIV measurement [11]:

(1) Model processing: It is to ensure that at least two faces of the model are transparent in order to arrange the camera and the laser. The optical divergence of the model should be eliminated as much as possible, and the inner circle and outer circle should be used for processing and manufacturing.
(2) Choice of the inquiry area: The interrogation area should be small enough so that the captured velocity vector can reflect the actual flow at that point.
(3) Selection of the tracer particles: It is important to try to balance the following movement of the particles and the intensity of scattered light. Particles of 1–10 μm should be selected in air, and particles of 1–20 μm should be selected in water.
(4) Check the number of particles in the region: In order to obtain effective and accurate data, the particles in each inquiry area should be between 3 and 20 pairs, and the particles should be clearly visible and slightly overexposed. This parameter depends on the particle concentration, the size of the interrogation area, and the time interval between the laser pulses.
(5) Particle displacement: The maximum particle displacement should generally be one-fourth of the size of the inquiry area, which can increase the percentage of particle pairs in the inquiry area. When the displacement of the particle image increases, the probability of only one particle image in the inquiry area increases, and the percentage of particle pairs decreases. This parameter can also be adjusted by changing the size of the interrogation area and the pulse interval. The minimum particle image displacement should be greater than twice the particle image diameter. When the particle moves less than one diameter, the image is a single elliptical particle image rather than two spherical particle images. This can generally be improved by increasing the pulse interval.

### 5.3.4 PIV Testing Instruments and System

The commercial PIV system of DANTEC company in Denmark is used in the PIV experimental results presented in this chapter. The velocity measurement range of the PIV system is 0–1000 m/s and has high accuracy. It can measure the two-dimensional and three-dimensional flow fields and can also analyze the temporal and spatial stability of the flow field structure. The main devices and equipment of the system are shown in Fig. 5.21, which include sheet light source lighting system, 38a7503 high-energy dual cavity laser, 80 × 39 optical arm, flow sensing Eo2mccd camera, 80N77 programmable synchronizer, and Dell workstation to collect the data, etc. Among them, the CCD camera has an FPGA chip, which is highly integrated and has hundreds of programmable functions. In addition, the built-in image processing system is installed, and the exposure time can be precisely controlled within 1 μs through programmable exposure control. Also, the software Dynamic studio v3.41 is installed on the Dell workstations to analyze and process the data. The synchronizer is connected to the Ni terminal of the computer through the high transmission data line and then connected to the CCD camera and pulse laser through the control cable at the output end to coordinate the trigger and shooting time between the camera and the laser and control the synchronous operation of the camera and the laser by sending a synchronization signal.

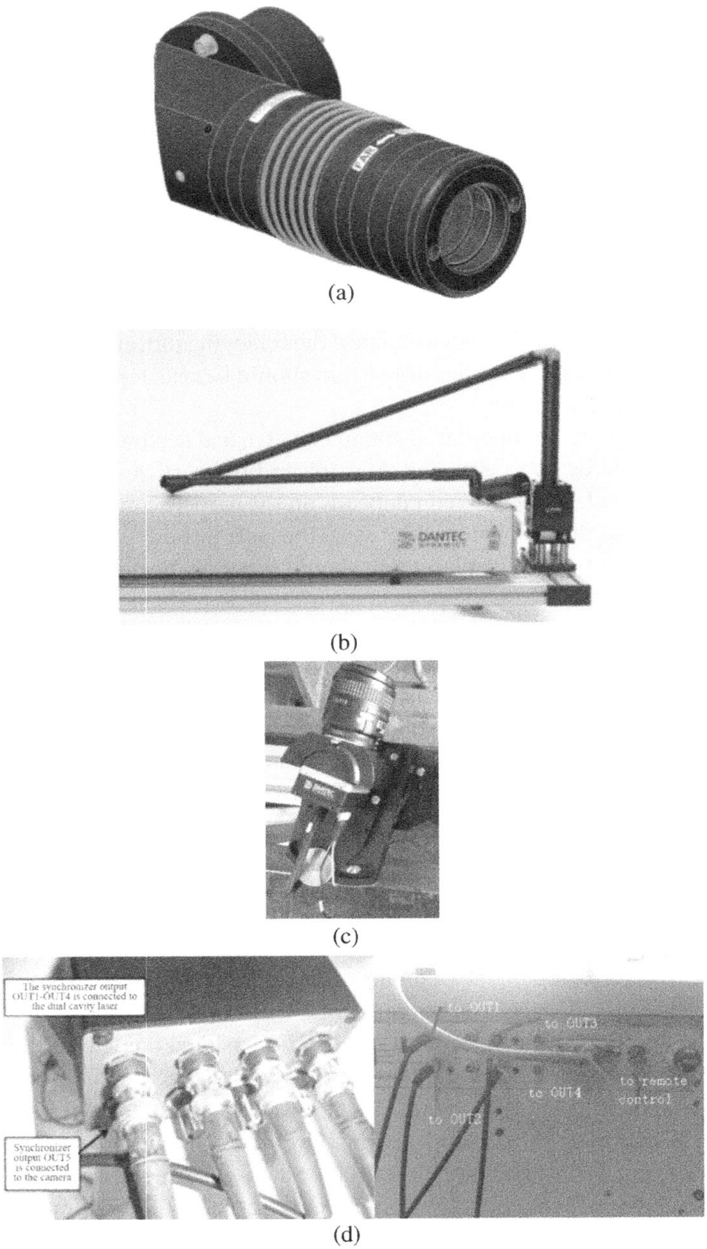


**Figure 5.21** PIV system and equipment: (a) sheet light source, (b) double cavity laser and long optical arm. Source: Dantec dynamics A/S, (c) CCD camera, and (d) programmable synchronizer.

### 5.3.5 PIV Test Method

The PIV test method involves the following steps.

#### 5.3.5.1 Selection and Addition of Tracer Particles

PIV test has certain requirements for the tracer particles. In addition to the stable chemical properties and nontoxicity, the particles should have good astigmatism and the ability to follow the flow. These factors should be considered when selecting the particles. Generally, if the medium is air, the smoke consisting of small particles is selected; if the medium is water, the hollow glass beads with a density close to that of water and organic materials such as polyethylene are selected so that the CDD camera can clearly capture the tracer particles in the area of inquiry to ensure the measurement accuracy of the test as much as possible. Therefore, after considering various factors such as the fluid medium and the cost, under the condition of ensuring the concentration and clarity of the particles, the hollow glass spheres with a density $1.05\,g/cm^3$ and diameter 20–60 μm are selected as tracer particles in the experiment reported in this chapter.

After the appropriate tracer particles are selected, the particles are added to the test pipeline. Since it is a closed test bed, the procedure of adding the particles is more complex; the particles need to be sent into the pipeline by employing differential pressure. First, the pipeline is filled with water, and then the device is started to make the water circulate. At this stage, in order to make the pressure stabilize in the tank, all external valves are closed, and the vacuum pump is opened to extract all the air from the tank. Now, the PIV tracer particles are added to the container filled with water in order to fully mix with water. Finally, the container is connected to the external valve of the test bench to send the particles to the testing device at atmospheric pressure. The distribution of tracer particles in the mixed-flow pump during PIV testing is shown in Fig. 5.22.

#### 5.3.5.2 Calibration Apparatus

The testing results are very much influenced by the calibration accuracy of the PIV test. Therefore, in order to improve the test accuracy as well as its feasibility, and to avoid difficulties in the disassembly of the pump, a calibration tank whose half shape is made the same as the organic glass chamber is used for calibration of the PIV test data, as shown in Fig. 5.23. The tank is open on the top for the purpose of filling water and placing a ruler.

Before the test, the following procedures are performed. The calibration tank is adjusted horizontally after it has been placed on a fixed bracket at the beginning of the test. The external side of the camera for the tank is thereafter adjusted to the vertical plane of the surface of the impeller chamber located on the same side. Afterward, the ruler is placed on the top surface of a base with the same curvature as the calibration tank. The end surface of the ruler containing scales is then adjusted in accordance with the shooting plane. The camera is finally moved in the same horizontal plane as the ruler. A schematic of the layout is shown in Fig. 5.24.

#### 5.3.5.3 Lens Group, Camera Fixing, and Adjustment

During the experiment, the base of the lens group support is fixed on the lifting table and is parallel to the axis, the lifting rod of the support is perpendicular to the base, the rod head is horizontal, and a lens group fixing device is hung. The positional relationship between the camera and the lens group is shown in Fig. 5.25. The device ensures the free movement and fixation of the lens group in the $X$, $Y$, and $Z$ directions, as shown in Fig. 5.26. The camera is installed on a special support, as shown in Fig. 5.27, which can also move and adjust freely in $X$, $Y$, and $Z$ directions.

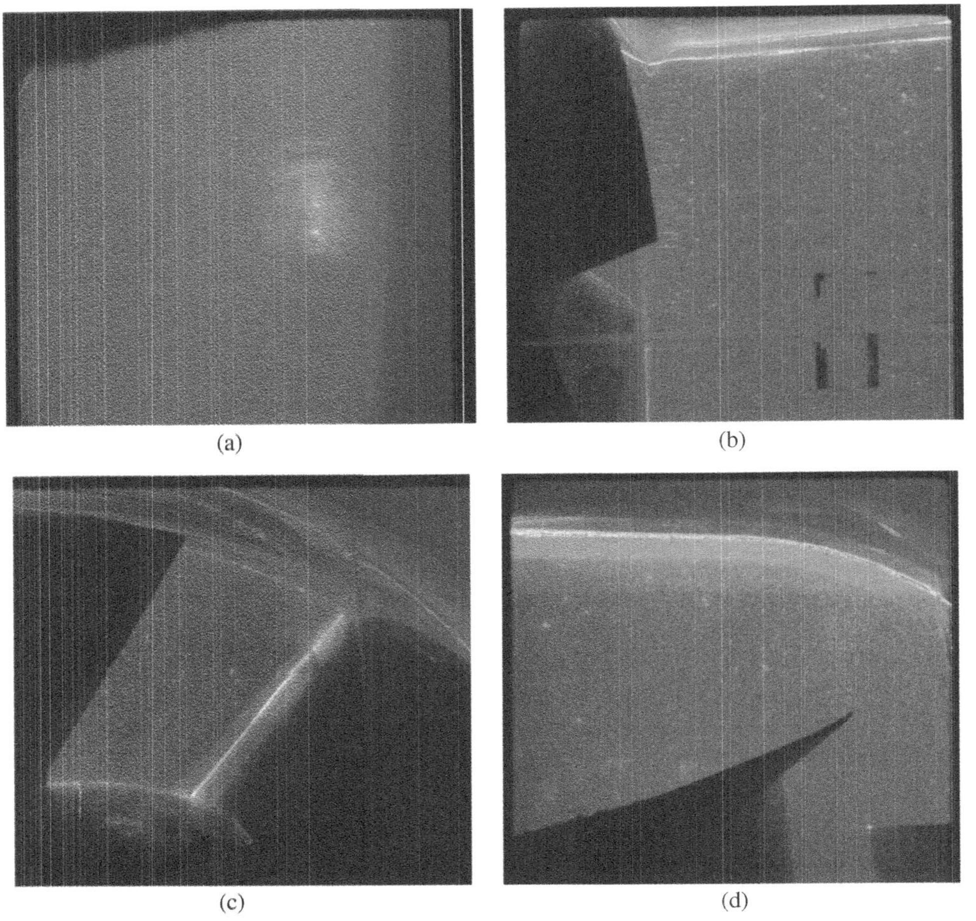


**Figure 5.22** Effect diagram of PIV tracer particles: (a) inlet vertical axis section, (b) inlet shaft section, (c) clearance between impeller and guide vane, and (d) inside guide vane.

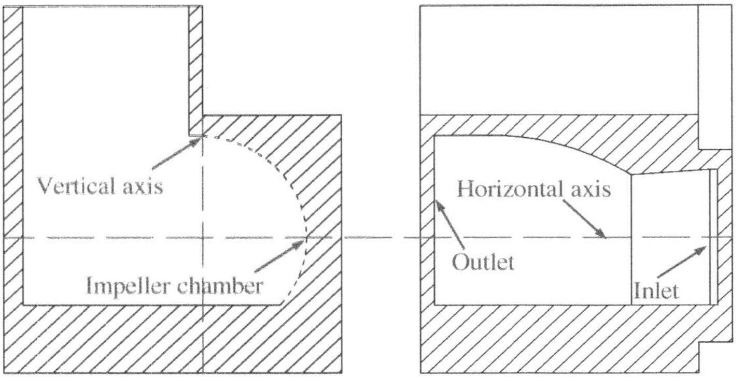


**Figure 5.23** Calibration tank.

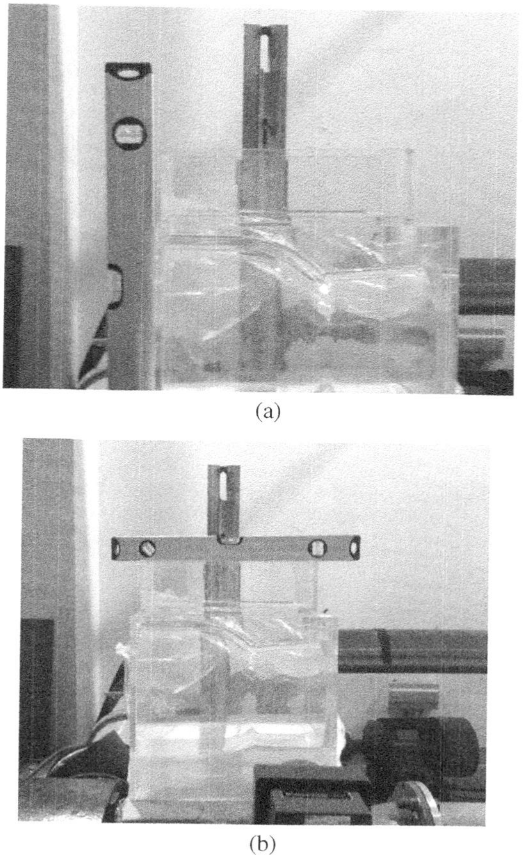
(a)
(b)

**Figure 5.24** Calibration process: (a) adjustment of the side surfaces in the same vertical plane and (b) adjustment of the end surface in the horizontal plane. Source: (a) Wei et al. (2019)/Sage Publications, Inc.

**Figure 5.25** Fixing of lens set and camera.

**Figure 5.26** Lens bracket diagram.

**Figure 5.27** Camera bracket.

#### 5.3.5.4 Shooting Section of the Dynamic and Static Interference Flow Field

During the PIV experiment, the flow field structure is photographed in the same section of the guide vane and the impeller at different phases. The shooting section, which passes through the rotation axis of the pump, is perpendicular to the horizontal plane. Since the blades used in the pump model are distributed uniformly in the circumferential direction, the total time for one blade of the impeller to rotate through the meridian plane is selected as the shooting time (i.e., for the impeller turning 90°, the time is about 10.34 ms). In total shooting time, at every turn when the impeller rotates 15° is set as a shooting section, the time interval is about 0.1724 ms, and each section shoots 500 frame images of the flow field. There are six shooting sections in total, as shown in Figs. 5.28 and 5.29; the phase positions in these figures are denoted as $1T$, $2T$, $3T$, $4T$, $5T$, and $6T$ in the following analysis and a critical phase is selected for the analysis. When the shooting is completed, the self-correlation process is conducted for 500 frame images of tracer particles' absolute velocity field using the Insight 3G software to obtain the instantaneous relative velocity vector of each image, and then it is imported into the Tecplot software to perform the numerical processing to obtain the relative velocity field and other flow variables in each instantaneous flow time. The requirements and experimental procedure of a PIV experiment are given in Ref. [12].

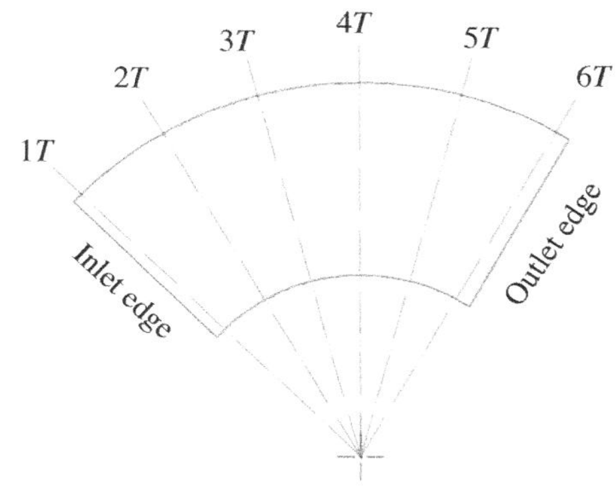


**Figure 5.28** Shooting sections of experiment.

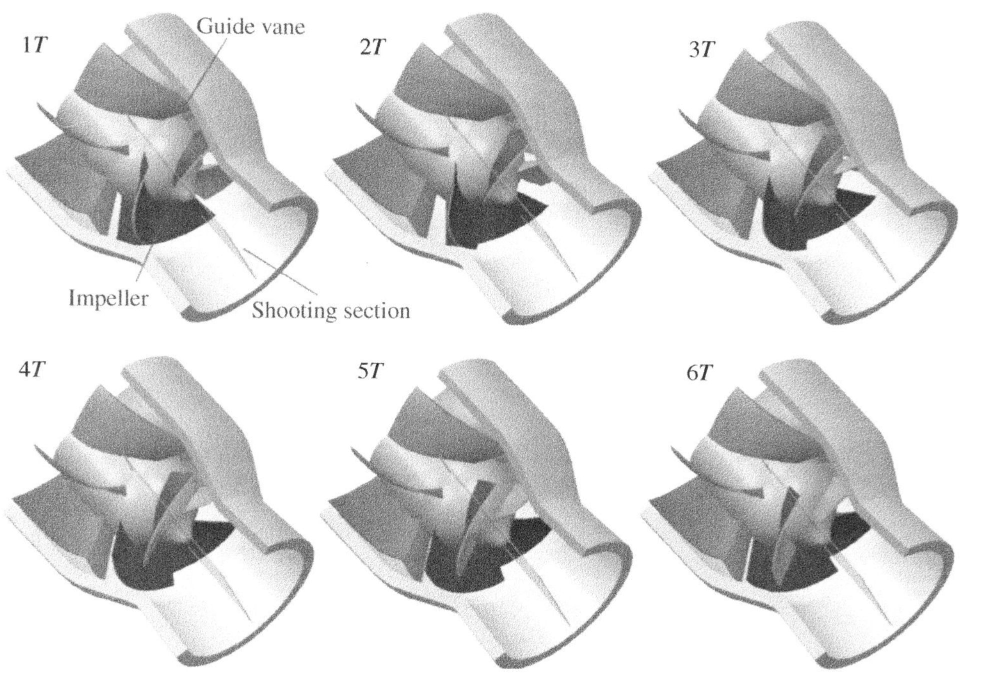


**Figure 5.29** Relative position of the guide vane and the impeller in different phases.

#### 5.3.5.5 Positions of the Equipment and the Shooting Sections

During the experiment, the laser plane generated by the laser generator is perpendicular to the camera as shown in Fig. 5.30. The laser generator and the camera are installed on special equipment that can be moved or fixed in the $X$, $Y$, and $Z$ direction. The shooting sections are shown in Fig. 5.31.

In order to obtain the inlet flow field of the pump, a 90° elbow is added to the experimental pipeline. The position of the elbow is set 90 mm away from the inlet of the mixed-flow pump, which reduces the effect of the elbow on the flow field in the pump while meeting the requirement for the cameras to capture the qualified images of the tracer particles in the pump. The elbow is designed with a window facing the inlet of the impeller. The experimental layout is shown in Fig. 5.31.

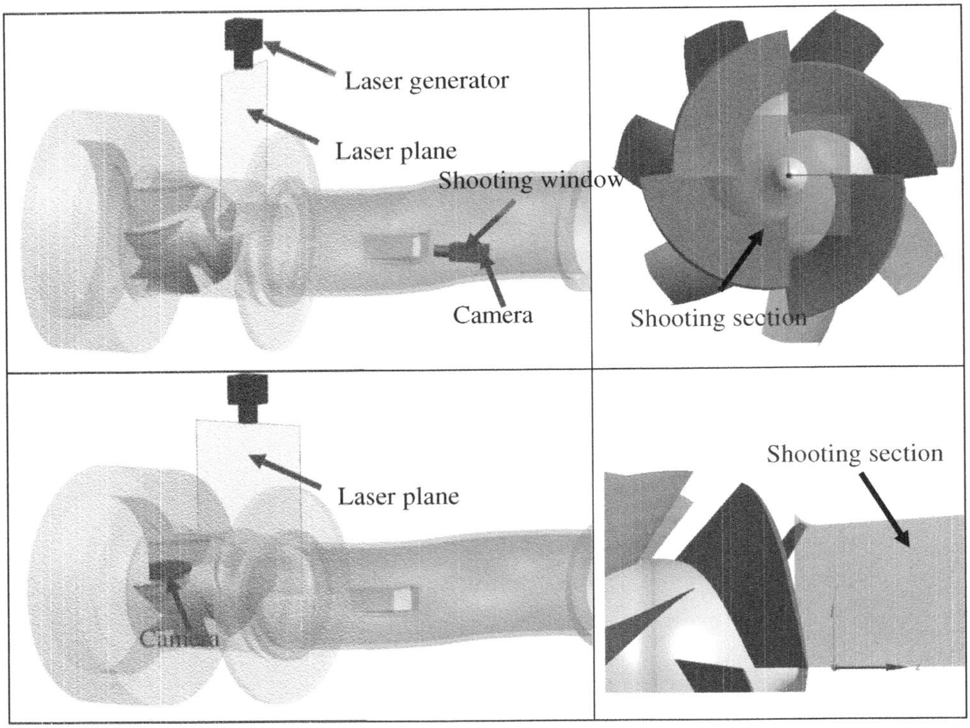


**Figure 5.30** Position of the laser and camera.

**Figure 5.31** Inlet layout for PIV tests.

### 5.3.6 Analysis of the PIV Results for the Flow Field of the Front Shaft Section of the Impeller Inlet

The flow field distribution in the front shaft section of the impeller inlet is shown in Fig. 5.32; the photographed section is the meridional plane passing through the axis. In this figure, the abscissa $X$ is the axial distance of the shooting section, and the ordinate $Y$ is the radial distance of the shooting section. As shown in this figure, the velocity distribution in the inlet flow field of the pump shows almost similar features in the three operating conditions, and the incoming flow at the impeller inlet is along the axial direction. This figure shows that there is a vertical high-speed area near the impeller inlet according to the PIV test results, and the high-speed area increases from the hub to the rim caused by the blade shape. Since the outer edge of the blade inlet is closest to the incoming

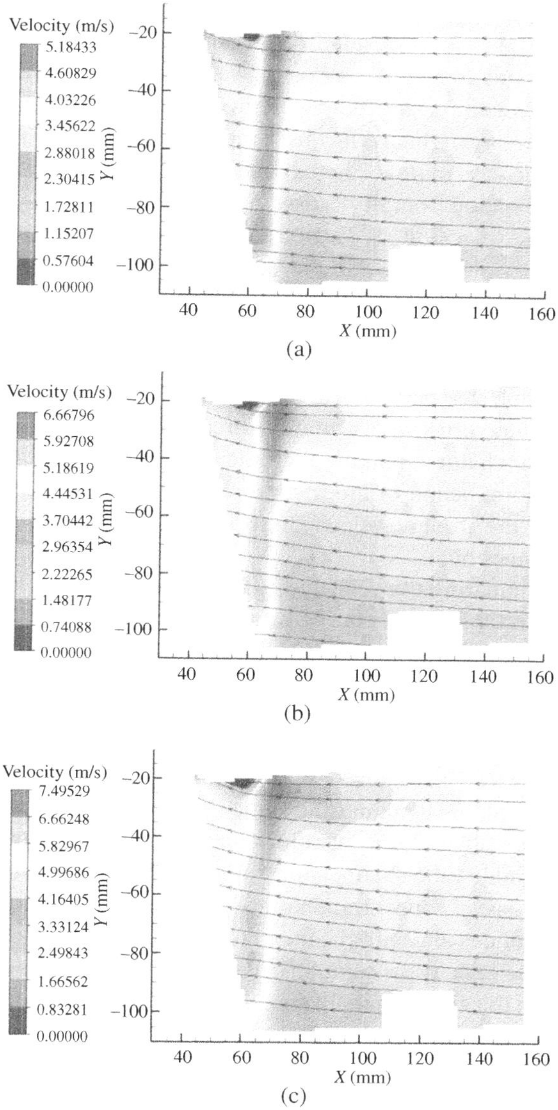


**Figure 5.32** PIV measurement results of inlet axial section: (a) $0.8Q_{des}$, (b) $1.0Q_{des}$, and (c) $1.2Q_{des}$.

flow direction and has the largest radius, the circumferential velocity of the outer edge of the blade inlet is the largest at the same speed, which has great significance on the flow field, which includes the prerotation and acceleration effects. Therefore, there is a high-speed area increasing from the hub to the rim. At the same time, there is also a low-speed area in the end wall region of the outer edge of the inlet, which may be due to the low-speed vortex formed by the leakage flow in the flange gap or the incorrectly measured velocity field in this area due to the light refraction caused by the transition fillet between the inlet section and the impeller chamber in the plexiglass section.

Comparing the impeller inlet velocity field distribution at various flow rates, it can be seen that the impeller inlet velocity increases with an increase in the flow rate, and its maximum value increases from 5.18 to 7.49 m/s. According to the flow and velocity triangle at flow rate of $1.2Q_{des}$, the circumferential velocity and axial velocity at this point are 6.32 and 4.06 m/s, respectively, and the synthetic velocity is about 7.51 m/s. The theoretical analysis results are basically consistent with the experimental results, which can be considered reliable. In addition, there is a velocity gradient from hub to rim in the existing vertical high-speed region, and the velocity contours distribution shows a decreasing trend with the upper left corner as the center of the circle. The low-speed region A increases, which is inconsistent with the empirical judgment that the leakage flow decreases with an increase in flow. Therefore, it is inferred that this region is more due to the incorrect flow field information from PIV caused by the light refraction.

The velocity field distribution in the clearance flow field between the impeller outlet and the guide vane is shown in Fig. 5.33. This figure shows that for various flow rates, the velocity distribution in the axial clearance between the impeller and the guide vane is not orderly, and there is a large velocity gradient due to the interference between the impeller and the guide vane. At the edge of the end wall, the velocity is small due to the influence of the wall boundary layer, and near the hub side of the guide vane inlet, PIV test results for the three operating conditions capture an obvious vortex whose rotation direction is counterclockwise. The vortex blocks part of the impeller outlet channel, making the fluid after the impeller outlet deflects toward the outer edge. Under the joint action of the friction resistance on the outer edge wall and the vortex blockage at the hub, the absolute velocity in the middle of the axial gap between the impeller and the guide vane becomes high.

Comparing the velocity field distribution of the clearance flow field between the impeller outlet and the guide vane at various flow rates, it can be seen that the clearance flow field is most disorderedly in the operating condition with $0.8Q_{\text{des}}$ flow rate, and the streamlines distribution is very uneven, and the reflux vortex A occupying more than one-third of the clearance flow field appears on the hub side of the guide vane inlet which blocks nearly one-half of the flow channel together with the low-speed area in the end wall region. With the increase in flow rate, the vortex in the low-speed area of the end wall region and the hub side of the guide vane inlet is significantly reduced, the dynamic and static interference effect is weakened, and the internal flow tends to be more stable. At the operating condition of $1.2Q_{\text{des}}$ flow rate, the inlet flow in the guide vane enters the guide vane channel approximately parallel to the end wall, the streamlines are smooth, and a high-speed concentration area is formed in the middle of the axial gap between the impeller and the guide vane with the maximum speed of 4.21 m/s.

The velocity distribution in the flow field inside the guide vane is shown in Fig. 5.34. It can be seen from this figure that, like the flow field in the clearance area, the fluid forms vortex A at the guide vane inlet due to the dynamic and static interference between the impeller and the guide vane. The vortex reduces the flow area at the inlet of the guide vane, increases the exclusion coefficient, and blocks part of the channel, and with the regularity of the flow in the end wall and guide vane channel, the liquid flow forms a high-speed flow area near the end wall boundary with the flow velocity distributed between 3.29 and 3.95 m/s. The velocity distribution in the impeller outlet flow in the circumferential direction shows an axial downward movement. At the same time, due to the influence of the outlet annular cochlear chamber, a large-scale vortex structure appears in region B at the connection between the guide vane outlet and the annular cochlear chamber, resulting in energy loss.

With increase in the flow rate, the energy due to dynamic pressure in the mainstream flow in the guide vane channel increases continuously, the high-speed area at the outer edge of the guide vane moves downstream, the internal high-speed area gradually expands, the dynamic and static interference effect is weakened, the vortex structure at the inlet of the guide vane gradually disappears, and the scale of vortex structure at the connection between the guide vane outlet and the annular volute chamber also decreases. For the operating condition of $1.2Q_{\text{des}}$ flow rate, the regularity of

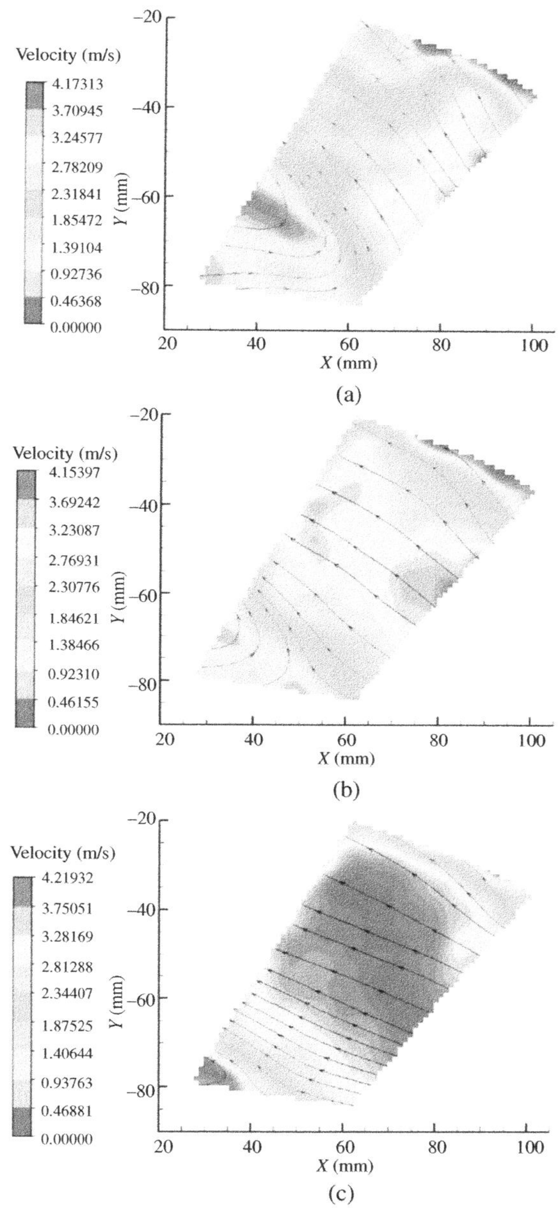


**Figure 5.33** PIV measurement results of the section between impeller and guide vane: (a) $0.8Q_{des}$, (b) $1.0Q_{des}$, and (c) $1.2Q_{des}$.

the flow in the guide vane is fully reflected. The flow is very stable and there is no obvious vortex flow. At this time, the average velocity of the fluid in the test plane is about 3.0 m/s.

The relative position of the impeller and the guide vane affects the flow field structure near the impeller, which has great significance for the design of the impeller. It is therefore important to note here that due to the instability of the flow under the low flow rate conditions, the impeller inlet reflux is inclined to shed vortices from the blade in the part-loading conditions. Therefore, this chapter also focuses on the analysis of the interference flow field with different phases in part-loading conditions. For this purpose, the flow fields shot in the shooting sections which are at

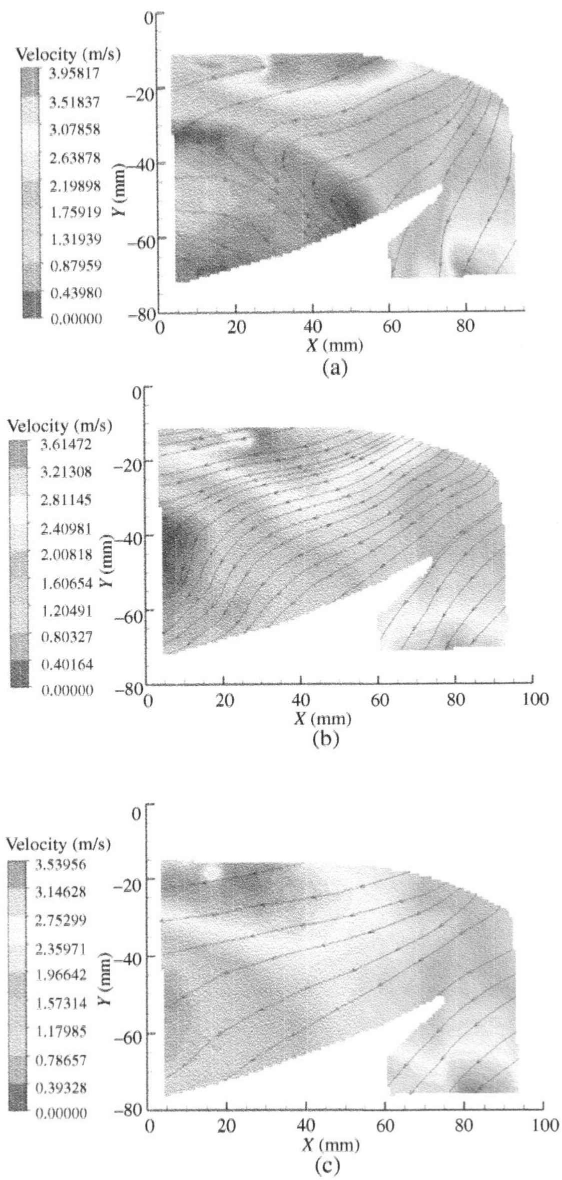


**Figure 5.34** PIV measurement results of the section in guide vane: (a) $0.8Q_{des}$, (b) $1.0Q_{des}$, and (c) $1.2Q_{des}$.

different phase positions noted as $1T$, $2T$, $3T$, $4T$, $5T$, and $6T$ at a very low flow rate of $0.2Q_{des}$ are shown in Fig. 5.35. These selected six phases are in just one rotation period of the impeller, so that the changes in the interference flow field of the impeller in one rotation period can be predicted.

As shown schematically in Fig. 5.35, due to the shadow that prior impeller blades had cast on the shooting sections and the influence of phase change during the rotation process, which changes as the impeller goes from $1T$ to $6T$ in one rotation, the flow field areas in the shooting sections first increase and then decrease. In the coherent region between the impeller outlet and the guide vane inlet, due to the influence of RSI, in the part-loading conditions, it can be noted that the internal flow in the interaction region is disorderly, and the changes in the impeller reflux and the vortex

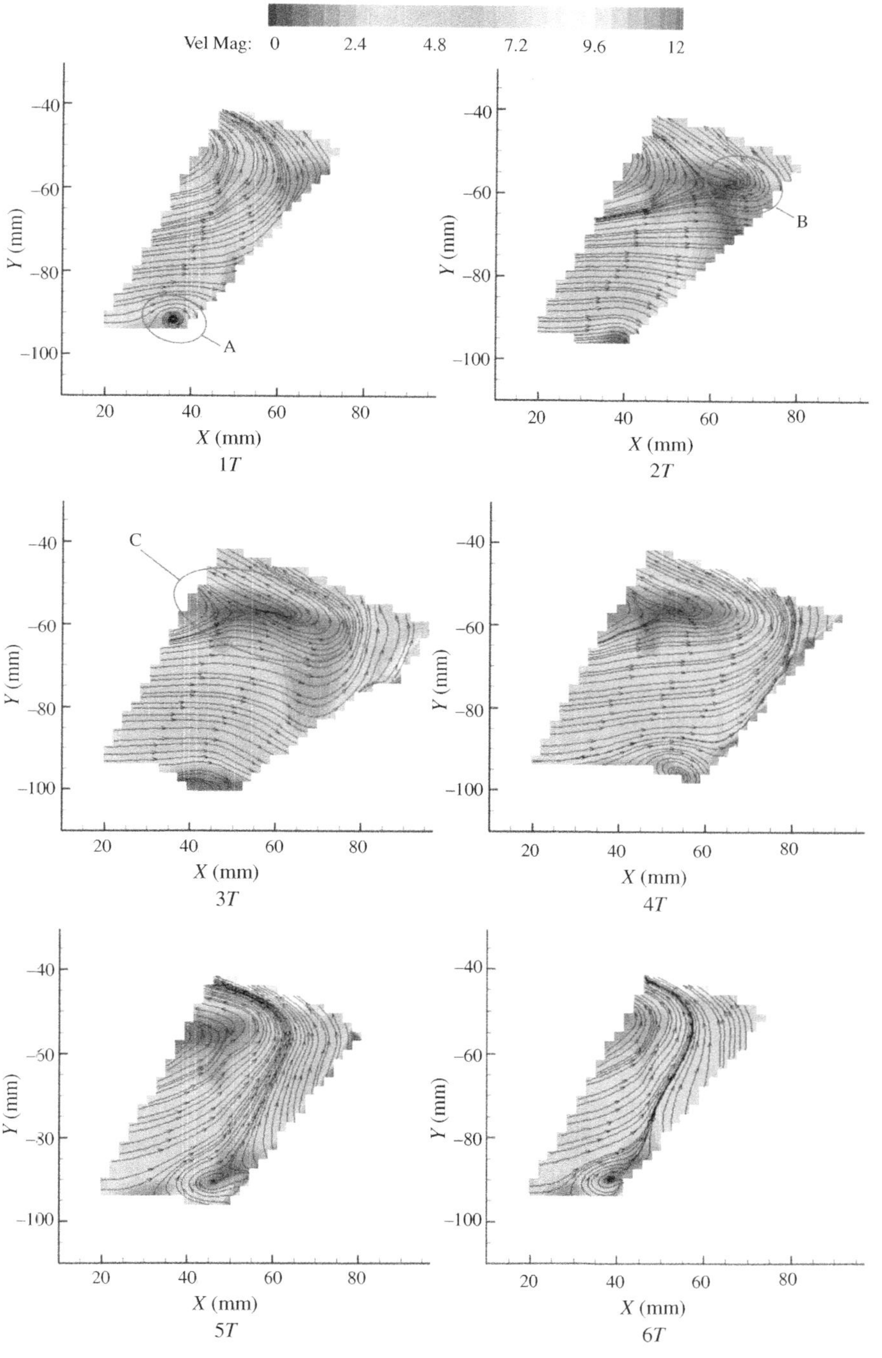


**Figure 5.35** Velocity distribution of flow field at different phases under part-loading condition.

flow can be seen in different phases. When the impeller is in the $1T$ phase, the inlet side of the impeller blade crosses the shooting section and since the flow rate is small and the internal flow in the vanes is unstable, there is fluid reflux from the inner vane to the interference flow field in the interaction region near the hub. It is affected by the main flow from the impeller outlet and flows along the radial direction. When the fluid is close to the end wall, the fluid flows along the main direction. At the same time, since the flow leaves from the prior blade suction surface, it forms a small-scale vortex (as shown in picture A) in the interaction region near the hub, and the coordinates of the vortex center are approximately (38, −93). When the impeller is in the $2T$ phase (which is the time when the inlet side of the impeller blade crosses the shooting section at a 15° phase angle), it can be found in the interference region that the impeller reflux in the guide vanes inlet is still present. Affected by the interference caused by the trailing edge of the prior blade, the impeller reflux area is stretched near the impeller outlet, and a low-speed vortex attachment region is formed (as shown in picture B). Meanwhile, the small-scale vortex center which is in the interference area close to the hub, moves about 5 mm toward the hub along the radial direction compared to the $1T$ phase. When the impeller is in the $3T$ phase, the shooting section is near the middle of the blade, and the shooting area reaches the maximum value. The flow field goes through the obvious disturbance caused by the RSI and forms the vortex structure caused by the impeller outlet stream, and the impeller reflux in the vanes increases further, as shown in picture C. The scale of the vortex structure, which is at the suction side of the prior blade, has increased, and the vortex center has moved along the hub toward the impeller inlet. When the impeller is in the $4T$ phase, the overall trend of the flow velocity distribution is almost identical to that in the $3T$ phase; the shooting section passes the middle of the blade, and the low-speed area near the interference flow field of the guide vane inlet contracts toward the inlet. Also, the small-scale vortex structure, which is near the hub of the impeller outlet, is reduced, and the vortex center position moves about 10 mm in the positive $X$-axis direction compared to that in the $3T$ phase section. When the impeller is in the $5T$ phase, the interference section becomes smaller with the work of the impeller increasing, and the flow velocity that exists near the end wall and impeller blades outlet is comparatively larger with closer streamlines. This results in a subsequent decrease in the low-speed reflux zone near the guide vane inlet and in the small-scale vortex structure near the hub. When the impeller is in the $6T$ phase, the shooting section is almost near the outlet side of the impeller blades, and the velocity distribution is very similar to that in the $5T$ phase; the low-speed area in the guide vane inlet reduces further rotating counterclockwise in the form of an ellipse and transforming gradually into $1T$ phase.

Vorticity, a common physical flow quantity that is used to describe the vortex structure, can be used to measure the magnitude and direction of the vortex. It is defined as the curl of the velocity field, and its value is used to define the intensity of the vortex and its positive and negative value indicates the direction of the movement of the vortex (counterclockwise direction is defined as positive). The vortical contours shot in the shooting sections at different phase positions $1T$, $2T$, $3T$, $4T$, $5T$, and $6T$ in the low flow rate $0.2Q_{des}$ operating condition are shown in Fig. 5.36.

From Fig. 5.36, in the six shooting phases, it can be noted that both the positive and negative vortex structures exist in all interference flow fields. The positive vortex structures are always near the end wall region of the interference flow field and when the shooting phase becomes close to the trailing edge of the blade, the intensity of the positive vortex structures becomes higher. Furthermore, there always exist a few negative vortex structures near the hub in the interference flow field, and as the shooting phase becomes close to the trailing edge of the blade, the strength of the vortex structure first decreases and then increases. In the $1T$ phase, due to the influence of the mainstream flow at the blade inlet and impeller reflux in the guide vane inlet, the positive vortex structure with

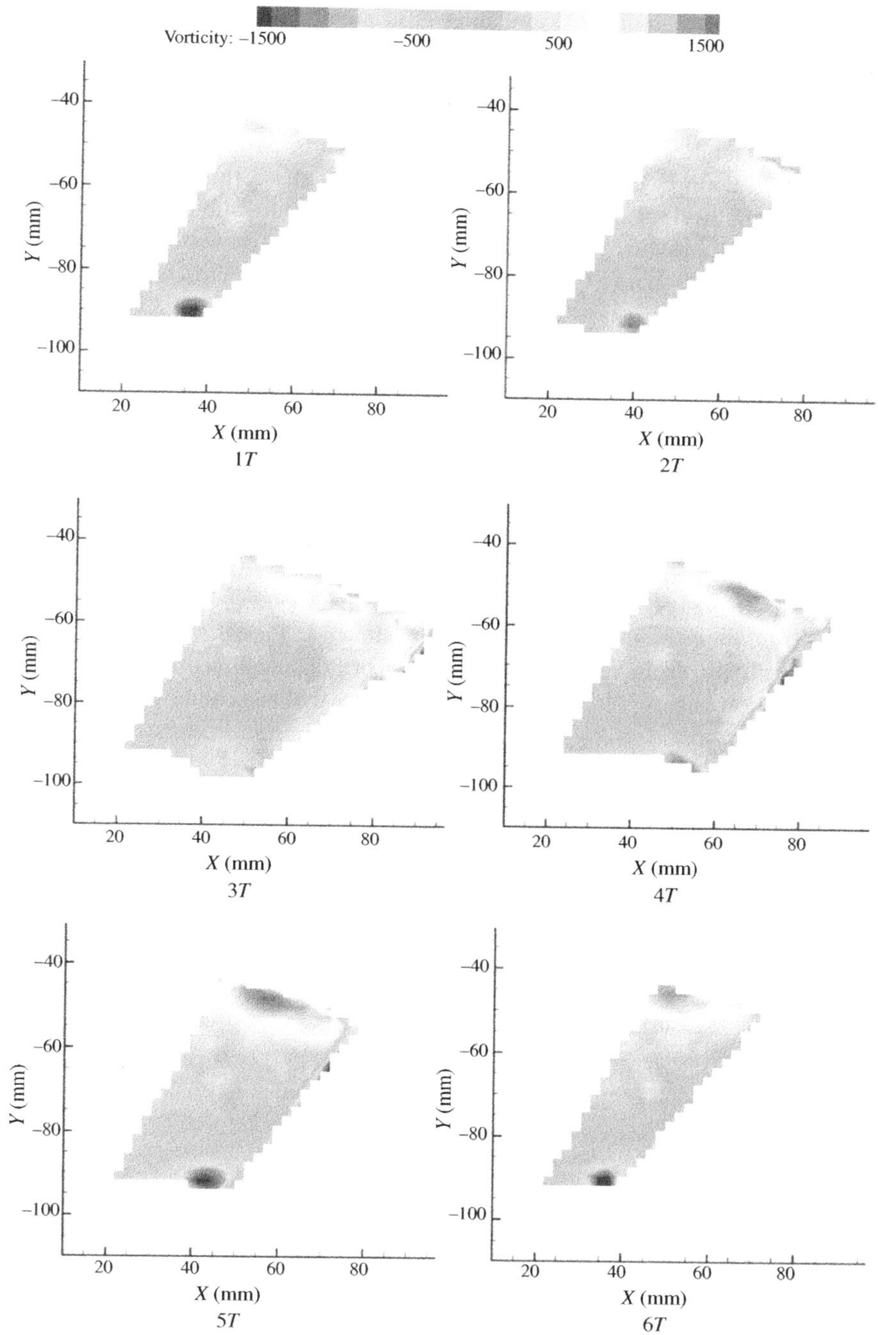


**Figure 5.36** Vorticity distribution of flow field at different phases under part-loading condition.

low intensity appears near the end wall of guide vane inlet. Also, since the flow detaches from the prior blade suction surface, the negative vortex appears near the hub with high intensity. In the $2T$ phase, there is a positive vortex structure near the end wall of the impeller outlet and a weak negative vortex structure near the hub. In the $3T$ phase, the intensity of the positive vortex structure in the end wall region increases significantly, while the intensity of the corresponding negative vortex further weakens. When the shooting section is located at the $4T$ or $5T$ phase, the intensity of both the positive vortex near the end wall region and the negative vortex near the hub increases significantly. When the impeller rotates to the $6T$ phase, the position of the positive vortex structure in the interference flow field is close to that in the $1T$ phase, but its intensity becomes higher.

### 5.3.7 Relative Velocity Distribution at the Monitoring Lines in Different Phases at the Impeller Inlet and Outlet Under the Part-loading Conditions

In order to quantitatively analyze the velocity magnitudes at different shooting sections at $0.2Q_{des}$ flow rate, three monitoring lines, which are parallel to the pump axis at $Y = -90$, $Y = -100$, and $Y = -110$, are selected at different positions of different shooting sections. These three monitoring lines are in the flow field at the impeller outlet. In addition, the other three monitoring lines are selected at the impeller outlet, which are parallel to the axis at $Y = 1.75X - 127.5$, $Y = 1.75X - 145$, and $Y = 1.75X - 162.5$. The spatial locations of these six monitoring lines are shown in Fig. 5.37. The relative velocity distributions at each monitoring line are shown in Figs. 5.38 and 5.39. Since the two-dimensional PIV only shoots the velocity field in the two orthogonal directions, the relative velocity of the shooting section only represents the particle velocity that is a combination of both the axial and radial directions.

The relative velocity distributions at monitoring lines which are at $Y = -90$, $Y = -100$, and $Y = -110$ in different phases of the mixed-flow pump inlet for $0.2Q_{des}$ flow rate, are shown in Fig. 5.38. As can be seen from this figure, the overall trends in the velocity distributions at these three monitoring lines are almost the same. When $X > 150$ which is the inlet segment region,

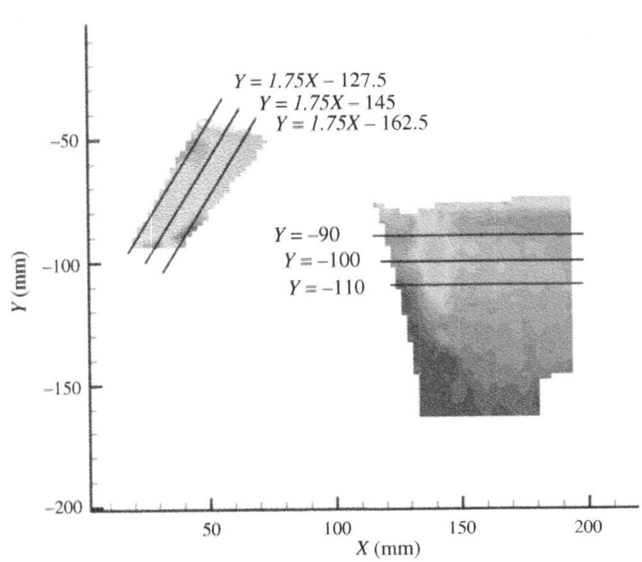


**Figure 5.37** Monitor lines.

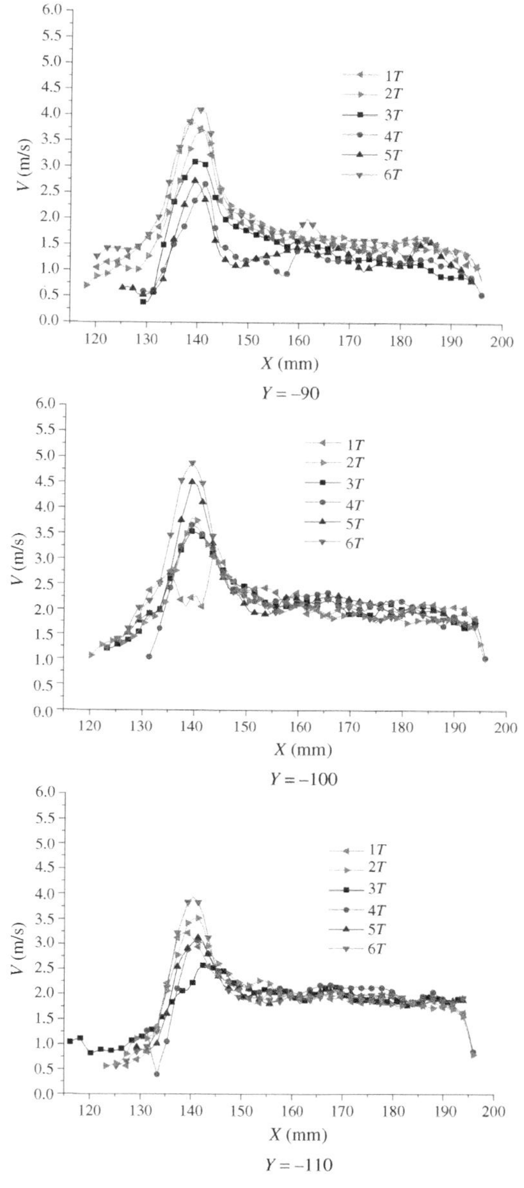


**Figure 5.38** Relative velocity distributions at monitoring lines which are at $Y = -90$, $Y = -100$, and $Y = -110$ in different phases of the mixed-flow pump inlet for $0.2Q_{des}$ flow rate.

the relative velocity is stable, and the fluid slightly accelerates the flow into the impeller inlet. In the $X = 35$–$150$ inlet area, due to the influence of the impeller rotational acceleration, the fluid velocity increases rapidly, reaching a peak on the inlet side of the blade at $X = 140$, and the impeller entrainment effect is significant. When $X < 135$, the impeller does work on the fluid, so that the axial velocity of the fluid gradually becomes the axial velocity component perpendicular to the shaft surface, and the axial velocity decreases. The relative velocity of the fluid decreases by about 1 m/s in the $X < 135$ region for each stage compared to the inlet region for $X > 150$. The relative velocities near the impeller inlet vary considerably at the three monitoring lines, with the maximum velocity occurring in the $6T$ stage. For comparing the velocity distributions at the

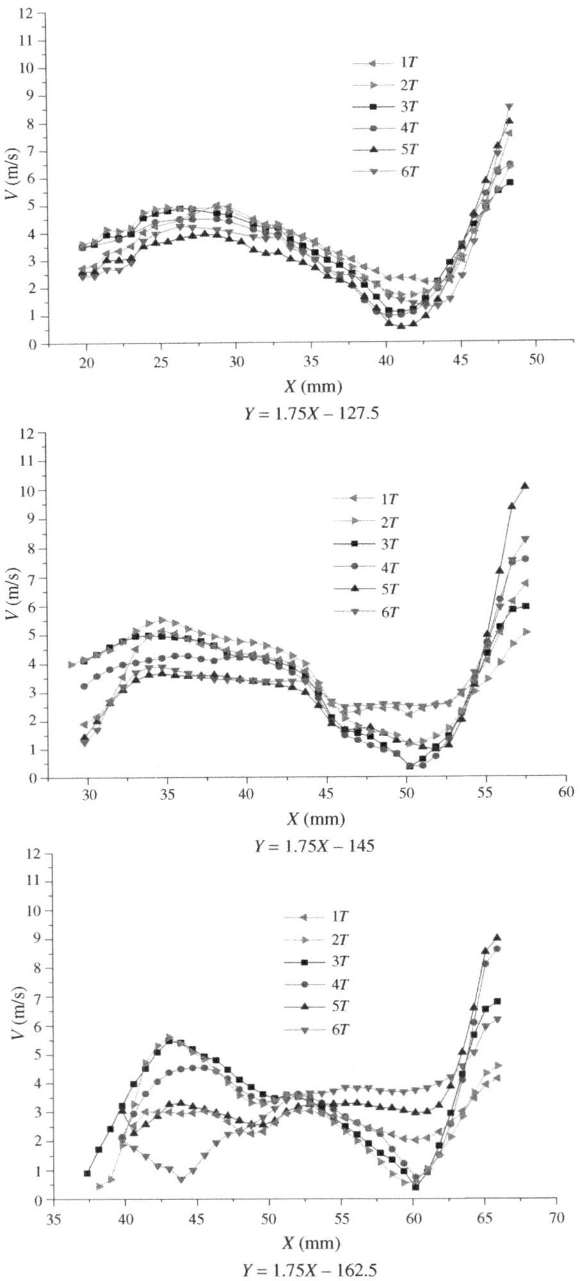


**Figure 5.39** Relative velocity distributions at different phases of the pump for $0.2Q_{des}$ flow rate at monitoring lines near the impeller whose locations are $Y = 1.75X - 127.5$, $Y = 1.75X - 145$, and $Y = 1.75X - 162.5$.

three monitoring lines, it should be noted that the $Y = -100$ straight line has more distance from the boundary layer than the $Y = -90$ straight line and is much closer than $Y = -110$ straight line to the outer edge of the impeller. Therefore, the relative velocity at $Y = -100$ straight line is slightly higher than that at the other two monitoring lines. The relative velocity at the $Y = -90$ monitoring line is relatively low due to the boundary layer effect.

Figure 5.39 shows the relative velocity distributions at different phases of the pump for $0.2Q_{des}$ flow rate at monitoring lines near the impeller whose locations are $Y = 1.75X - 127.5$, $Y = 1.75X - 145$, and $Y = 1.75X - 162.5$. It can be seen from this figure that the relative velocity distributions at the three monitoring lines (which are in the interference region of the impeller inlet) have significant differences from each other. The relative velocity distributions near the monitoring lines at the impeller inlet are stable and similar in each phase, which indicates that the guide vanes have some rectifying effect. But farther from the guide vane inlet, the velocity distribution is more disorderly in each phase due to strong RSI. At the $Y = 1.75X - 127.5$ monitoring line, the relative velocity distributions of each phase along the negative $X$ line decrease first and then increase, while near the hub, they show a slight decrease. Combining these results with the velocity distributions in Fig. 5.36, it can be noted that at $X = 41$ near the end wall, due to the presence of the reflux low-velocity zone of guide vane inlet, all the relative velocity distributions of each phase in this region have a minimum value. In the $1T$ shooting phase, there is the highest relative velocity while in the $5T$ shooting phase, there is minimal relative velocity distribution. This is consistent with the distribution of the relative velocity field in each phase. At the $Y = 1.75X - 145$ monitoring line, the trend of relative velocity distribution in each phase is similar to that at the $Y = 1.75X - 127.5$ monitoring line; the overall trend first decreases and then increases along the direction of the negative $X$-axis, while near the hub shows a slight decrease. However, because of dynamic and static interference, there is a significant difference between the relative velocities of the fluid in the central position, in the low reflux region, and in the end wall region; the extremes in the relative velocity occur in different stages and in the reflux low-velocity region along the $X$-axis. At the $Y = 1.75X - 162.5$ monitoring line, since the flow is much closer to the impeller inlet, the RSI is more significant than in case of the other two monitor lines. The relative velocity distribution of each phase has a remarkable difference; at the same location in different phases, the relative velocity difference can be up to 5 m/s. The small-scale vortex near the hub and the low velocity region near the end wall region vary with the phase angle, showing different scales and locations resulting in a disturbed fluid velocity distribution on the monitoring line. Thus, the dynamic and static interference under partial loading conditions is the source of the unstable characteristics of the flow field in the impeller outlet and guide vane inlet gap region.

## 5.4 Orbit of Shaft Centerline

### 5.4.1 Measurement of the Spindle Axis Trajectory

Since the rotation error in the rotor of the mixed-flow pump is generally much greater than the manufacturing error, the shape error in the roundness of rotor shaft can be neglected. In the test reported here in this section, the two-way dynamic measurement technique [13] is used to measure the spindle axis trajectory as it rotates. Since the motion of the rotor spindle is a two-dimensional planar motion, at least two sensors are required to collect data in a cross-section of the spindle at the same time and then process the displacement data of each sensor to obtain the axis trajectory of the spindle as shown in Fig 5.40. The Bentley 408 data acquisition system is employed to quantify the vibration of the rotor shaft in the design flow condition, as shown in Fig. 5.41. The system consists of 408 dynamic signal processing equipment (DSPi) and ADRE Sxp software.

The experiment is conducted with a 3300 XL 8 mm eddy current displacement sensor consisting of a probe, a proximal wire, and an extension wire. The diameter of the probe is 8mm, the type of probe is 330130-040, the type of proximitor is 330180-50, the extension of the cable is 4 m, and

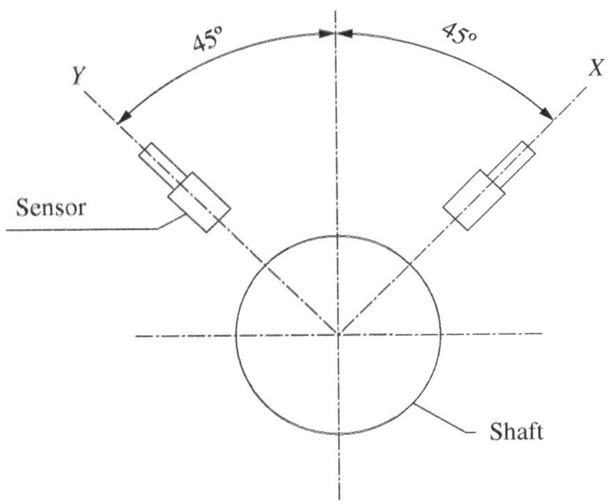


Figure 5.40 Position of the eddy current displacement sensors.

Figure 5.41 Bentley 408 data acquisition system.

the metal for calibration is No. 45 steel. The gap voltage of the sensor is −10 V, the sensitivity of No. 45 steel is 7.87 V/mm, and the range is −2 to +2 mm. The sensor is installed at the midpoint of the line so that the measurement range is equal in positive number and negative number. A digital multimeter is installed at the OUT terminal of the proximitor. It is important to observe the output voltage of the OUT and COM terminal when the probe is screwed into the hole, and the sensor should be fixed when the voltmeter is at −10 V so that the sensor is installed at the middle point of the line. The key phase sensor used in this experiment is an infrared sensor. This sensor is only used to stick the luminous on the rector and its probe is installed perpendicular to the rector. In order to exclude the influence of the driving device, the power supply and the ground wire of the eddy current displacement sensor are independent and not shared with the electric motor. The electromagnetic interference is also considered; therefore, the drive device power of the electric motor is more than 10 m away from the eddy current displacement sensor. The double-stranded double-layer shielding cable is adopted in the eddy current displacement sensor, which can doubly protect the measuring signal.

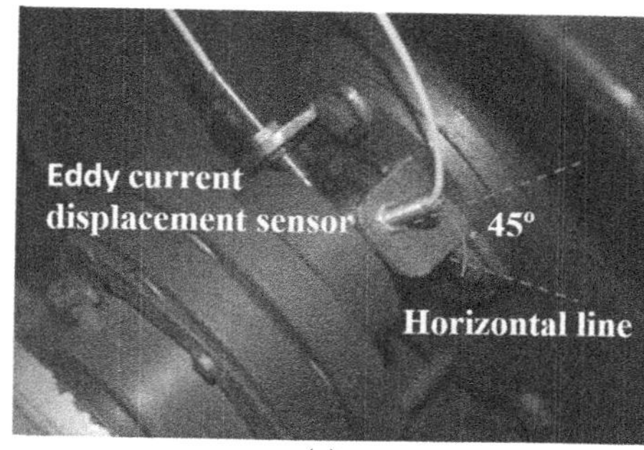


(a)

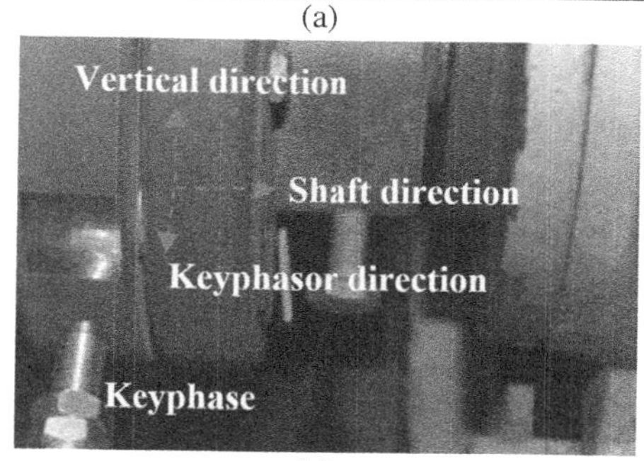


(b)

**Figure 5.42** Measurement of rotor axis locus: (a) position of eddy current displacement sensor and (b) position of key phase.

The vibration of the rotor is measured by the bidirectional dynamic measurement in the experiment [12]. According to the experimental requirements, the data acquisition and detection system is connected, and the monitoring points of the axis orbit are selected on the side of the connection of the rotor shaft coupling and the electric machinery. The eddy current displacement sensor is installed with the key phase as required and is connected to the Bentley 408 data acquisition system. The eddy current displacement sensor is vertically intersected with a distance of 1 mm between the probe and the rotor. The key phase is mounted vertically on the rotor shaft with 5 mm between the reflective paper and the rotor shaft. The measurement method and the experimental arrangement of the axis orbit are shown in Fig. 5.42.

### 5.4.2 Characteristics of Vibrations of the Shaft System

The vibration of the rotor is measured. The magnitude of the rotor vibration is represented by the axis orbit of the rotor. The shaft track is an important graphic representation of the rotor vibration state, which contains a large amount of fault information and is an effective way to visually obtain the actual motion of the rotor [14]. The time-domain diagram of the axis orbit can provide the change in the vibration quantity with time, and the frequency spectrum can reflect the frequency component of the complex signal.

In the processing of the axis orbit signal, the Butterworth filter is used to filter the original signal and obtain the 1$X$ and 2$X$ frequency of the rotor axis orbit. The characteristic of the Butterworth filter is that the frequency response curve in the pass frequency band is maximally flat without fluctuation, while the frequency response curve in the blocked frequency band gradually decreases to zero. The relationship between the square of the amplitude of the Butterworth filter and the frequency can be expressed as follows:

$$G^2(\omega) = |H(j\omega)|^2 = \frac{G_0^2}{1 + \left(\frac{\omega}{\omega_c}\right)^{2n}} \tag{5.17}$$

where $G$ represents the magnification, $H$ represents the transfer function, $j$ is the imaginary value of unity, $\omega$ is the angular frequency of the signal in rad/s, $n$ is the order of the filter, $\omega_c$ is the cut-off frequency (frequency is about −3 dB) in rad/s and $G_0$ is the DC gain (gain at frequency of 0).

Using the Laplace transform and setting $s = \sigma + j\omega$ in the two-dimensional complex plane, it can be shown that

$$|H(s)|^2 = H(s)\overline{H(s)} \tag{5.18}$$

where $H(s)$ is the transfer function of the complex variable $s$.

From the general nature of the Laplace transform, it can be shown that $s = j\omega$ and $|H(-j\omega)| = \overline{H(j\omega)}$. Therefore, Eq. (5.19) can be obtained by the analytic continuation as

$$H(s)H(-s) = \frac{1}{1 + \left(\frac{-s^2}{\omega_c^2}\right)^n} \tag{5.19}$$

where $H(-s)$ is the transfer function of the complex variable $(-s)$.

The poles of the function in Eq. (5.19) are equally distributed on the circle whose radius is $\omega_c$. Therefore, Eq. (5.20) can be obtained as:

$$\frac{-s^2}{\omega_c^2} = (-1)^{\frac{1}{n}} = e^{\frac{j(2k+1)\pi}{n}} \tag{5.20}$$

where $k = 0, 1, 2, \ldots, n-1$. Thus, Eq. (5.21) can be obtained from Eq. (5.20) as:

$$s_k = \omega_c e^{\frac{j\pi}{2}} e^{\frac{j(2k+1)\pi}{2n}} \tag{5.21}$$

where $s_k$ is the value of the complex variable at the $k$th pole.

The transfer function can be represented by these poles, and Eq. (5.22) is obtained as follows:

$$H(s) = \frac{1}{\prod_{k=1}^{n}(s - s_k)/\omega_c} \tag{5.22}$$

where the denominator of the Eq. (5.22) is the Butterworth polynomial.

### 5.4.3 Original Axis Trajectory Diagram and Time-domain Diagram

The original rotor-axis-center trajectory and the time-domain diagram of the rotor at $\Delta t = 0.3$ seconds for various flow conditions are measured as shown in Fig. 5.43. Examining the overall trend, the axis-center trajectory measured in the test for various flow conditions is not ideally smooth and circular but has nearly the shape of the figure of "8" with many serrated sharp angles; however, the shape and size of the axis-center trajectory shows good repeatability indicating that the rotor whirl is relatively stable. According to the rotating machinery fault research literature [15], this feature shows that there is a typical misalignment fault in the rotor. Since the test monitoring point is very close to the bearing, it is likely that the alignment between the rotor and the support bearing is out of the design tolerance. For various flow rates, the trajectory of the rotor axis center is accompanied by the changes in the serrated sharp angles, and there are many "burrs," indicating that either there is inevitable friction between the rotor and the bearing pair or there are other high-order harmonic vibrational components during the operation of the pump such as the hydraulic vibration and the noise, which would have a considerable impact on the axis-center trajectory.

By observing the time-domain diagrams of the axis-center trajectory in the $X$ direction and the $Y$ direction at various flow rates, it can be seen that at various flow rates, the time-domain diagram in the $X$ direction is a periodic distorted sine wave, and there is a second frequency with a secondary

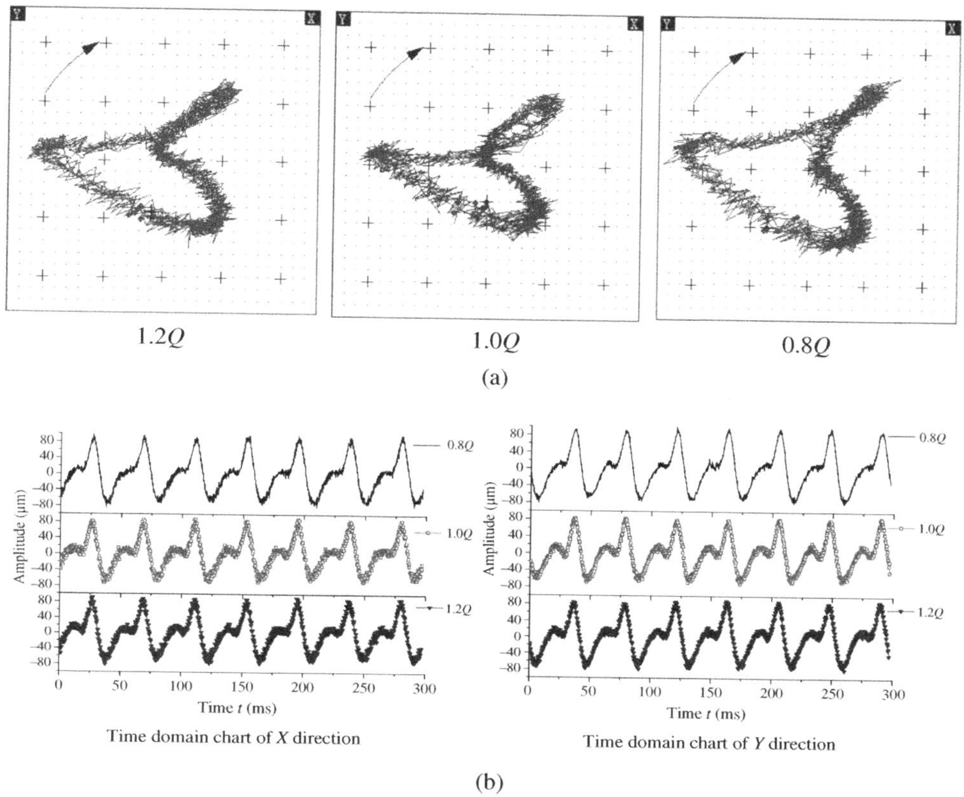


**Figure 5.43** Rotor axis locus chart and time-domain chart of the original rotor axis locus: (a) original rotor axis locus chart and (b) time-domain chart of the original rotor axis locus.

peak on the sine wave of the power frequency; on the other hand, the time-domain diagram in the *Y* direction is a periodic distorted cosine wave, and the phase difference between the time-domain diagrams in *X* direction and *Y* direction is just one-fourth cycle.

Comparing the time-domain diagrams for the three flow rates, it is found that the change in the flow rate has a significant impact on the rotor axis-center trajectory, and the fluctuation amplitude is at a low level at the design flow rate. From the high flow rate to the design flow rate with decrease in the flow rate, the peak value of the time-domain waveform in the *X* direction and the *Y* direction decreases, and the amplitude of the axis-center trajectory also decreases; it shows that the influence of the hydraulic excitation on the shaft vibration decreases with decrease in the flow rate at this stage. During the change from the design flow rate to smaller flow rates, the peak value of the time-domain waveform in the *X* direction and the *Y* direction continues to increase, and the amplitude of the axis-center trajectory continues to increase, which indicates that the vibration condition of the shaft system continues to deteriorate as the flow rate of the pump shifts toward smaller flow rates. This is consistent with the empirical observation that it is easy to cause vibrational instability in the rotor system when the hydraulic machinery operates under eccentric, nonaligned shaft conditions. The vibration inducements at different frequencies are not analyzed and presented in this chapter. The reader is recommended to refer to recent research papers for this information.

### 5.4.4 Decomposition and Refinement of the Rotor Axis-centered Trajectory

In order to obtain an accurate rotor axis-centered trajectory as well as to accurately analyze the influence of the rotor vibration, whirl, eccentricity, and other design and manufacturing faults on the axis-centered trajectory for various flow rates, the original axis-centered trajectory is decomposed and refined by excluding the ultra-high harmonic signals such as noise and electromagnetic interference, and only the main factors such as power frequency and double frequency are retained.

The first harmonic axis-centered trajectory and the time-domain diagram of the mixed-flow pump rotor are obtained from the experiment, as shown in Fig 5.44. This figure shows that the shaft center track diagram of frequency doubling for the three flow rates is approximately circular, and with the decrease in flow rate, the shaft center tracking contour diagram of the rotor changes from large to small and then to large. Similar to the variation in the trend of the axis-center tracking diagram of frequency doubling for the three flow rates, the peak value of the axis-center tracking time-domain diagram of frequency doubling in the $X$ direction and the $Y$ direction first decreases and then increases with a decrease in the flow rate, and the peak value is the smallest at the design flow rate.

Since the rotor axis-centered trajectory diagram of the frequency doubling is approximately circular, and the circle is drawn in the counterclockwise direction opposite to the direction of the rotor rotation, according to the rotating machinery fault research literature [16], it can be concluded that the difference in the direction of the bearing stiffness is very small and the influence

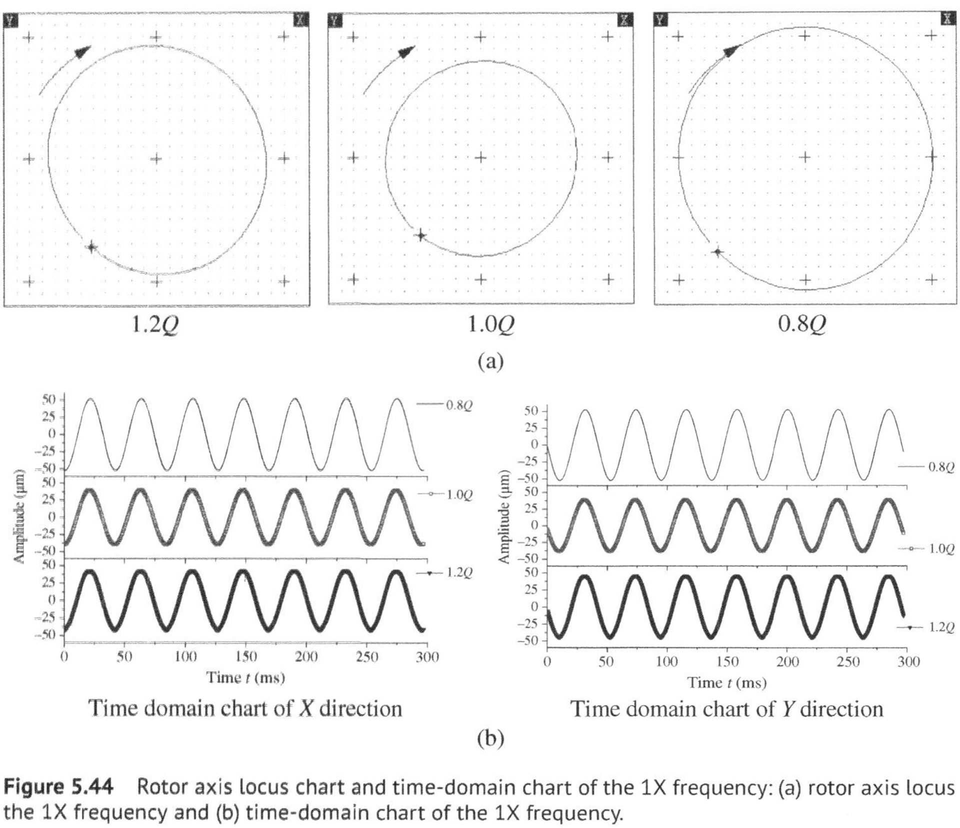


**Figure 5.44** Rotor axis locus chart and time-domain chart of the 1X frequency: (a) rotor axis locus chart of the 1X frequency and (b) time-domain chart of the 1X frequency.

of bearing related factors on the frequency doubling can be ignored. At the same time, it can also be concluded that there is reverse (precession accurate) rotary whirl in the operation of the pump, i.e., there is a power frequency vibration caused by the rotor imbalance. In the process of going from high flow rate to design flow rate, the rotary whirl decreases gradually with decrease in the flow rate, indicating that the power frequency vibration caused by the rotor imbalance decreases gradually with decrease in the flow rate from the design flow rate. In the process of changing from the rated flow rate condition to smaller flow rate condition, this rotary whirl increases gradually with decrease in the flow rate, indicating that the power frequency vibration caused by the rotor imbalance increases gradually with the decrease in the flow rate. In addition, the peak value of the power frequency vibration appears at flow rate of $0.8Q_{des}$, and the fluctuation amplitude increases by 23% relative to $1.2Q_{des}$ flow rate in the $X$ direction and 30% relative to the design flow rate. In the $Y$ direction, the fluctuation amplitude increases by 17.5% relative to $1.2Q_{des}$ flow rate and 36.9% relative to the design flow rate or 1.35 times relative to the design flow rate. It can be seen that when the flow rate drops to smaller flow rate, multi-scale vortices from backflow and secondary flow appear in the mixed-flow pump. The medium and high-frequency excitation force caused by these unstable flow fields are the main reason for the deterioration of the power frequency vibration. Thus, in order to improve the reliability and service life of the rotor system, the mixed-flow pump should operate as far away as possible from the low flow rate conditions.

In a similar fashion, the original axis trajectory is decomposed and refined into the second harmonic axis trajectory diagram and the time-domain waveform diagrams in the $X$ direction and the $Y$ direction, as shown in Fig. 5.45. It can be seen from this figure that the axis trajectory of the frequency doubling is an ellipse whose long axis is approximately in the horizontal direction and has a flat shape. The length of the long axis of the ellipse is much larger than that of the short axis, and the axis trajectory of the frequency doubling almost coincides with the short axis direction in the design flow condition. As the flow deviates from the design operating point, the shape of the axis trajectory remains a flat ellipse with a gradual increase in the length of the long and short axes. At the same time, the amplitude of the waveform under double frequency increases gradually with a change in the flow rate. By observing the double frequency axis trajectory diagram and the time-domain diagram, since the double frequency axis trajectory diagram is an ellipse with significant difference between the long and short axes and the long axis of the ellipse is located in the horizontal direction, according to the rotating machinery fault research literature [17], it can be concluded that this is caused by the misalignment of the mixed-flow pump rotor and this misalignment has significant impact on the vibration in the horizontal direction. It is one of the main factors that lead to the outer "8" shape in the original axis trajectory and the existence of double frequency subpeak in the waveform. By comparing the double frequency axis-center track diagram and the time-domain diagram for different flow rates, it can be seen that the vibration amplitude reaches the maximum value with the increase in hydraulic load under low flow rates because the vibration due to misalignment is more sensitive to the change in the rotor load.

### 5.4.5 Flow Spectrum Analysis

Flow spectrum diagrams of the flow measuring points in the $X$ direction and the $Y$ direction for various flow conditions are drawn, as shown in Fig. 5.46. Since the rated speed of the pump is 1450 r/min, the rotor speed frequency $f$ can be calculated to be 24.17 Hz. It can be seen from Fig. 5.46 that the energy is concentrated in the spectrum with one- and two-times the frequency accounting for the main vibrational energy, while the three and four times the frequency account

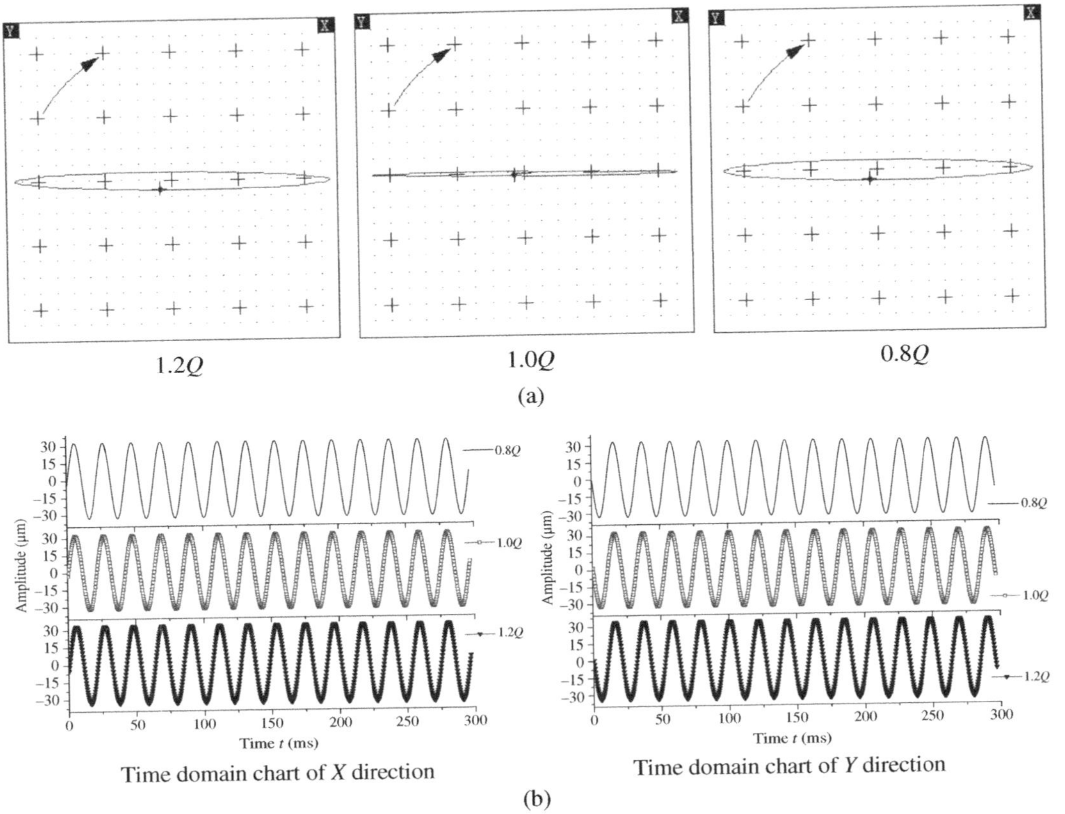


Figure 5.45 Rotor axis locus chart and time-domain chart of the 2X frequency: (a) rotor axis locus chart of the 2X frequency and (b) time-domain chart of the 2X frequency.

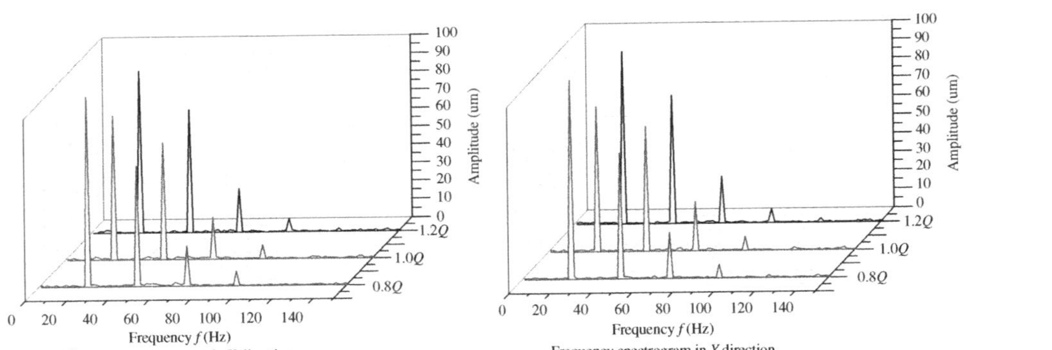


Figure 5.46 Frequency spectrogram in X and Y direction.

for a small portion of energy. Comparing the flow spectrum at different flow rates, it can be seen that the amplitude of the first harmonic frequency is quite different, and the amplitude of the second harmonic frequency is slightly different; however, the amplitude of other harmonic frequencies is basically similar at various flow rates. The amplitude of the first harmonic frequency and the second harmonic frequency at the design flow rate is the smallest, while the amplitude of the first harmonic frequency and the second harmonic frequency for flow rates different from the design flow rate increases. Combining the first harmonic axis trajectory diagram and the time-domain diagram for various flow rates, it can be concluded that when the mixed-flow pump deviates from the design flow point, the imbalance in the rotor increases, resulting in the continuous increase in the vibration amplitude of the rotor with impact on vibration of the entire rotor system. Combining the second harmonic axis trajectory diagram and the time-domain diagram for various flow rates, it can be concluded that as the difference in the flow rates from the design flow rate increases, the misalignment of the rotor increases. The measurements show that as the flow rate deviates from the design flow rate, the radial offset and the misalignment of the rotor increase, and the vibration of the rotor shaft system intensifies. The power frequency vibration is the main component in vibration deterioration. The double frequency also significantly affects the vibration of the shaft system; however, other double frequencies have little effect on the vibration of the shaft system. The contribution of imbalance and misalignment to the shaft vibration achieves a peak at very small flow rates.

### 5.4.6 Effect of the Rotor–Stator Interaction on the Shaft Vibration

In order to study the characteristics of pressure fluctuations caused by the interaction between the guide vane and the impeller in the mixed-flow pump at design flow rate, the experiment focused on the pressure fluctuation characteristics at the impeller inlet (monitor position A), the middle of the impeller (monitor position B), the impeller outlet (monitor position C), the inlet of the guide vane (monitor position D), the outlet of the guide vane (monitor position E) and the pump unit outlet (monitor position F). Thus, there are six pressure monitoring sensor that are separately attached in different parts of the pump to capture the pressure pulsations. The arrangement of the monitoring points is shown in Fig. 5.47. The monitoring point F is placed at the outlet of the pump unit, and all other monitoring points are on the axis in the horizontal plane. The monitor position F is not shown in Fig. 5.47.

Based on the time and frequency domain diagrams of the pressure pulsation in different positions, it can be noted that due to the rotation of the impeller and the RSI between the impeller and the guide vane, periodic pressure fluctuations in the pump are generated. Therefore, in the process of impeller rotation, the fluctuating pressure acting on the rotor has a periodic force on the rotor shaft of the pump.

Since the effect of the fluid flow on the rotor system is mainly reflected on the impeller, therefore the influence of the fluctuating pressure in the impeller area on the vibration of the shaft system is analyzed. At monitoring point B, the amplitude of the pulsation has four peaks in one rotating cycle of the impeller and there are one or two secondary peaks that occur simultaneously. This indicates that the impeller is not only affected by the RSI but also is affected by the secondary flow such as leakage flow and backflow in the end wall region. As a result, the periodic flow vibrations on the rotor system are generated, and the constantly changing zigzagged sharp corners appear on the original axis orbit.

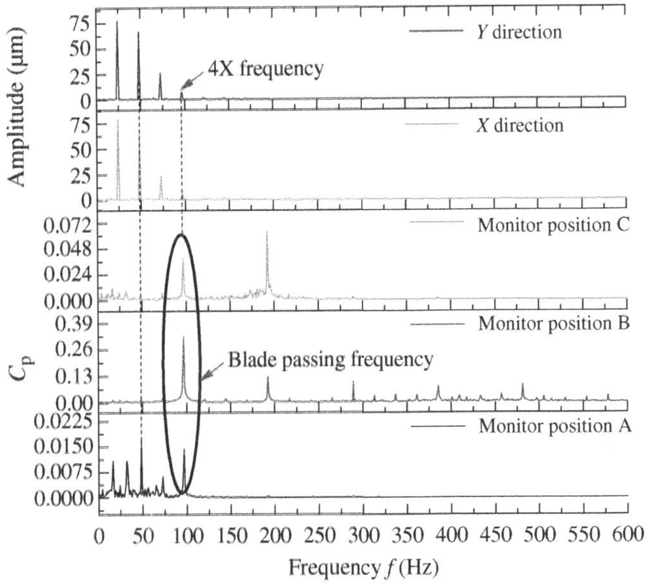


**Figure 5.47** Frequency domain diagram of pressure fluctuation and axis orbit.

## 5.5 Summary

This chapter describes the experimental methods used in the data collection and analysis of the flow field and performance characteristics of a mixed-flow pump under different operating conditions. The details of the energy performance test, PIV measurements of the flow field, pressure pulsation monitoring and data collection, and axis orbit signal collection are provided. The experimental results presented in this chapter can be summarized as follows:

(1) The change in the tip clearance can affect the head and efficiency of the mixed-flow pump. Considering various tip clearances, when the tip clearance increases from 0.2 to 1.1 mm, the head and efficiency of the pump decrease by 18.1% and 11.6%, respectively, at design flow rate. When the tip clearance is 0.2 mm or less than 0.5 mm, the pump head or efficiency does not change much. However, when the tip clearance is 0.8 mm or more than 1.1 mm, the pump head or efficiency decreases significantly. Therefore, in order to ensure that the mixed-flow pump operates at a higher head and efficiency, the tip clearance of the impeller needs to be kept at a small level.
(2) The amplitude and intensity of pressure pulsation at the impeller inlet are mainly affected by the change in the tip clearance. With the increase in tip clearance, the peak-to-peak value of the pressure pulsation curve and the degree of fluctuation in the pressure pulsation show a decreasing trend first and then an increasing trend. Furthermore, the change in the tip clearance makes the time-domain features of the impeller outlet flow field very complex. At the impeller inlet, the low-frequency pulsation with energy concentration is the main disturbance frequency, and with the increase in the tip clearance, not only does the high-value region of the wavelet spectrum expand in the low-frequency direction, but it is also easy to form second-order peaks in the time-averaged wavelet curve. At the impeller outlet, affected by the BPF and RSI, the extreme value of the time-averaged wavelet curve also decreases first and then increases with the increase in tip clearance, indicating first the decrease in the high-frequency disturbance

area of RSI and then its increase. The wavelet coherence at different positions indicates a higher similarity in the low-frequency pulsation at the impeller inlet in the circumferential direction, demonstrating that the low-frequency disturbances at the impeller inlet flow field are more stable than in other positions, but the low-frequency disturbances in the impeller inlet still affect the low-frequency pulsation in the middle of the impeller. Thus, the sensor distance is also one of the most important factors in coherence analysis.

(3) Under the influence of dynamic and static interference, the velocity distribution in the axial gap between the impeller and the guide vane at various flow rates is relatively disordered. There is a counterclockwise rotating vortex near the hub on the inlet side of the guide vane, which deflects the fluid at the outlet of the impeller to the outer edge. The dynamic and static interference between the impeller and the guide vane further affects the flow field at the inlet of the guide vane and forms a vortex structure at the inlet of the guide vane resulting in the channel blockage. At the same time, a large-scale vortex structure is formed at the guide vane outlet due to the influence of the outlet annular cochlear chamber. The velocity distributions in the flow field at different phases indicate that in the interference region between the impeller outlet and the guide vanes inlet, the RSI affects the internal flow of the interaction field in part-loading condition, and as a result, the internal flow is disorderly and the flow field has remarkable reflux and vortex flow in different phases. With the phase angle changing from the leading edge to the trailing edge of blade, the positive intensity of the vortex structure gradually increases while the negative intensity of the vortex structure intensity first decreases and then increases. The RSI in part-loading condition is the root cause for unsteadiness in the interference flow field.

(4) The change in the flow rate has a significant impact on the axis trajectory, and its fluctuation amplitude is at a low level at the design flow rate. With the change in the flow rate, the shaft vibration continues to deteriorate, which is consistent with the empirical knowledge that it is easy to cause vibrational instability in the rotor system when the hydraulic machinery operates in off-design conditions. The rotor system of the mixed-flow pump has the power frequency vibration caused by imbalance, and the power frequency vibration peak appears at small flow rates. The medium and high-frequency excitation force caused by the unstable flow is the main reason for the deterioration of the power frequency vibration. The analysis of the time-domain diagram of the double frequency axis trajectory shows that the horizontal vibration caused by the misalignment of the rotor shaft is the main factor leading to the external "8" shape of the original axis trajectory. It is also responsible for the double frequency secondary peak in the waveform diagram, and the vibration amplitude reaches the peak with the increase of hydraulic load at small flow rates. As the flow rate deviates from the design flow rate, the radial offset and the misalignment of the rotor shaft increase, and the vibration of the shaft intensifies. The hydraulic imbalance at the impeller rotating frequency has the main influence on the vibration of the shaft, and the RSI effect induced by the BPF has a significant but secondary effect on the vibration of the shaft.

## References

**1** Li, W., Liu, M., Ji, L. et al. (2023). Study on the trajectory of tip leakage vortex and energy characteristics of mixed-flow pump under cavitation conditions. *Ocean Engineering* 267: 113225.

**2** Ji, L., Li, W., Shi, W. et al. (2020). Experimental study of pressure pulsation in a mixed-flow pump with different tip clearances based on wavelet analysis. *Shock and Vibration* 2020: 1–15.

3 Li, W., Ji, L.L., Shi, W.D. et al. (2018). Vibration of shaft system in the mixed-flow pump induced by the rotor-stator interaction under partial load conditions. *Shock and Vibration* 1–12.

4 Li, Y.L., Yuan, S.Q., Tang, Y., and Yuan, J.P. (2012). Calculation of tracer particle following in PIV measurement of flow field in centrifugal pump. *Journal of Drainage and Irrigation Machinery Engineering* 30 (01): 6–10.

5 Adrian, R.J. (2005). Twenty years of particle image velocimetry. *Experiments in Fluids* 39 (2): 159–169.

6 Wereley, S.T. and Meinhart, C.D. (2010). Recent advances in micro-particle image velocimetry. *Annual Review of Fluid Mechanics* 42: 557–576.

7 Adrian, R.J. and Westerweel, J. (2010). *Particle Image Velocimetry*. Cambridge: Cambridge University Press.

8 Raffel, M., Willert, C.E., and Kompenhans, J. (1998). *Particle Image Velocimetry: A Practical Guide with 24 Tables*. Berlin: Springer.

9 Zhao, B.J. (2007). Numerical simulation and PIV test of unsteady three-dimensional turbulence in double channel pump. Thesis, Jiangsu University.

10 Yuan, J.P. (2008). Internal flow PIV test and unsteady full flow numerical simulation of centrifugal pump under multiple design schemes, Thesis, Zhenjiang, Jiang University, viewed 9 June 2008, Research Direct database.

11 Westerweel, J. (1997). Fundamental of digital particle image velocimetry. *Measurement Science and Technology* 8 (12): 1379–1392.

12 Li, W., Zhou, L., Shi, W.W. et al. (2017). PIV experiment of the unsteady flow field in mixed-flow pump under part-loading condition. *Experimental Thermal and Fluid Science* 83: 191–199.

13 Li, W., Ji, L.L., Shi, W.D. et al. (2018). Fluid-structure interaction study of a mixed-flow pump impeller during startup. *Engineering Computations* 35 (1): 1–17.

14 Li, W., Ji, L.L., Shi, W.D. et al. (2016). Influence of different flow conditions on rotor axis locus of mixed-flow pump. *Transactions of the Chinese Society of Agricultural Engineering* 32 (4): 91–97.

15 Liu, Y.B. and Tan, L. (2019). Spatial–temporal evolution of tip leakage vortex in a mixed-flow pump with tip clearance. *Journal of Fluids Engineering* 141 (8): 081302.

16 Hao, Y. and Tan, L. (2018). Symmetrical and unsymmetrical tip clearances on cavitation performance and radial force of a mixed-flow pump as turbine at pump mode. *Renewable Energy* 127: 368–376.

17 Shi, L., Zhang, D.S., Zhao, R.J. et al. (2017). Effect of blade tip geometry on tip leakage vortex dynamics and cavitation pattern in axial-flow pump. *Science China Technological Sciences* 60 (10): 1480–1493.

# 6

# CFD Simulations of a Mixed-flow Pump Using Various Turbulence Models

## 6.1 Comparison and Validation of Numerical Results from Three Two-equation Turbulence Models (SST $k$–$\omega$, $k$–$\omega$, and Standard $k$–$\varepsilon$)

### 6.1.1 Comparison of Energy Performance

The numerical results for head and efficiency of the pump are presented using the RANS equations with the three turbulence models (shear stress transport [SST] $k$–$\omega$, $k$–$\omega$, and standard $k$–$\varepsilon$) and are compared with the experimental results as shown in Fig. 6.1.

It can be seen from the flow-head curve that the computed head is in good agreement with the experimental head for flow rates from $0.4Q_{des}$ to flow rates larger than the design flow rate. However, there is large error between the experimental heads and the computed heads at flow rates lower than $0.4Q_{des}$. The error between the predicted head using the three turbulence models and the experimental value is less than 7% in the vicinity of the design operating point. The computed efficiency is also in good agreement with the experimental efficiency with an error of less than 5% at the design flow rate. The computed efficiency results using the SST $k$–$\omega$ turbulence model are closer to the experimental values compared to those obtained from the other two models.

It is found that each performance curve has a positive slope; therefore, the pump exhibits a phenomenon of rotating stall. As the flow rate gradually decreases from the design point, the head of the pump increases gradually. When the flow rate decreases below $0.52Q_{des}$, the head of the three computed curves using the three different turbulence models begins to decrease slightly. At this stage, the $0.52Q_{des}$ flow rate is determined to be the stall inception point and the $0.5Q_{des}$ flow rate is determined to be the deep stall point. The hump zone calculated by the SST $k$–$\omega$ turbulence model is more prominent compared to that computed by the $k$–$\omega$ model and the standard $k$–$\varepsilon$ model; however, the flow rate condition at the lowest point of the hump zone has little difference in the results computed from the three turbulence models.

### 6.1.2 Comparison of CFD Results with Phase-Averaged Velocity Predictions from PIV

As described in Chapter 5, the particle image velocimetry (PIV) measurements inside the model mixed-flow pump were conducted in detail at the Research Center of Fluid Machinery Engineering and Technology at Jiangsu University, China, to understand the features of its complex flow field and evaluate its performance at various flow rates (lower and higher than the design flow rate). Here, one impeller position is considered to compare the experimental and computed results in

*Mixed-flow Pumps: Modeling, Simulation, and Measurements*, First Edition.
Wei Li, Leilei Ji, Ramesh Agarwal, Weidong Shi, and Ling Zhou.

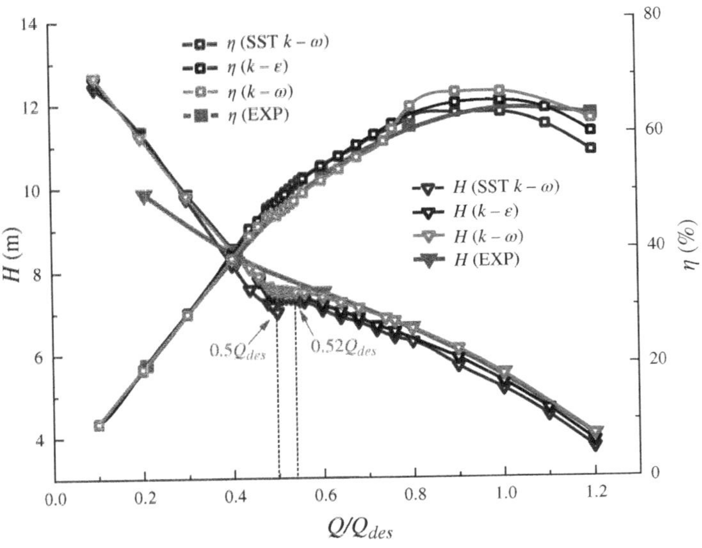


**Figure 6.1** Comparison of experimental results with simulations using the three turbulence models (SST $k$–$\omega$, $k$–$\omega$, and $k$–$\varepsilon$).

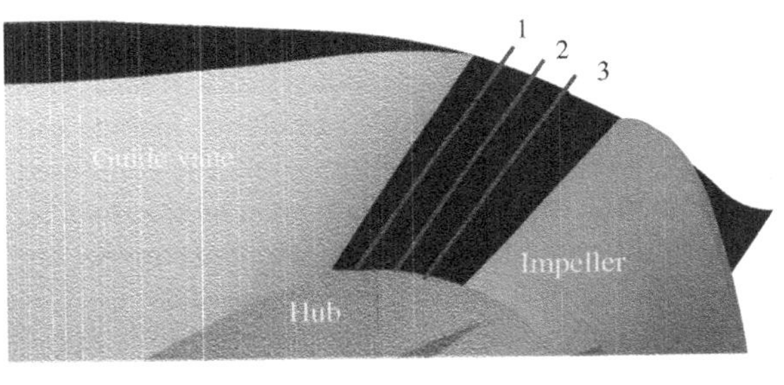


**Figure 6.2** Positions of the target area and monitoring lines in the PIV experiment.

detail using the three turbulence models; for this purpose, three monitoring lines in the target flow area are created. The monitoring lines and the initial relative position between the impeller and the guide vane are shown in Fig. 6.2.

Figure 6.3 compares the phase-averaged velocity contours and streamlines obtained by PIV measurements and the numerical simulations at 0.4 times the design flow rate $Q_{des}$ at the same scale. It can be noted from Fig. 6.3 that the trend in the streamlines' distribution using the three turbulence models is basically the same. Due to the deviation from the design flow rate, at lower flow rates, a large backflow vortex is formed at the inlet of the guide vane, making the impeller outlet fluid to flow upward, creating the high-velocity flow region in the end wall area of the runner chamber. The high-velocity region also exists on the blade pressure surface (PS). Overall, the simulation results are in good agreement with the experimental results. Comparing the backflow vortex structure using the three different turbulence models, the vortex core of the backflow is not yet visible in the computation with the SST $k$–$\omega$ model, but the position of the vortex is close to the PIV results. However, the vortex size in the computation with the $k$–$\omega$ model and standard $k$–$\varepsilon$ model is larger and the position of the vortex core deviates from the wall to the middle region of the guide vane inlet. Therefore, the results of the SST $k$–$\omega$ model can be considered more accurate based on the analysis of the vortex position and the vortex size.

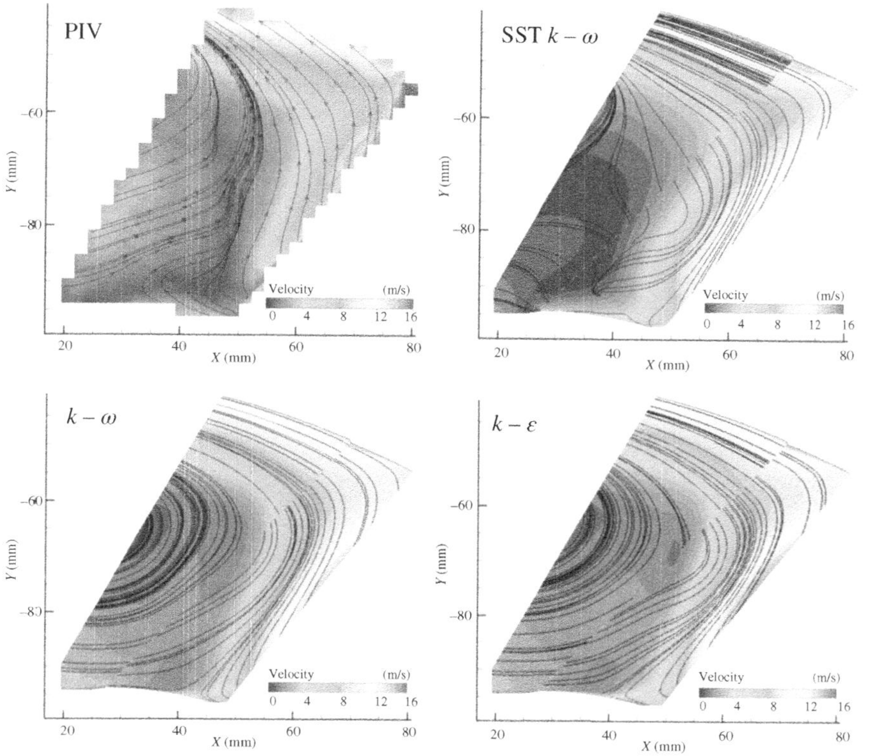


**Figure 6.3** Velocity contours and streamlines in the target area at $0.4Q_{des}$ flow rate.

Figure 6.4 compares the computational fluid dynamics (CFD) and PIV results for the velocity at the three monitoring lines. Overall, the velocity predicted by the simulations is in close agreement with the PIV measurements. The SST $k$–$\omega$ model is known to produce somewhat larger turbulence levels in regions with large normal strain, therefore the velocity results have more fluctuations and more changes in the velocity gradient. On monitoring line 1, the simulation results using the $k$–$\varepsilon$ model and $k$–$\omega$ model are identical. The velocity values obtained with the SST $k$–$\omega$ model near the hub region are closer to the experimental results, while the $k$–$\varepsilon$ model and $k$–$\omega$ model show higher accuracy in the region $X = 32$–$44$. The simulation results from the three turbulence models are basically consistent in the region $X > 44$. Overall, the SST $k$–$\omega$ model is closer to the experimental results. On monitoring line 2, the simulation results using the three turbulence models are basically the same as the PIV results. At the same time, the simulation results using the standard $k$–$\varepsilon$ model and $k$–$\omega$ model are still very close. From the overall flow field trend, the performance of the SST $k$–$\omega$ model is in better agreement with the PIV results. On monitoring line 3, all models appear to give predictions that deviate substantially from the velocity measurements. By checking the images captured by the camera, this could be explained by noting that there is serious refraction of light close to the spherical end wall, which may lead to distortion in the outlet flow field of the impeller. As can be seen from Fig. 6.4, the simulation results using the standard $k$–$\varepsilon$ model are closest to the experimental results in the region $X = 38.5$–$56$, while the results from the SST $k$–$\omega$ model are in best agreement with the experimental data in the region $X > 56$. Therefore, based on this detailed analysis of results from the three turbulence models, it can be concluded that the SST $k$–$\omega$ turbulence model gives the highest accuracy among the three turbulence models in predicting the flow field behavior.

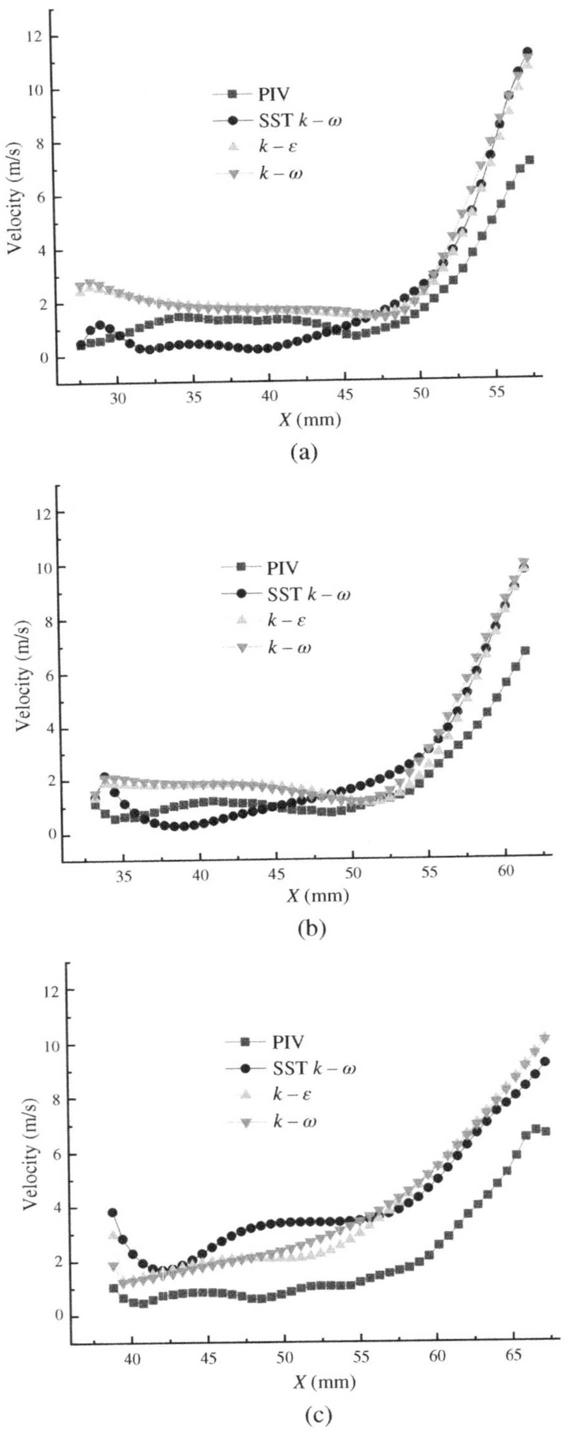


**Figure 6.4** Comparison of velocities from simulations using the three turbulence models (SST $k$–$\omega$, $k$–$\omega$ and $k$–$\varepsilon$) with PIV experiment along the three monitoring lines. (a) Monitoring line 1, (b) Monitoring line 2, and (c) Monitoring line 3.

### 6.1.3 Flow Field Under Incipient Stall Condition

The velocity distribution in the outlet section of the impeller for $0.52Q_{des}$ flow rate (incipient stall condition from Chapter 3) is computed using the three turbulence models as shown in Fig. 6.5. This figure shows the internal flow field of the impeller at the outlet section of the pump near the incipient stall condition.

It can be seen from Fig. 6.5 that the three vortex structures in the adjacent flow passages are captured by the SST $k$–$\omega$ model, which are all located near the outlet of the suction surfaces of the impeller blades. Additionally, the velocity in the region occupied by the vortices becomes the lowest in the flow path and the vortices block the flow passages. The other flow passage without vortices shows a slight distortion in the streamline at the outlet, where the velocity is obviously lower than that upstream. Similarly, the two vortex structures in the adjacent flow passages are captured by the $k$–$\omega$ model and the other two flow passages have distorted streamlines, but no vortices have yet formed. However, no vortex structure is seen in the four flow passages using the standard $k$–$\varepsilon$ model, and there is only streamline distortion. When the flow rate is $0.52Q_{des}$, the flow

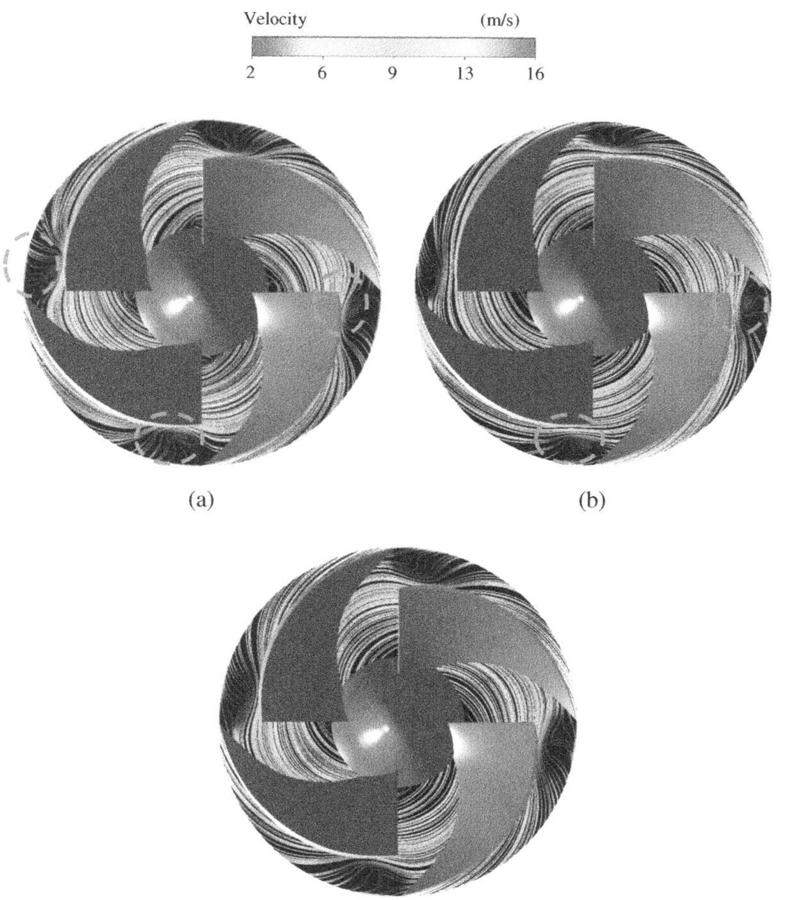


(c)

Figure 6.5 Velocity distributions at the section of the impeller outlet using the three turbulence models (a) SST $k$–$\omega$, (b) $k$–$\omega$, and (c) $k$–$\varepsilon$.

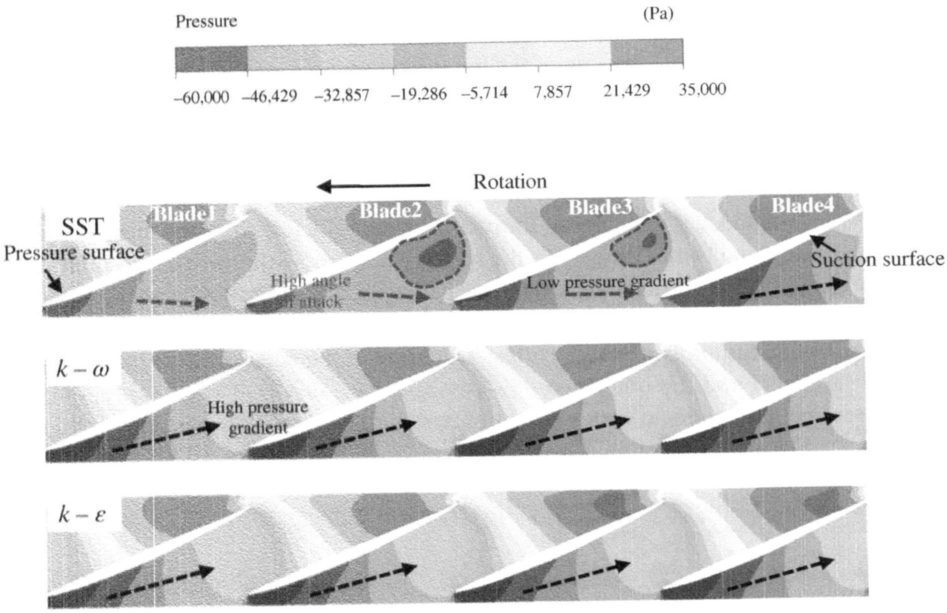


**Figure 6.6** Pressure distributions in the blade passages in the plane at 50% span using the three turbulence models (a) SST $k$–$\omega$, (b) $k$–$\omega$, and (c) $k$–$\varepsilon$.

distribution with SST $k$–$\omega$ and $k$–$\omega$ turbulence models is almost the same in the macro-structure; however, a difference can be noted in capturing the flow loss which is shown by the head-flow rate characteristics in Fig. 6.1. According to the classical stall theory [1], stall vortex should appear in the individual flow passages at the beginning of the stall; again, the SST $k$–$\omega$ model appears to be more suitable to conduct the simulations.

In the initial stage of the stall, the pressure distribution in the impeller passage is uneven due to the influence of the initial stall. Figure 6.6 shows the pressure distribution in the blade passages in the plane at 50% span for $0.52Q_{des}$. It can be seen from this figure that the pressure field calculated by the SST $k$–$\omega$ model differs significantly from that calculated by the other two models; it is mainly manifested by the spatial distribution of the low-pressure area and changes in the pressure gradient. From the results obtained with the SST $k$–$\omega$ model, it can be seen that there are two annular low-pressure regions on the suction surface near the tailing edge of blade #2 and blade #3. Examining the characteristics of the pressure contours, it can be noted that the low-pressure region is caused by the stall vortex, which can also be verified by the characteristics of the previous streamlines. In the two flow passages of the stall, the inlet flow is hindered by the stall vortex, resulting in the deflection of the inflow angle, which increases the attack angle of the incoming flow to the next blade and forms a flow separation bubble on the suction surface of the blade leading edge. The low-pressure zone corresponding to the inlet of the blade in this figure reflects the strength of the flow separation. Simultaneously, the positive pressure gradient in the blade passage decreases when the stall occurs; it has also been found in the study of the instability mechanism in a mixed-flow pump by Hagiya et al. [2] using the large eddy simulation (LES). This is different from the pressure gradient distribution in each passage without stall calculated by the standard $k$–$\varepsilon$ model and $k$–$\omega$ model. Again, the SST $k$–$\omega$ model shows better prediction capability for the uneven spatial distribution of the low-pressure area and the change in pressure gradient due to initial stall.

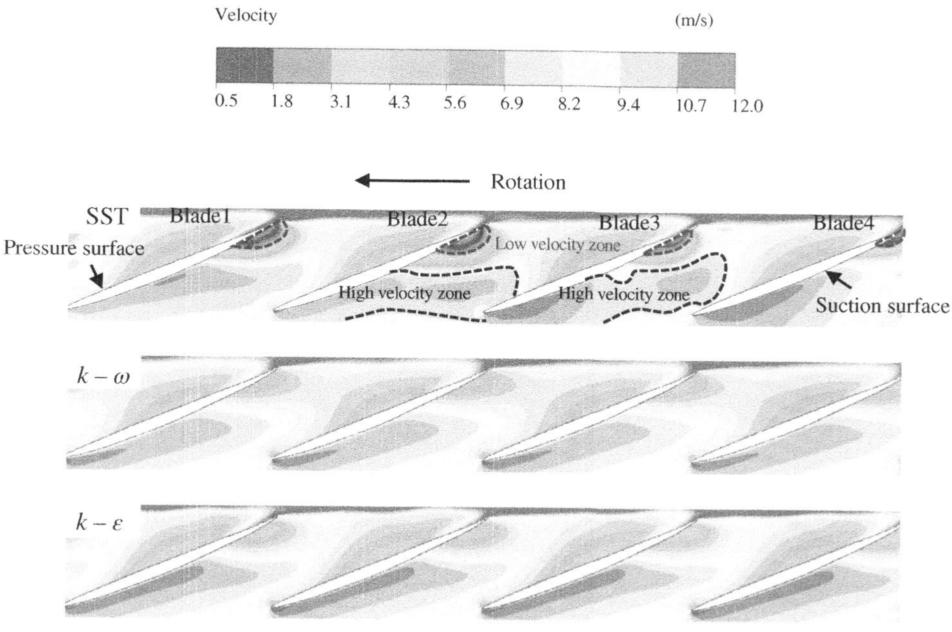


**Figure 6.7** Velocity distributions in the blade passages in the plane at 50% span using the three turbulence models (a) SST $k$–$\omega$, (b) $k$–$\omega$, and (c) $k$–$\varepsilon$.

Figure 6.7 shows the velocity distributions in the blade passages in the plane at 50% span for $0.52Q_{des}$ flow rate. It can be seen from this figure that the results of the SST $k$–$\omega$ model are clearly different from those of the other two models, and the velocity field predicted by the standard $k$–$\varepsilon$ model and $k$–$\omega$ model are also different. The differences in the velocity fields are shown in two aspects. The first feature is the size and position of the high-velocity zone at the leading edge of the blade, and the second feature is the existence of a low-velocity zone near the trailing edge of the suction surface. The results of the SST $k$–$\omega$ model show these two features better than those from the other two models. The high-velocity zone at the entrance of the two flow passages behind blades #2 and blades #3 extends from the leading edge of the blades to the PS of the next blade, which is due to the blockage caused by the stall cells in the flow passage resulting in the reduction of the local flow area. The high-velocity zone is depicted by the blue dotted line in the figure. The low-velocity area at the trailing edge of the blade is caused by the decrease in the flow rate, which results in a decrease in the axial velocity and creates flow separation due to a lack of sufficient momentum for the fluid to reach the outlet of the suction surface of the blade. The first feature has also been observed in the computations by using the standard $k$–$\varepsilon$ model and $k$–$\omega$ model. The low-velocity zone is depicted by the red dotted line in this figure. In addition, there is no difference in the velocity fields among the flow passages calculated by the standard $k$–$\varepsilon$ model and $k$–$\omega$ model. Furthermore, the flow separation has not yet formed the stall vortex. As to the second feature, both the standard $k$–$\varepsilon$ model and $k$–$\omega$ model fail to capture the low velocity region near the trailing edge of the suction surface. Thus, the SST $k$–$\omega$ model shows better prediction capability compared to the other two models in calculating the unstable flow in the impeller of the mixed-flow pump in the incipient stall condition.

The three-dimensional streamlines in a flow passage of the impeller for $0.52Q_{des}$ flow rate obtained by using the three turbulence models are shown in Fig. 6.8. It can be seen from this

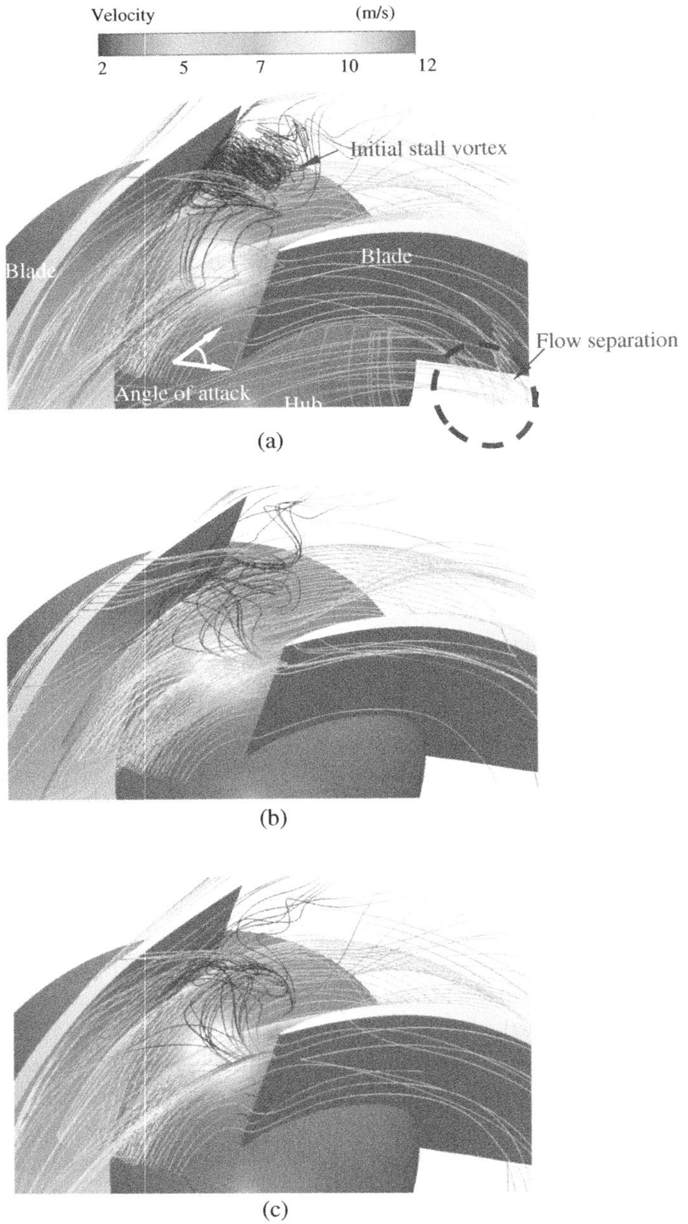


**Figure 6.8** Initial stall vortex structure in the impeller flow passage.

figure that a large vortex at the outlet of the impeller is captured by the SST $k$–$\omega$ model. The vortex stays near the outlet of the impeller and blocks almost half of the flow passage. This blocking of the passage causes higher turbulence dissipation and, consequently, higher energy loss in the impeller. At the same time, the vortex flows from the pressure side to the suction side of the current blade and propagates to the next flow passage, which leads to an increase in the angle of attack at the inlet of the blade. With the rotation of the impeller, the flow separation occurs in the

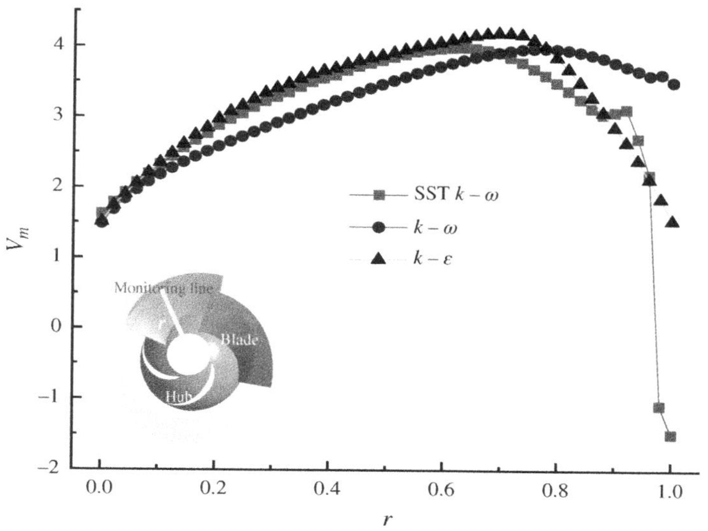


**Figure 6.9** Axial velocity in the impeller inlet under the initial stall condition using the three turbulence models (SST $k$–$\omega$, $k$–$\omega$, and $k$–$\varepsilon$).

next flow passage, and the incipient stall vortex is formed, which moves in the direction opposite to the impeller rotation. However, it should be noted that the $k$–$\omega$ model and the standard $k$–$\varepsilon$ model do not capture the vortex structure in the flow passage.

The axial velocity distributions at the impeller inlet in the incipient stall condition obtained using the three turbulence models are shown in Fig. 6.9. A dimensionless number $r$ for the radial position is defined as:

$$r = (R - R_h)/(R_c - R_h) \tag{6.1}$$

where $R$ is the radius of any point on the inlet monitoring line, $R_h$ is the hub radius, and $R_c$ is the rim radius.

As can be seen from Fig. 6.9, the velocity distributions obtained with the three turbulence models are basically the same; velocity gradually increases from the hub to the rim. The peak value of the velocity appears at 60–80% of span, which then gradually decreases. It should be noted that the velocity computed using the SST $k$–$\omega$ model is negative near the end wall region, which further proves that the flow capacity of the fluid in the inlet rim region decreases, and there is a backflow vortex. The tip leakage flow is the main reason for the induced stall, which is consistent with the previous flow field analysis. Thus, the SST $k$–$\omega$ model can best simulate the incipient stall in the mixed-flow pump compared to the other two turbulence models.

The velocity circulation distribution at the impeller inlet in the incipient stall condition is obtained by using the three turbulence models as shown in Fig. 6.10. The abscissa $r$ represents the blade height position of any point on the inlet monitoring line. The velocity circulation at the impeller inlet is obtained by the integration of the fluid velocity along the path of a closed curve and is given by,

$$\Gamma = 2\pi R_1 v_u \tag{6.2}$$

where $R_1$ is the impeller inlet radius, and $v_u$ is the circumferential velocity at the impeller inlet.

Due to the work done by the blades, the velocity circulation distribution at the impeller inlet using the three turbulence models increases from the hub to the rim. The peak value appears at 78–88%

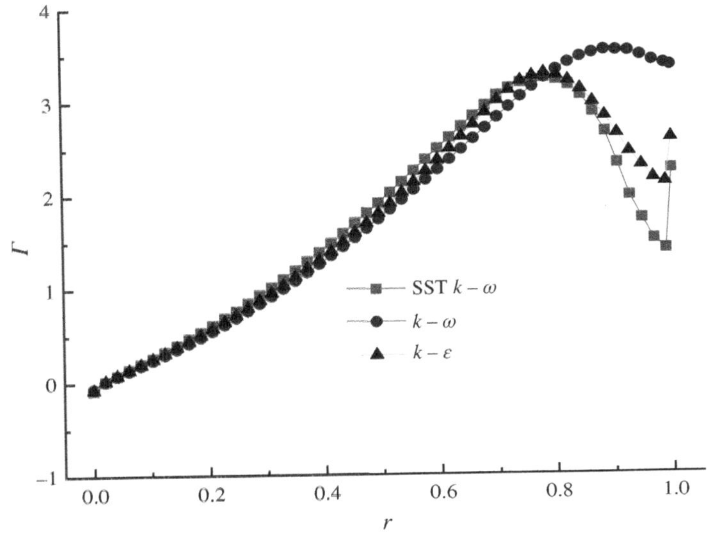


**Figure 6.10** Velocity circulation distribution in the impeller inlet in the initial stall condition.

of span, which then gradually decreases. It should be noted that the velocity circulation calculated by the SST $k$–$\omega$ model and the standard $k$–$\varepsilon$ model also increases near the end wall region, which implies that the backflow vortex in the end wall region further reduces the flow capacity in the flow passage and the velocity circulation in the end wall region is affected by the leakage flow. Comparing the results using the three turbulence models, the velocity circulation value computed using the SST $k$–$\omega$ model is higher than that obtained with the other two turbulence models at 10–70% of span, indicating that the inlet flow field calculated by the SST $k$–$\omega$ model is more likely to result in the flow separation. At the same time, the change in the velocity circulation using the SST $k$–$\omega$ model near the end wall region is larger than that obtained from the other two models. Overall, these results show that the SST $k$–$\omega$ model can accurately predict the occurrence of rotating stall compared to the other two turbulence models, making the inlet flow field more unstable.

## 6.2 Application of Wray–Agarwal (WA) One-Equation Turbulence Model

In Section 6.1, the relative accuracy of the three most widely used turbulence models, namely the two-equation SST $k$–$\omega$, $k$–$\omega$, and standard $k$–$\varepsilon$ in conjunction with RANS equations, was discussed for computing the flow fields inside the mixed-flow pump for various flow rates (below and above the design flow rate) by comparing the computations with the PIV data. It was found that overall, the use of the SST $k$–$\omega$ model gave the best results in capturing most of the flow features as well as the performance parameters of the pump. In recent years, a recently developed one-equation turbulence model called the WA model has shown lot of promise in both accuracy and efficiency in a wide variety of complex 2D and 3D flows including those in the hydraulic machinery.

The model is listed on the NASA Turbulence Modeling Resource (TMR) (https://turbmodels.larc.nasa.gov/wray_agarwal.html). Therefore, the application of WA model for computing the flow in a mixed-flow pump is considered in this section.

WA model is a one-equation turbulence model which behaves like the $k$–$\omega$ model near the wall and like $k$–$\varepsilon$ model away from the wall in the freestream region; a switching function is used for this purpose (see Chapter 3 for details). In many applications, it exhibits similar behavior near the wall and in the freestream as the SST $k$–$\omega$ model. However, compared to the SST $k$–$\omega$ model, it is equally or more accurate and more efficient being a one-equation model. In the following sections, the results obtained with the WA model in computing the flow field in a mixed-flow pump are presented and compared with those obtained with the SST $k$–$\omega$ model.

### 6.2.1 Energy Performance Comparison

In Section 6.1, it has been demonstrated that the simulations using the SST $k$–$\omega$ model can better capture the backflow in the end wall region, separated flow, stall vortex, the uneven spatial distribution of pressure in the low-pressure region, and the change in pressure gradient due to initial stall in the mixed-flow pump compared to $k$–$\omega$ and $k$–$\varepsilon$ models. However, in some cases, if accurate energy characteristics of the pump and its internal flow field results need to be obtained quickly and efficiently, the WA model has an advantage over the SST $k$–$\omega$ model, as described below.

Figure 6.11 shows the comparison of the energy performance curves of the mixed-flow pump obtained by numerical simulations using the WA model and the SST $k$–$\omega$ model and their comparison with experimental data. It can be observed that the trends in both the pump head and efficiency with flow rate are well predicted by both the WA model and the SST $k$–$\omega$ model. However, at majority of the computed points for the pump head and efficiency, the values of both the head and efficiency predicted by the WA model are slightly higher than the experimental data,

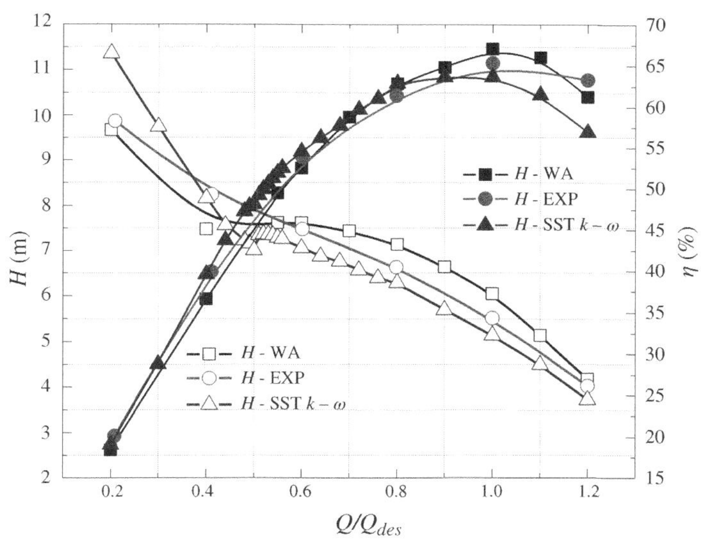


**Figure 6.11** Comparison of simulations using the SST $k$–$\omega$ and WA turbulence models and experiment for pump head and efficiency.

while the results obtained using the SST $k$–$\omega$ model are lower than the measured results. For the pump efficiency, the results obtained using both models are higher than the experimental values for most flow rates, except that the results using the SST $k$–$\omega$ model are slightly lower than the experimental values at flow rates higher than the design flow rate. By comparing the error in computed efficiency with experimental value at the design flow rate, it can be noted that the accuracy of the computed results using the WA model is slightly higher than that obtained with the SST $k$–$\omega$ model; however, the error in computed head is slightly higher than that obtained with SST $k$–$\omega$ model. Considering the whole range of flow conditions, it is found that the pump efficiency computed by using the WA model is closer to the experimental results, while the pump head is also within the acceptable engineering accuracy. At $0.4Q_{des}$ flow rate which is close to the rotating stall condition, the head curve shows a "saddle area" using the SST $k$–$\omega$ model, which is not predicted by the WA model and the experiment. This result again shows that the WA model can better predict the unsteady and high Reynolds number flow in the mixed-flow pump. At the lower flow rate of $0.2Q_{des}$, the computed error in the head compared to experimental data using the WA model is smaller than the error obtained with the SST $k$–$\omega$ model, which again indicates that the WA model can achieve higher accuracy at lower flow rates. Thus, for most of the operating flow rate conditions of the mixed-flow pump, the WA model shows the ability to accurately and efficiently predict the energy characteristics of the mixed-flow pumps.

### 6.2.2 Comparison of the Flow Field in the Rotor–Stator Interaction Zone Using Different Turbulence Models

In this section, in order to further demonstrate the accuracy of the WA model in calculating the flow field in a mixed-flow pump, the flow field of the rotor–stator interaction zone is computed by using the SST $k$–$\omega$ and WA models and is compared with the experimental results. The selected section of the rotor–stator interaction zone is shown in Fig. 6.12, and the comparisons of computations with the experiment results are shown in Fig. 6.13(a)–(c). The entire flow fields simulated by using the SST $k$–$\omega$ model and the WA model are quite similar, and both the low flow velocity zone and the vortex induced by the backflow from the guide vane channel are well predicted by both models. With an increase in the flow rate, the computed low flow velocity zone shrinks like in the PIV data. Furthermore, at $0.8Q_{des}$ flow rate, the range of low flow velocity zone predicted by the WA model has higher accuracy than that predicted by the SST $k$–$\omega$ model compared to the PIV data.

According to the results shown in Section 6.1, the SST $k$–$\omega$ model has significant advantages over $k$–$\omega$ and $k$–$\varepsilon$ models in predicting the internal and external flow characteristics of mixed-flow pumps under multiple working conditions, which implies that the WA model can also inherit this advantage to great extent being quite similar to SST $k$–$\omega$ model in many physical and theoretical aspects. Therefore, in terms of the computational accuracy, WA can meet the engineering

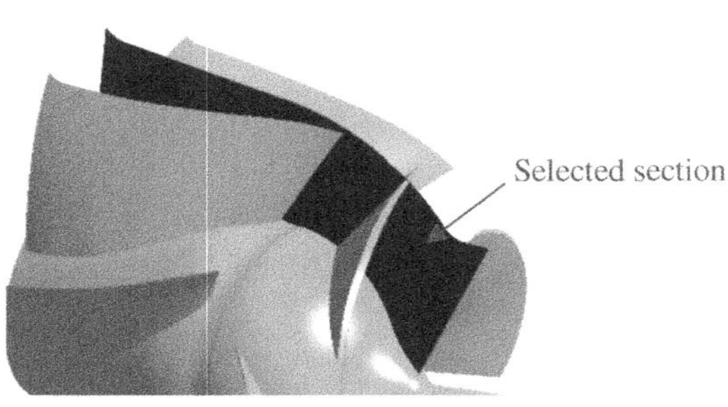


**Figure 6.12** Selected section in the mixed-flow pump model.

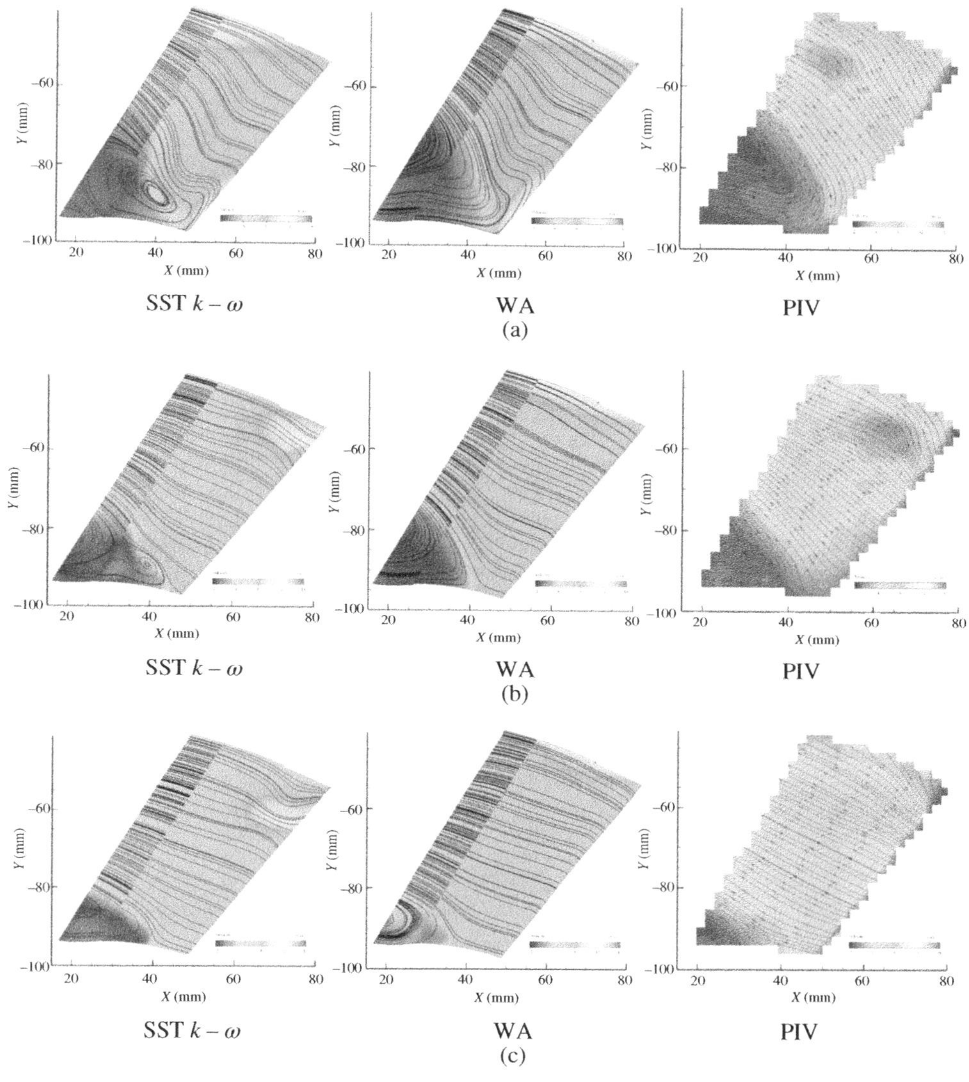


**Figure 6.13** Velocity comparisons between simulations using the SST $k$–$\omega$ and WA turbulence models and experiment in the rotor-stator interaction zone. (a) $0.8Q_{\text{des}}$, (b) $1.0Q_{\text{des}}$, (c) $1.2Q_{\text{des}}$.

requirements of the mixed-flow pump near the design and off-design conditions at twice the computational efficiency of SST $k$–$\omega$ model.

### 6.2.3 Comparison of Eddy Viscosity Variable $R = k/\omega$ in the Pump Using the Four Turbulence Models (SST $k$–$\omega$, $k$–$\omega$, $k$–$\varepsilon$, and WA)

In this section, in order to reveal the differences in the numerical results calculated by using the four turbulence models with RANS equations, the variable $R = k/\omega$ which is closely related to turbulent eddy viscosity in all the models, is selected for calculation in various parts of the pump in order to understand its role in energy dissipation. Figure 6.14 shows the comparison of $R$ distribution in the

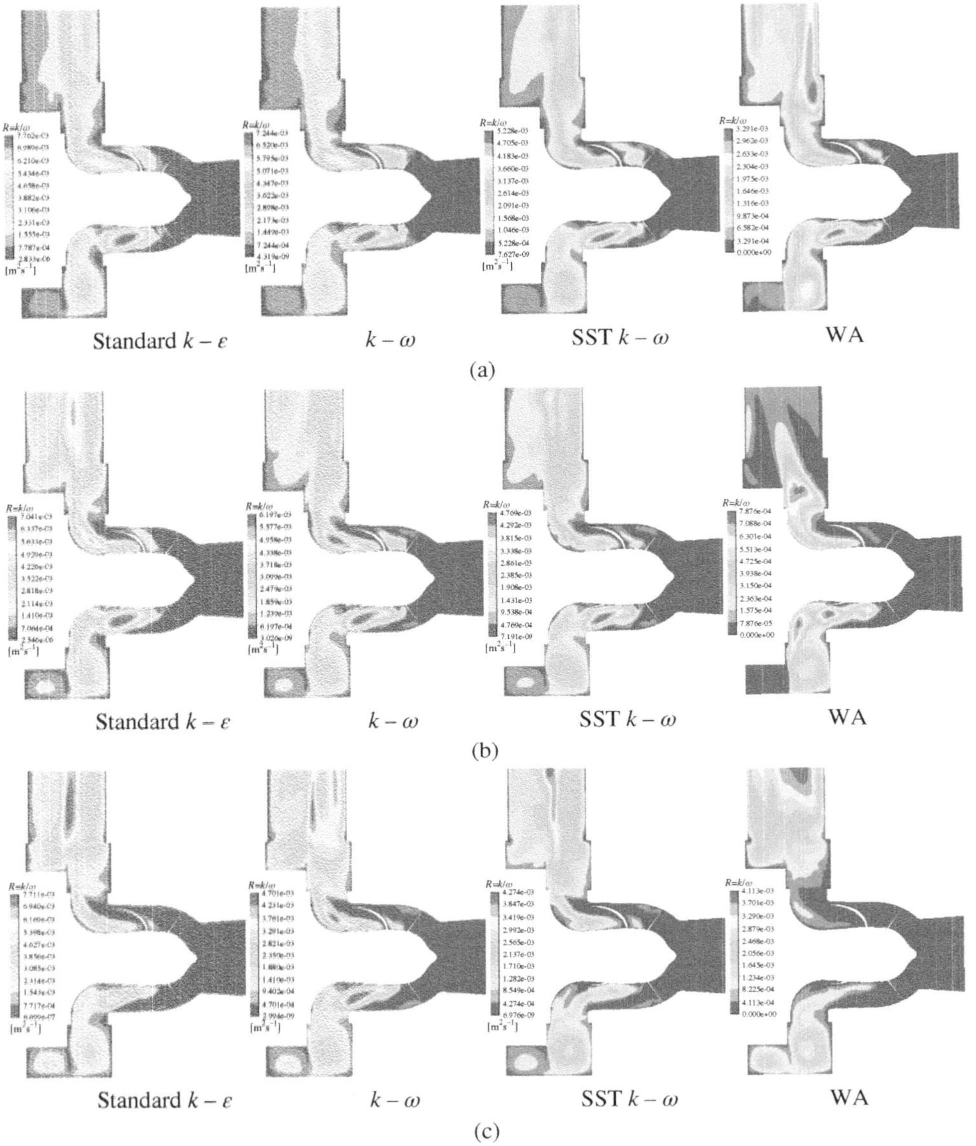


**Figure 6.14** $R$ distribution in the middle cross-section of the mixed-flow pump using the four turbulence models (SST $k$–$\omega$, $k$–$\omega$, and $k$–$\varepsilon$, WA). (a) $0.8Q_{des}$, (b) $1.0Q_{des}$, (c) $1.2Q_{des}$.

middle section of the whole mixed-flow pump for the three flow rates ($0.8Q_{des}$, $1.0Q_{des}$, and $1.2Q_{des}$). It can be noted from the computed results from all four turbulence models that the flow fields from the guide vane to the annular volute have higher $R$-value which is due to the vortices in the high Reynolds number flow. With a decrease in the flow rate to $0.8Q_{des}$, the range of high $R$-value zone increases in the guide vane flow channel, which implies that more vortices are generated when the pump operates at the design flow rate. In the high $R$-value regions, the flow energy dissipates into internal energy and potential energy. Thus, if judged by this criterion, all four turbulence models can predict this trend in $R$ reasonably well. However, for the two flow rates ($0.8Q_{des}$ and $1.2Q_{des}$)

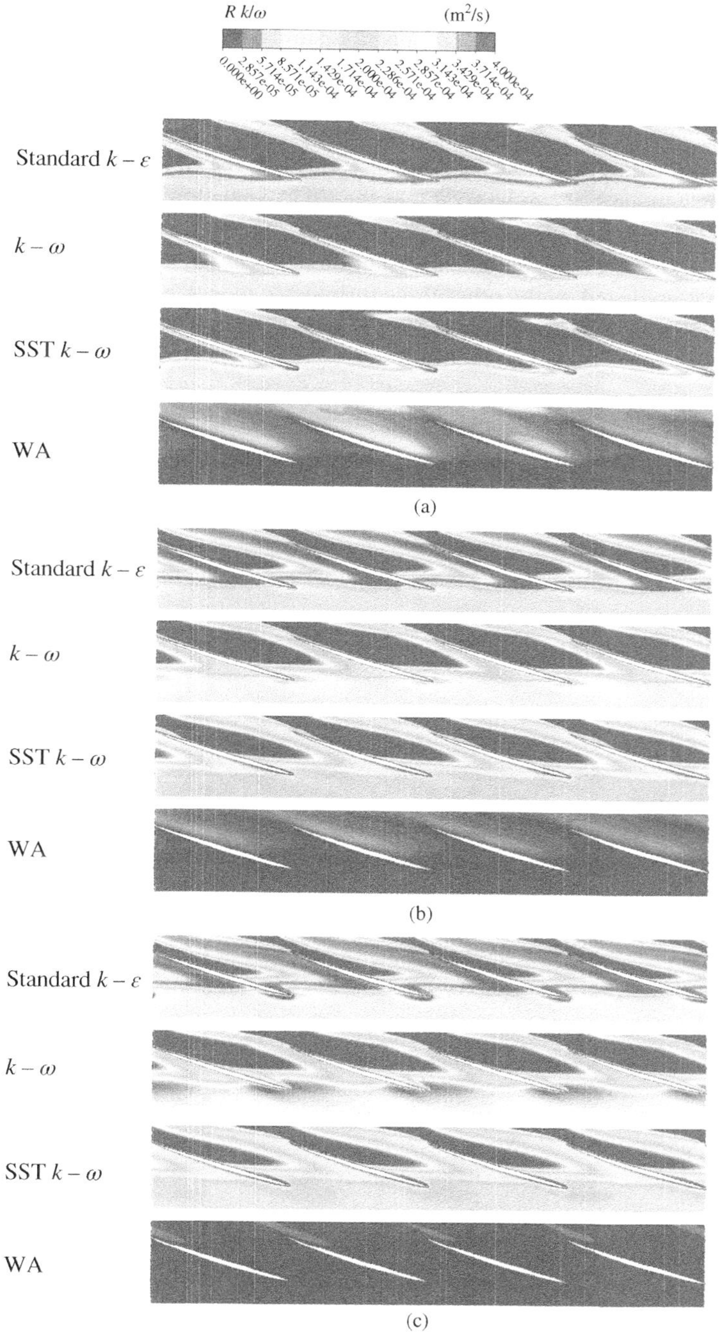


**Figure 6.15** $R$ distribution on the turbo surface of the impeller at 90% span using the four turbulence models (SST $k$–$\omega$, $k$–$\omega$, and $k$–$\varepsilon$, WA). (a) $0.8Q_{des}$, (b) $1.0Q_{des}$, (c) $1.2Q_{des}$.

other than the design flow rate, the $R$ distributions predicted by the WA model appear to be weaker compared to that obtained from the other three models; this result is similar to that reported for a centrifugal pump [3]. This is due to the fact that the undamped eddy viscosity variable $R$ $(= k/\omega)$ in the WA model does not overestimate the turbulent kinetic energy and turbulent eddy frequency obtained with the other three models. The $R$ distribution on the turbo surface at span = 0.9 in the impeller is also compared for the three flow rates, as shown in Fig. 6.15. From the computed results obtained using the four turbulence models, it can be seen that the tip leakage flow (TLF) would cause a higher $R$-value near the blade rim region, and the higher $R$-value zone becomes larger with the decrease in flow rate ($0.8Q_{des}$). However, the $R$-value distribution predicted by the WA model has lower values compared to that predicted by the other three turbulence model; again, it is due to the fact that the WA model does not overestimate the turbulent kinetic energy. Thus, since the

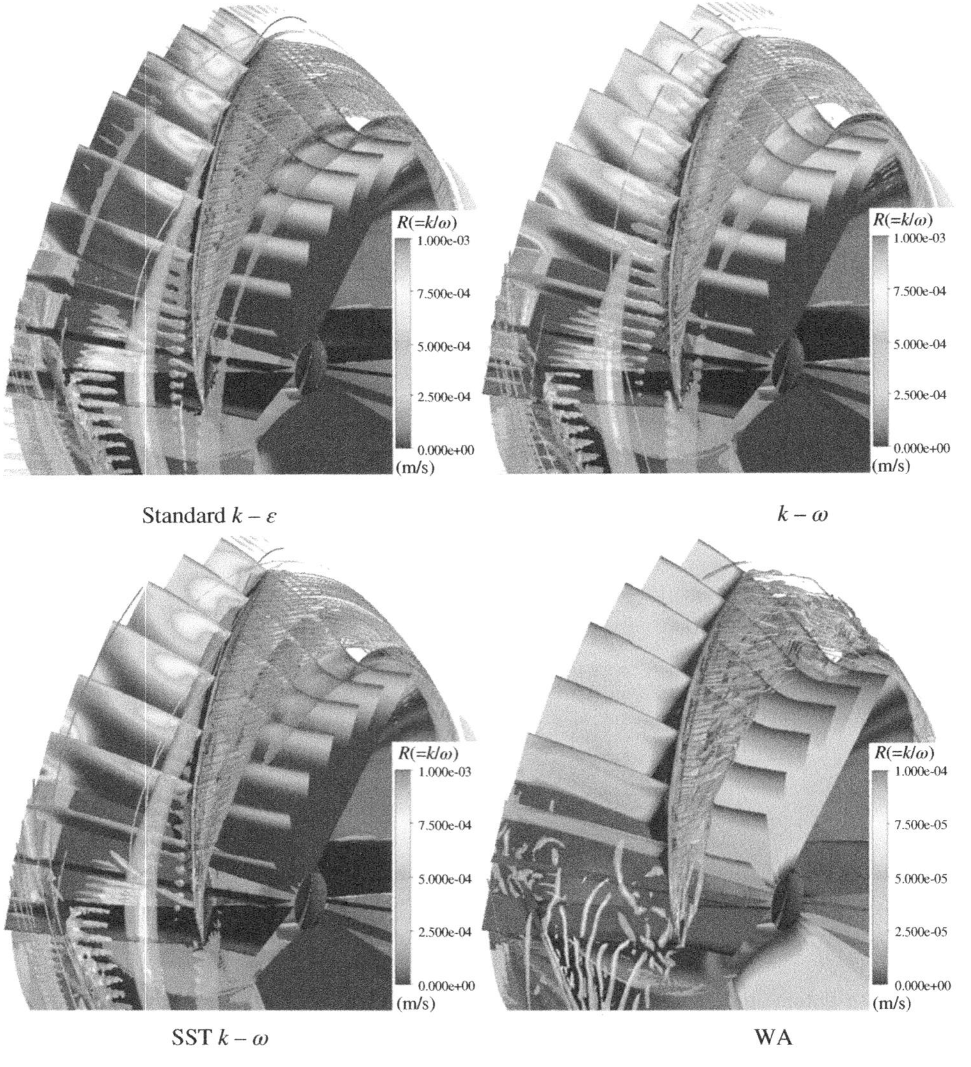


**Figure 6.16** $R$ distribution and tip leakage vortex in the impeller at the design flow rate using the four turbulence models (SST $k$–$\omega$, $k$–$\omega$, and $k$–$\varepsilon$, WA).

turbulent kinetic energy is not estimated higher by the WA model compared to that predicted by the other three models, the pump head and efficiency predicted by the WA model are closer to the experiential data, and therefore, the model has the advantage in accurate prediction of both the internal and external flow characteristics of the mixed-flow pump.

Since the tip leakage vortex (TLV) is the one of the main unsteady flow structures in the mixed-flow pump impeller, therefore the $R$ distribution in the impeller and the flow field near the blade tip are computed and compared at design flow rate using the four turbulence models. It can be seen from Fig. 6.16 that the flow direction and the main vortex structure of the TLF are well predicted by the four turbulence models. The vortex core region is marked by the $\lambda_2$ criterion, and the threshold level is 0.02. The TLV creates a higher $R$-value region along the main flow direction, since the higher dissipation of turbulent kinetic energy occurs in these areas. However, the tails of the TLV structures predicted by the different turbulence models have some differences. The TLV calculated by the WA model breaks into several strip vortex structures at the tail, which is different from that obtained with the other three models. This is again due to the contribution of the undamped eddy viscosity $R$, which is also consistent with the energy transfer process of a large vortex breaking into smaller vortices and eventually disappearing. Furthermore, in case of the WA model, not only the whole $R$-value distribution is lower than that of the other models, but also the relative $R$-value difference between the main flow and the TLV region is lower, which indicates that the turbulent kinetic energy generated by the TLV and the turbulent kinetic energy generated by the main flow are within one order of magnitude as predicted.

## 6.3 Summary

The results presented in this chapter can be summarized as follows:

(1) The computed pump performance curves obtained using the standard $k$–$\varepsilon$, $k$–$\omega$, and SST $k$–$\omega$ models with RANS equations in ANSYS/CFX CFD flow solver show a positive slope in a small region of flow rates (lower than the design flow rate) implying that the pump will exhibit a phenomenon of rotating stall. The pump efficiency computed with the SST $k$–$\omega$ model is more accurate compared to that obtained with the other two models, as well as the internal flow field of the pump computed using the SST $k$–$\omega$ model is also in good agreement with the PIV experimental results. Furthermore, the simulations performed using the SST $k$–$\omega$ model show better capture of the backflow in the end wall region, separated flow, stall vortex and the uneven spatial distribution of the low-pressure area, and the change in pressure gradient due to initial stall in the pump. It has been demonstrated that among the three models, the SST $k$–$\omega$ model has greater potential for accurately predicting the performance of the pump in incipient stall condition.

(2) The energy performance curves calculated by the WA turbulence model show a similar trend in the results as predicted by SST $k$–$\omega$ model, but overall, it shows better accuracy compared to the SST $k$–$\omega$ model. For majority of the flow rates, the WA model shows better accuracy in predicting the head and efficiency. Furthermore, near the stall condition also, the WA model exhibits better flow field prediction capability. In addition to accuracy, for solving engineering problems, the WA model has a quicker convergence rate and lower computational requirements, being a one-equation turbulence model.

(3) The velocity fields in the middle cross-section of the rotor-stator zone of the pump predicted by the WA model for different flow rates agree best with the PIV experimental results. Comparing

the whole flow field and tip region of the mixed-flow pump, the values of the eddy viscosity variable $R = k/\omega$ predicted by the WA model are relatively lower than those obtained for the other three models. The main reason for better prediction of the external flow characteristics of the mixed-flow pump is better control of undamped variable $R$ in the WA model, which does not allow for the overestimation of the turbulent kinetic energy and turbulent eddy frequency.

## References

1 Emmons, H.W., Pearson, C.E., and Grant, H.P. (1995). Compressor surge and stall propagation. *Transactions of the ASME* 77 (3): 455–469.

2 Hagiya, I., Kato, C., Yamade, Y., et al. (2015). Clarification of performance curve instability mechanism using large eddy simulation of internal flow of a mixed-flow pump. *ASME/JSME/KSME 2015 Joint Fluids Engineering Conference, American Society of Mechanical Engineers.*

3 Ji, L.L., Li, W., Shi, W.D., and Agarwal, R.K. (2020). Application of Wray-Agarwal turbulence model in flow simulation of a centrifugal pump with semi-spiral suction chamber. *ASME Journal of Fluids Engineering* 143 (3).

# 7

# Tip Leakage Flow In a Mixed-flow Pump

## 7.1 Energy Characteristics

### 7.1.1 Comparison of Energy Performance of the Pump for Various Tip Clearances

Figure 7.1 shows the computed energy performance curves for various blade tip clearances; the head versus efficiency curves for six flow rates $Q/Q_{des}$ varying from 0.2 to 1.2 are shown in this figure, $Q_{des}$ being the design flow rate. These curves are smoothened by the Bessel method. It can be seen from this figure that when the flow rate changes from $0.4Q_{des}$ to $1.2Q_{des}$, both the pump head and the efficiency drop with the increase in the tip clearance and near the design point, the difference between the head and the efficiency of the pump is the largest for all tip clearances. It can also be noted that the head of the pump gradually increases with decrease in the flow rate, and the pump has the highest efficiency near the design point.

Furthermore, as the flow rate decreases, the differences in the efficiency of the pump gradually diminish with change in the tip clearance. It is because the losses caused by the unstable flow generated by the TLV have a lesser proportion of the vortex-induced energy losses in the impeller at smaller flow rates compared to the losses caused by the increase in the secondary flow in the pump. Thus, the effect of the increase in the tip clearance is closely related to the external characteristic performance of the pump, which is consistent with the conclusions of some other reports in the literature on TLF of an axial pump [1]. This is also the conclusion by Li et al. [2] that the tip clearance size has a strong relationship with the internal flow field of the pump and can cause a lot of losses. It is also verified in Fig. 7.1 that larger the tip clearance, greater are the losses.

In the mixed-flow pump, in order to evaluate the head loss in each part of the pump, the total pressure drop between the inlet and outlet of each component or domain is calculated. For the stationary domains (inlet domain, tapered joint domain, guide vane domain, annular volute domain, and outlet domain), Eq. (7.1) is employed to calculate the head-drop loss coefficient $h_{\Delta p}$. For the rotating domain (impeller domain), since the impeller keeps on inputting the power into fluid, the head loss is computed as the difference between the total rotational input work and output work obtained from the impeller, which can be calculated by Eq. (7.2):

$$\text{Stationary domain: } h_{\Delta p} = \frac{\int p_1 d\dot{m} - \int p_2 d\dot{m}}{P_{tol}} \tag{7.1}$$

$$\text{Rotating domain: } h_{\Delta p} = 1 - \frac{\int p_2 d\dot{m} - \int p_1 d\dot{m}}{P_{tol}} \tag{7.2}$$

*Mixed-flow Pumps: Modeling, Simulation, and Measurements*, First Edition.
Wei Li, Leilei Ji, Ramesh Agarwal, Weidong Shi, and Ling Zhou.

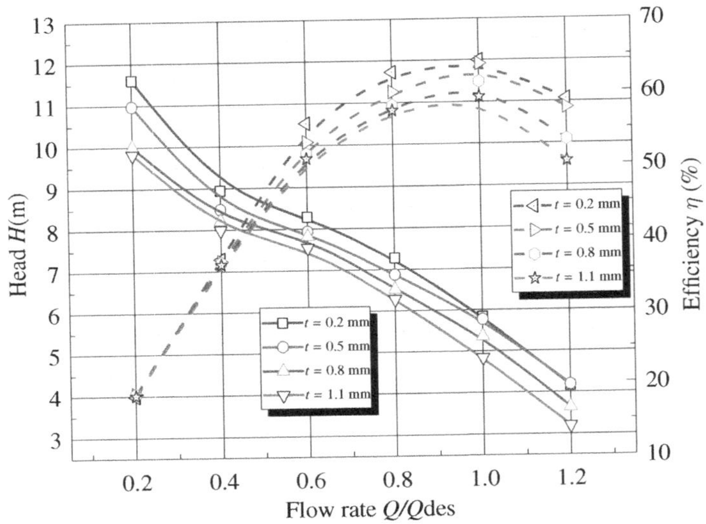


**Figure 7.1** Comparison of head and efficiency for various tip clearances at design flow rate.

In Eqs. (7.1) and (7.2), $p_1$ and $p_2$ represent the total pressure at the inlet and outlet of each domain, respectively, in Pa, and $P_{tol}$ represents the total input power W. $\dot{m}$ represents the mass discharged from each domain in $m^3/s$.

Figure 7.2 shows the head-drop loss coefficient at various flow rate rates in various components of the pump. It can be seen from this figure that under partial-loading condition, the head-drop loss coefficient of the combined impeller and guide vane accounts for nearly 54% to 80% for flow rates $0.8Q_{des}$ to $0.2Q_{des,}$ respectively. For these flow rates, the flow separation and unsteady flow features are prevalent in the impeller and the guide vane, which cause a great deal of losses in these domains. The losses incurred due to the change in tip clearances are not much evident for these flow rates. The most obvious flow rate conditions that affect the losses due to the change in the tip clearance are the design flow rate or near design flow rates. At the design flow rate, the head loss coefficient in the impeller is clearly impacted by the change in tip clearances; when the tip clearance increases from 0.2 to 1.1 mm, the head-drop loss coefficient increases 1.62 times in the impeller, which is due to enormous change in the tip region flow fields of the pump as the tip clearance increases. Thus, the energy loss in the impeller is highly dependent on the size of the tip clearance.

### 7.1.2 Distribution of the Total Entropy Production in the Pump for Various Tip Clearances

Figure 7.3 shows the total entropy production in the entire computational domain of the pump for different blade tip clearances and various flow rates. At larger flow rates, the total entropy production in the pump increases slightly with increase in the flow rate for the same tip clearance. At smaller flow rates, the total entropy production increases with decrease in the flow rate for the same tip clearance, indicating that the turbulence effects inside the pump and the hydraulic loss caused by the secondary vortex structure increase. For each flow rate, the influence of different tip clearances on the total entropy production of the pump is different as expected. At the design flow rate $Q_{des}$ and flow rates greater than $Q_{des}$, the total entropy production of the pump increases

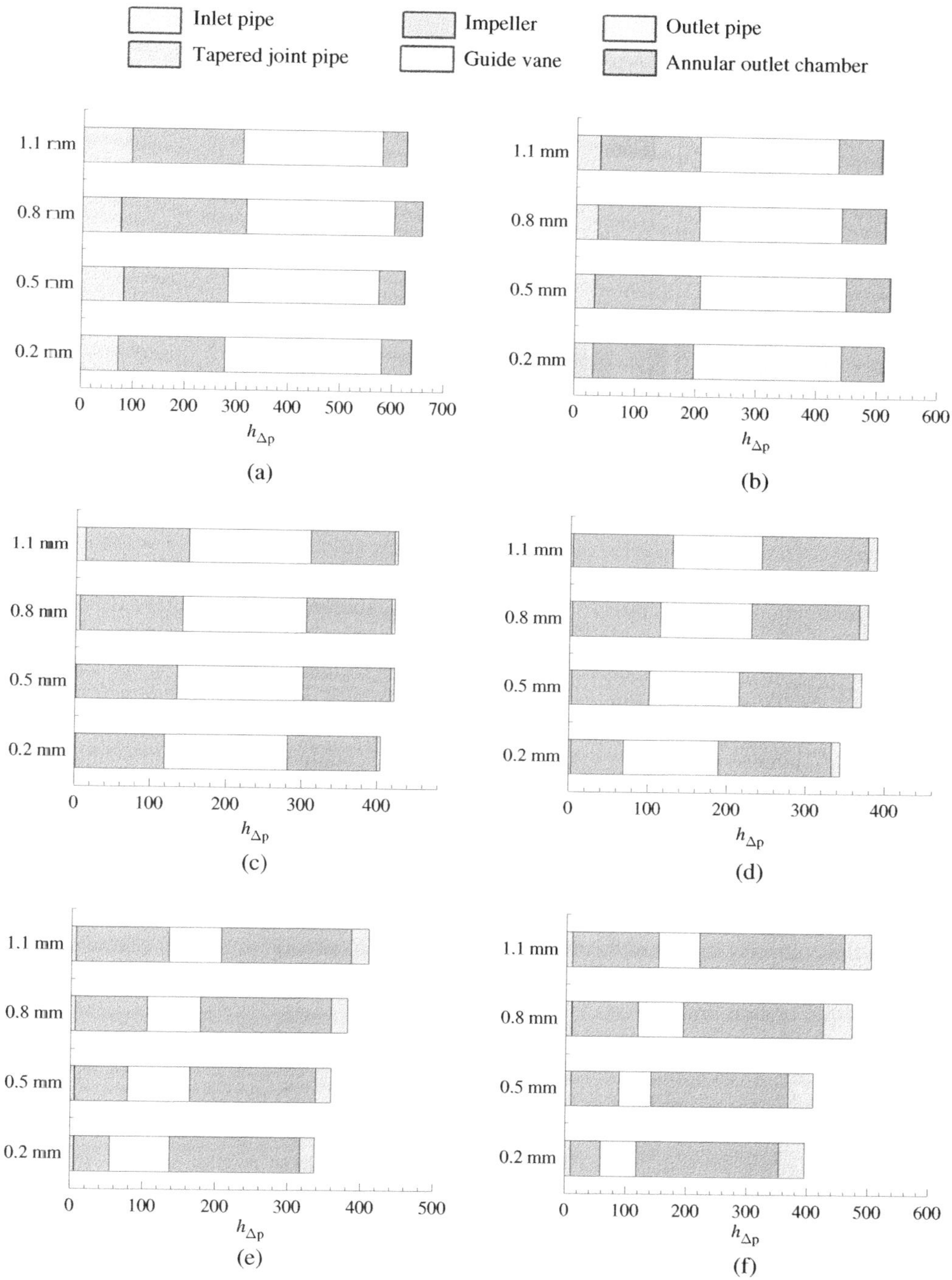


**Figure 7.2** The head-drop loss coefficient for various flow rates. (a) $0.2Q_{des}$, (b) $0.4Q_{des}$, (c) $0.6Q_{des}$, (d) $0.8Q_{des}$, (e) $1.0Q_{des}$, (f) $1.2Q_{des}$.

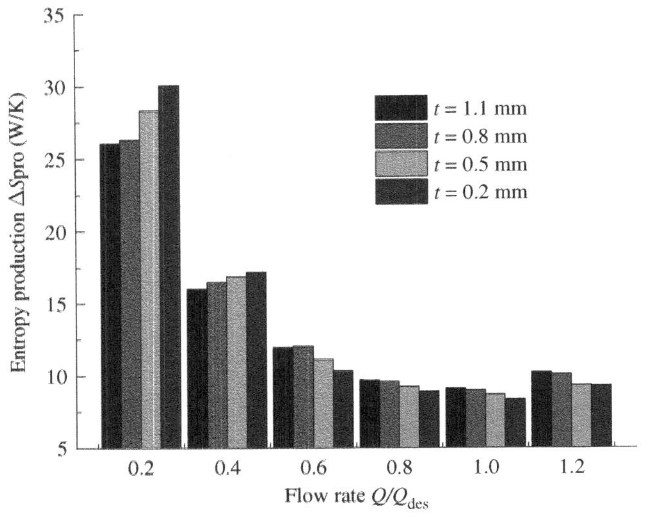


**Figure 7.3** Total entropy production in the mixed-flow pump for different tip clearances at various flow rates.

gradually with increase in the tip clearance. As the tip clearance increases from 0.2 to 1.1 mm, the total entropy production increases by 0.75 W/K. However, the increment rate of total entropy production is not linear with the tip clearance. When the tip clearance increases from 0.8 to 1.1 mm, the increase in total entropy production is significantly smaller compared to when the tip clearance increases from 0.2 to 0.5 mm. The difference in the total entropy production for change in the tip clearance from 0.2 to 1.1 mm is 0.11 W/K; however, the difference in entropy production for change in tip clearance from 0.8 to 1.1 mm is 0.32 W/K. Thus, it can be concluded that when the blade gap increases up to a certain value, the effect of tip clearance on the total entropy production of the pump is insignificant. Compared to the large flow rate conditions, the influence of tip clearance on the total entropy production of the pump is opposite to when the flow rates are small enough. The total entropy production decreases with increase in the tip clearance. At the same time, for the two large tip clearances ($t = 0.8$ mm and $t = 1.1$ mm), the difference in the total entropy production of the mixed-flow pump is less than the difference in the entropy production for tip clearances 0.2 and 0.5 mm. Thus, the influence of tip clearance on the total entropy production of the pump is constrained within certain limits.

Figure 7.4 shows the total entropy production in the entire computational domain of the pump obtained for various tip clearances and flow rates. When the tip clearance increases from 0.2 to 1.1 mm, the total entropy production in the impeller increases to 2.85 W/K, which is a 142% increase, whereas the total entropy production in the guide vane drops slightly. The total entropy production in the guide vane decreases by 21.8% to 1.61 W/K. These results indicate that the increase in the TLF increases the energy loss in the impeller but inhibits, to some extent, the hydraulic losses in the guide vane. Due to the influence of the guide vane, the high-energy fluid loses more kinetic energy to get more potential energy. Therefore, the total entropy production in the guide vane changes in a direction opposite to the changes in TLF. At smaller flow rates, with increase in the blade tip clearance, the total entropy production in the annular outlet chamber and the outlet pipe does not change significantly, and the total entropy production in the inlet pipe and the tapered joint pipe changes slightly indicating that the change in the blade tip clearance slightly

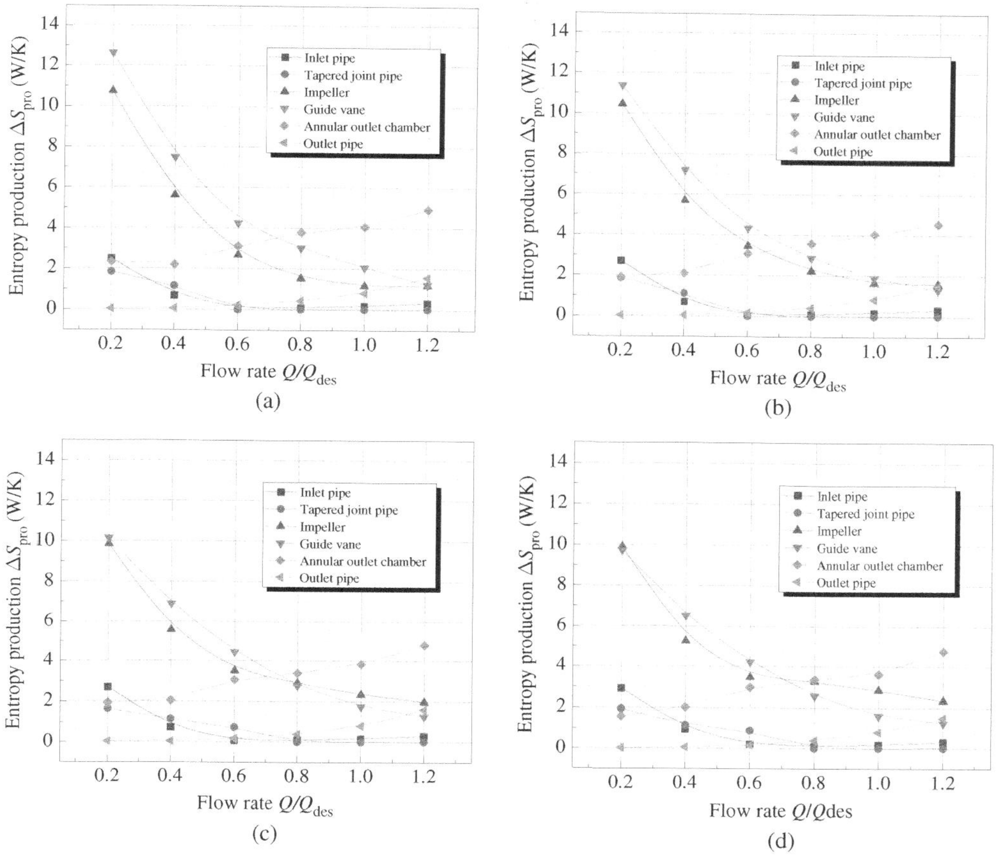


**Figure 7.4** Variation in the total entropy production with the flow rate in different domains of the pump for various tip clearances. (a) $t = 0.2$ mm, (b) $t = 0.5$ mm, (c) $t = 0.8$ mm, (d) $t = 1.1$ mm.

affects the inlet flow field of the of mixed-flow pump. Meanwhile, at smaller flow rates, the total entropy production in the impeller reduces slightly with increase in the tip clearance, while the total entropy production in the guide vane reduces more indicating that the TLF and TLV are not the main reason for the increase in hydraulic losses in the impeller at smaller flow rates; the losses are mostly impacted by the secondary vortex, flow separation, and the backflow. Furthermore, the local unsteady flow caused by the TLF near the blade rim at smaller flow rates is also not the main factor for entropy generation in the guide vane; however, it does account for the secondary vortices in the impeller, which reduce the flow rate of the mainstream and prevent some vortices from entering into the guide vane; thus the total entropy generation in the guide vane decreases at smaller flow rates with increase in the tip clearance.

In summary, the size of the tip clearance in a mixed-flow pump is highly correlated with the hydraulic losses in the pump, which not only enhances the hydraulic losses in the impeller but also reduces the flow velocity of the high-speed fluid in the impeller due to the TLF and TLV. The energy required for TLF to enter the guide vane is small, thus reducing the hydraulic losses in the guide vane. Therefore, in order to improve the energy performance of the mixed-flow pump, it is necessary to consider the scale of blade tip clearance and optimize the hydraulic design structure of the guide vanes to match the scale of the tip clearance.

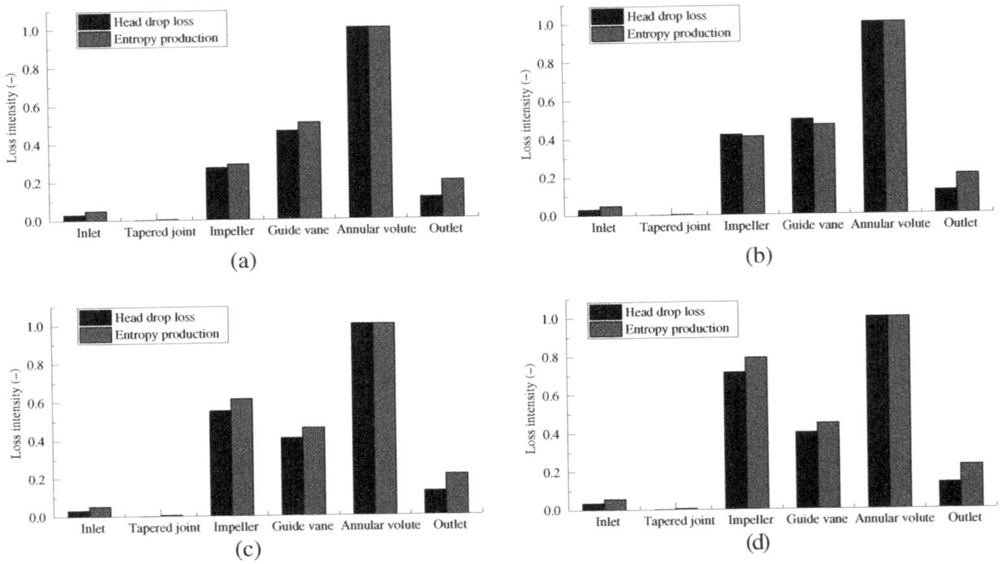


**Figure 7.5** Comparison of the head-drop loss and the entropy production at design flow rate in each component of the pump. (a) $t = 0.2$ mm, (b) $t = 0.5$ mm, (c) $t = 0.8$ mm, (d) $t = 1.1$ mm.

As mentioned above, the performance of the mixed-flow pump is most seriously affected at the design flow rate due to change in the tip clearance. Thus, it is important to determine the relationship between the head-drop loss coefficient and the entropy production. Figure 7.5 shows the head-drop loss and the entropy production at the design flow rate for each simulation domain; the loss is normalized by the maximum value of each variable for the same tip clearance. Furthermore, the entropy production has the same trend as the head-drop loss coefficient, which indicates that the entropy production method could be used to evaluate the head-drop loss in a mixed-flow pump. From this figure, it can be inferred that the losses in the impeller become greater at larger tip clearance. Thus, the TLF induced by the larger tip clearance generates more energy losses, which is the main reason for head drop at the design flow rate.

### 7.1.3 Local Entropy Production Rate in the Impeller

To analyze the local entropy production in the impeller, the impeller domain is divided into ten volumes based on the span coefficient, as shown in Fig. 7.6. It can be noted from this figure that with increase in the tip clearance, the increment in the average entropy production of each component near the tip region is not linear with the increment in the tip clearance. For example, the difference in the average entropy production among the four tip clearances is 909.34, 864.28, 503.01 W/m$^3$/K, respectively, from the small tip clearance to large tip clearance: at the largest tip clearance, the difference being zero. It indicates that the relationship between the average entropy production and the scale of the tip clearance is not a simple linear proportionality function. It can also be noted from the proportion of the total entropy production in different volumes for different tip clearances, as shown in Fig. 7.7. It can be noted from this figure that the larger proportion of the total entropy production occurs in the blade rim region (Volume 9 and Volume 10) of the impeller for different tip clearances; it corresponds to the regions where the unsteady flow occurs, and more

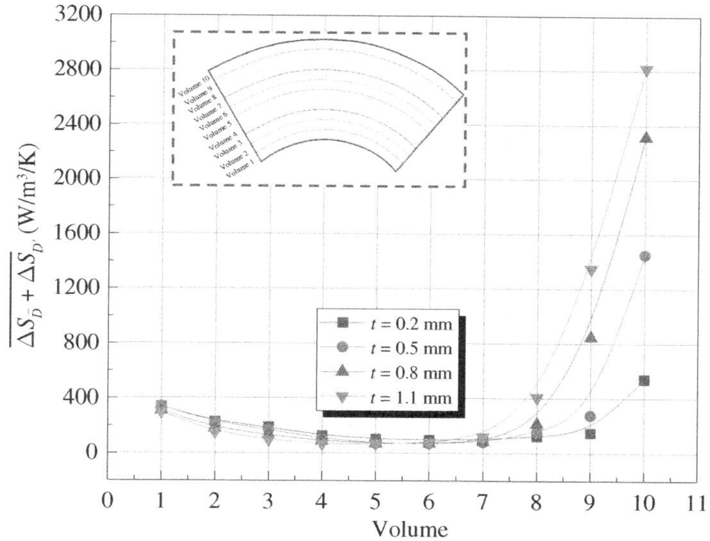


**Figure 7.6** Average entropy production in the different volumes of the impeller for various tip clearances.

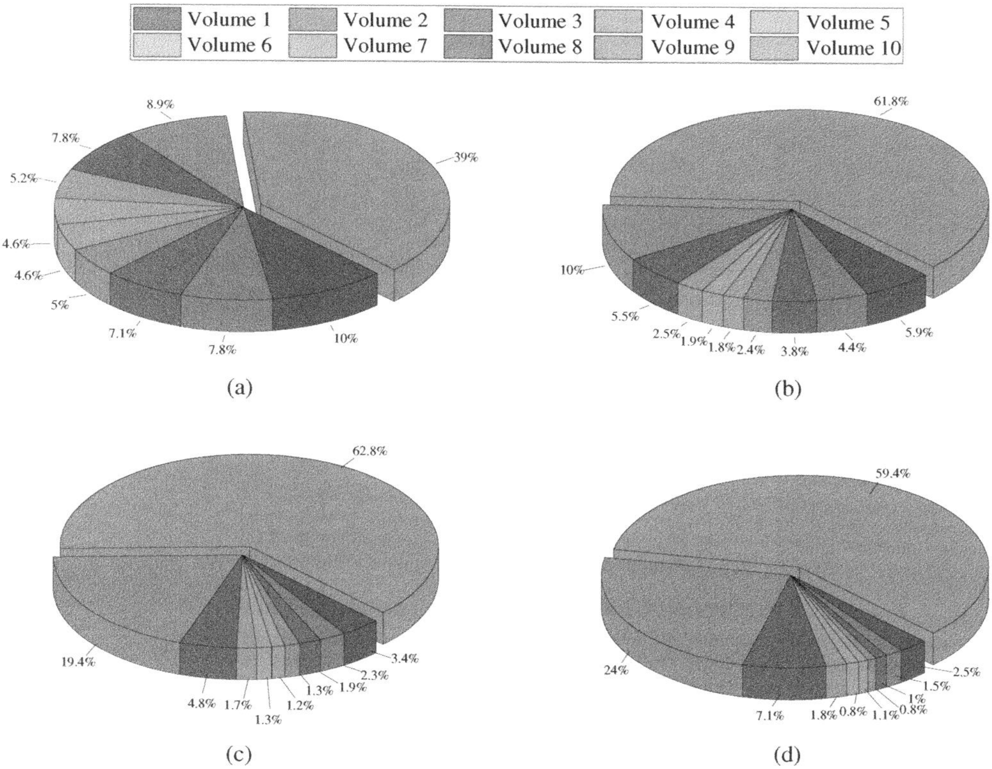


**Figure 7.7** Proportion of the total entropy production in each volume for various tip clearances. (a) $t = 0.2$ mm, (b) $t = 0.5$ mm, (c) $t = 0.8$ mm, (d) $t = 1.1$ mm.

energy losses are generated. It can also be noted that the proportion of the total entropy production is also relatively larger (although smaller than that in the blade rim region) in the hub region (Volume 1 and Volume 2) compared to that in the middle of the impeller, which indicates that the flow is relatively steady in the middle of the flow channel of the impeller. Furthermore, with the increase in the tip clearance, the proportion of the total entropy production in the middle region (Volume 5 and Volume 6) decreases a little, which means that the TLF also affects the flow field in the middle region of the impeller. Additionally, the proportion of the total entropy production in the hub region to the total entropy production in the impeller gradually decreases with increase in the tip clearance, while the proportion of the total entropy production in the blade rim region increases from 47.9% to 83.4% (sum of Volume 9 and Volume 10). It indicates that the TLF greatly increases the hydraulic loss in the impeller, and the scale of the gap can affect the strength of the TLF, thereby affecting the hydraulic losses inside the impeller. It can also be seen that when the tip clearance is 0.8 or 1.1 mm, the total entropy production difference in the blade rim region is about 1.2%, which indicates that the incremental acceleration of total entropy production in the rim region reduces significantly, and therefore there is a limiting value for the hydraulic losses as the tip clearance continues to increase. Thus, the TLF directly influences the overall flow field in the impeller and thereby affects the pump's efficiency.

Figure 7.8 shows the local entropy production on the blade-to-blade surfaces of the impeller at various span locations for different tip clearances at the design flow rate. As can be noted from this

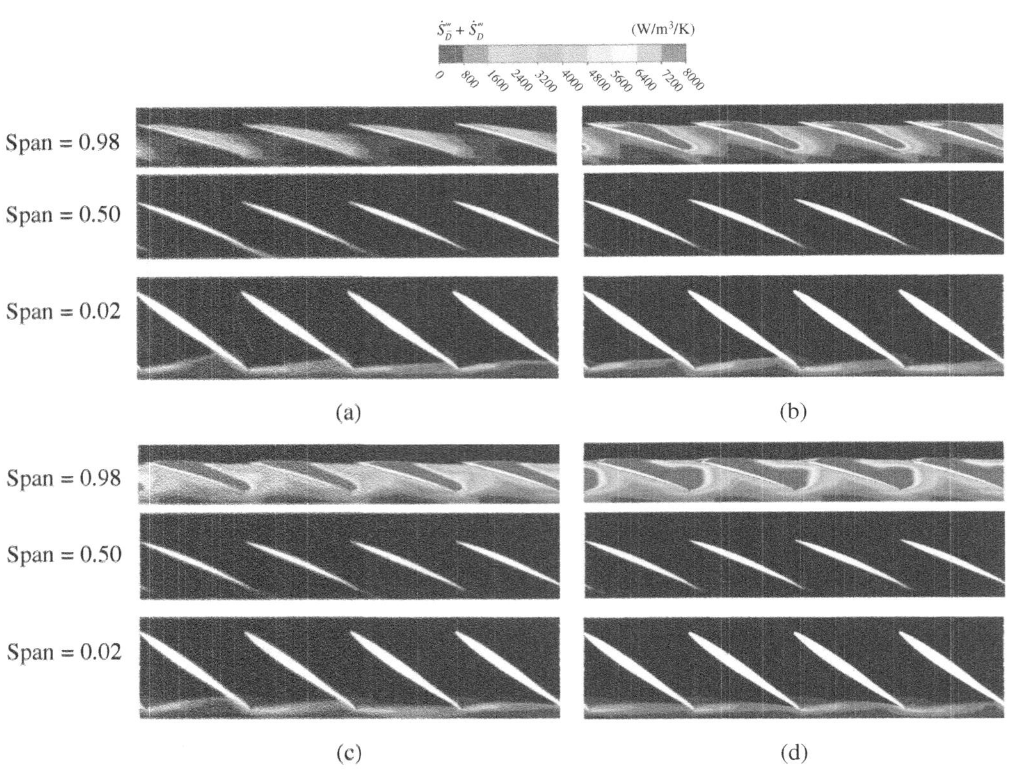


**Figure 7.8** Distribution of local entropy production on blade-to-blade surfaces at different Span locations for various tip clearances. (a) $t = 0.2$ mm, (b) $t = 0.5$ mm, (c) $t = 0.8$ mm, (d) $t = 1.1$ mm.

figure, the entropy production contours in the middle section of the impeller are essentially the same for each tip clearance, and the entropy production value is higher in the small area near the trailing edge of the blade close to the outlet. Near the vane hub of the pump, the entropy production contours for each tip clearance are basically similar, and the entropy production at the impeller outlet is higher. In the middle of the flow channel, the blade pressure surface (PS) and the suction surface (SS) converge near the blade trailing edge, forming a high-loss region. Near the impeller rim region, the difference in the entropy production contours for each tip clearance is large, which signifies that the entropy production in the blade rim region increases with increase in the tip clearance. At smaller tip clearance, the TLF-induced high entropy generation zone is close to the blade suction surface. From the blade inlet to the blade outlet, the high entropy production area gradually expands to the middle of the impeller flow channel, but it does not affect the next stage of the blade. With increase in the tip clearance, the local entropy generation in the high entropy production area near the suction surface of the blade increases, and the area of the high entropy generation also increases significantly.

### 7.1.4 Local Entropy Production Rate in the Guide Vane

Figures 7.9 and 7.10 depict the average entropy production in various sections of the guide vane and the local entropy production on different blade-to-blade surfaces, respectively. In the middle region of the guide vane, the average entropy production decreases with increase in the tip clearance. However, in the rim region of the guide vane, with increase in the tip clearance, the average entropy production first decreases and then increases. On the other hand, for smaller tip clearance, the average entropy production in the middle region of the guide vane is significantly larger than that in the hub region of the guide vane, whereas the average entropy production in the blade rim region of the guide vane for larger tip clearance is greater than that in the middle region of the guide vane.

It can also be seen that the local entropy production near the guide vane inlet is small when the tip clearance is small, and the entropy production in the guide vane inlet is significantly augmented

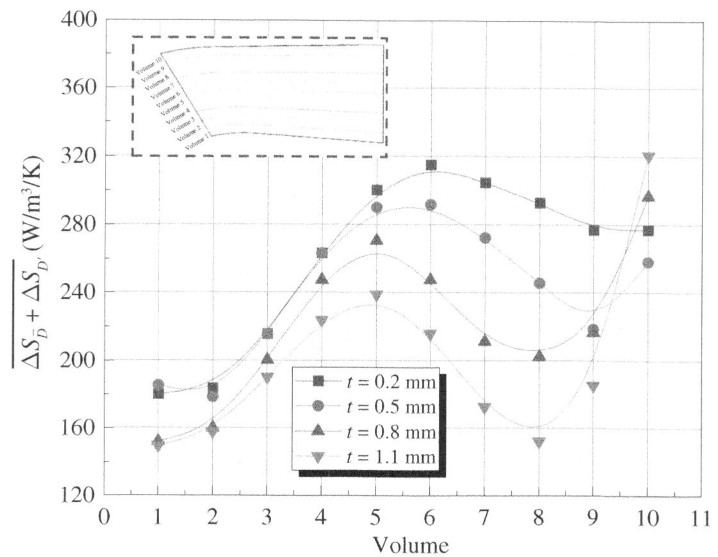


**Figure 7.9** Local average entropy production in different parts of the guide vane for various tip clearances.

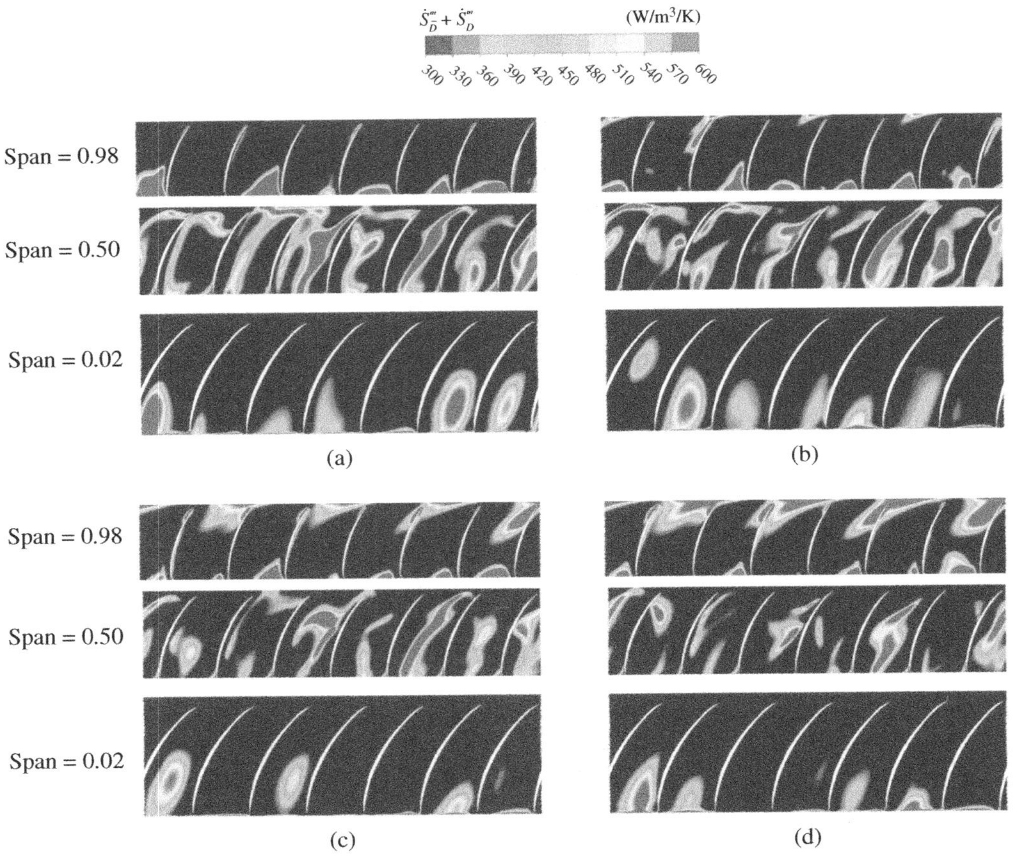


**Figure 7.10** Local entropy production on blade-to-blade surfaces for various tip clearances. (a) $t = 0.2$ mm, (b) $t = 0.5$ mm, (c) $t = 0.8$ mm, (d) $t = 1.1$ mm.

with increase in the tip clearance. This is because the energy of the TLV in the impeller has not completely dissipated, which affects the flow field in the guide vane. Furthermore, the larger the tip clearance, the greater the strength of the TLV. The TLV is not dissipated in the impeller and causes more hydraulic losses in the guide vane inlet. In the middle region of the guide vane, the local entropy production in the guide vane flow channel for each tip clearance is higher; however, it decreases with increase in the tip clearance. This is because the hydraulic loss in the impeller is small when the TLF is small, and therefore, the fluid carrying the high kinetic energy flow enters the guide vane. But, when the TLV is big enough to influence the mainstream flow in the impeller, it lowers the velocity of the mainstream flow, which causes the velocity of fluid that enters the guide vane to become smaller. Since in the mixed-flow pump, the guide vane is a hydraulic component that converts the kinetic energy of fluid into potential energy, and the vortex structures are formed in the guide vane to dissipate the kinetic energy of the fluid and to decrease the flow velocity. This is the reason that there are more hydraulic losses in the hub region and the middle region of the guide vane when the tip clearance is smaller. Near the guide vane hub, the influence of TLV is relatively weak, and therefore, there is prominence of turbulent flow in the guide vane channel and the recirculatory flow in the annular volute.

In summary, the TLF affects the blade tip region flow fields in the guide vane, and its influence becomes more prominent as the tip clearance increases. The size of the tip clearance determines the interference level of TLF with the main flow in the impeller. Furthermore, TLF also affects the velocity of fluid into the guide vane. When the energy conversion process of the fluid in guide vane is considered, the high-speed fluid needs more kinetic energy to dissipate and convert it into potential energy; however, this is not the case with low-speed fluid since it dissipates less energy. Thus, the hydraulic losses in the guide vane are negatively correlated with the size of the tip clearance.

## 7.2 Flow Structures

### 7.2.1 Vortex Patterns in the Blade Rim Region of the Pump for Various Tip Clearances

Figure 7.11 shows the morphology of TLF and the distribution of the local entropy production in the TLV for various tip clearances. It can be noted that, for each tip clearance, the starting point of the TLV does not start from the blade inlet but starts at a distance from the leading edge of the blade inlet. In the case of smaller tip clearance, the flow rate and the axial momentum of the TLF are small, and it is close to the end wall region of the impeller chamber due to the action of the main flow. Therefore, in the blade rim region, the high entropy production zone is in the end wall region along the direction of TLF. However, as the tip clearance increases, both the flow rate and

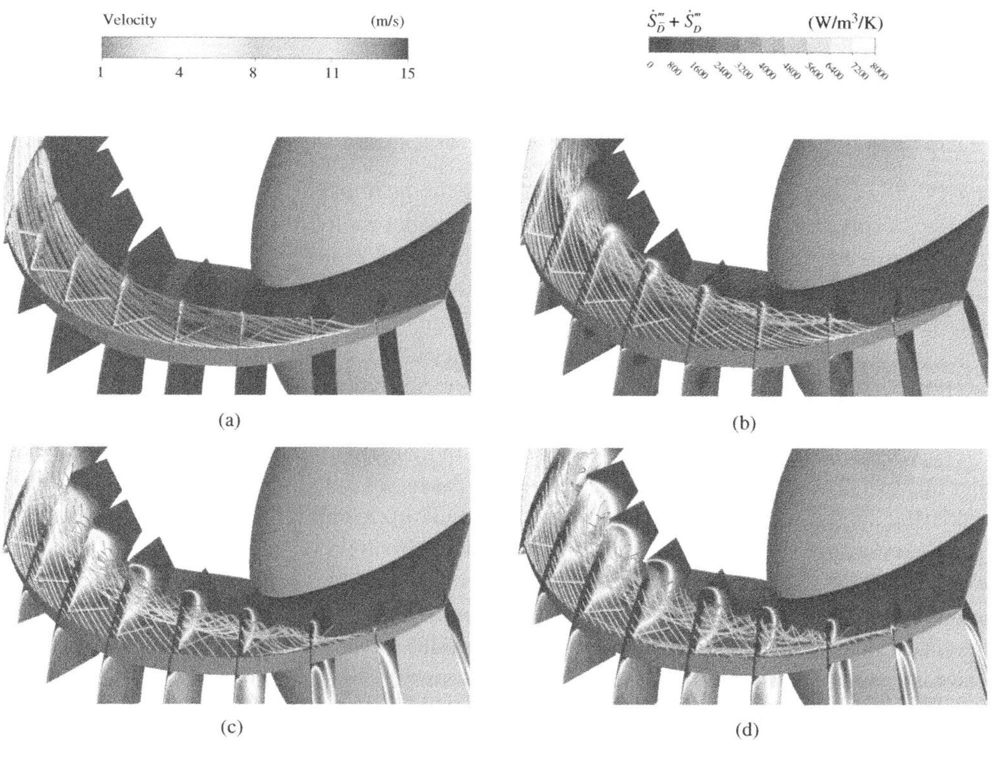


**Figure 7.11** Morphology of tip leakage vortex (TLV) and the distribution of the local entropy generation along the vortex for various tip clearances. (a) $t = 0.2$ mm, (b) $t = 0.5$ mm, (c) $t = 0.8$ mm, (d) $t = 1.1$ mm.

**Table 7.1** Comparison of rim leakage flow for various tip clearances.

| Clearance (mm) | Single-blade tip leakage flow (kg/s) | Relative to total leakage (%) |
|---|---|---|
| 0.20 | 0.374 | 1.42 |
| 0.50 | 0.905 | 1.72 |
| 0.80 | 1.222 | 4.65 |
| 1.10 | 1.469 | 5.59 |

the axial momentum of TLF increase as well as the strength and the scale of the TLV increase; as a result, the affected area also expands into the impeller flow channel.

Table 7.1 shows quantitatively the TLF of a single-blade clearance for different tip clearances and the relative leakage flow compared to the total leakage flow in the rim region of the four blades of the impeller at the design flow rate. The single-blade rim leakage flow can be obtained directly from the area fraction of the normal velocity pairs in the middle section of the tip clearance region. From this table, it can be noted that increase in the tip leakage clearance size and the increase in the TLF do not exactly have a linear relationship. The structure of the tip leakage vortex (TLV) also shows that when the tip clearance increases, the increase in the TLF is the main reason for increase in the size of the TLV; therefore, controlling the relative leakage flow is one of the key ways to weaken the TLV.

### 7.2.2 Tip Leakage Vortex (TLV) Intensity and the Vortex Core Distribution

Figure 7.12 shows the distribution of the vortex intensity in the impeller TLF ($Q$-criterion $>200\,s^{-1}$) and the line of vortex cores at different sections of the impeller for four tip clearances at the design flow rate. In this figure, the TLV core is selected as follows: it is near the suction surface of the blade and is in the middle of the swirling flow line through the maximum leakage flow vortex strength (called the swirling strength). As can be seen from Fig. 7.12, there is some vorticity loss in the impeller for each tip clearance, and there is a primary leakage vortex and a secondary leakage vortex in the rim area (red dashed area), and there are return vortex and separation vortex in the blade hub area (yellow dotted line). From the leakage vortex intensity in the tip area, it can be noted that the vortex intensity of the TLV tends to increase with increase in the tip clearance, while in the impeller hub exit area, the high vortex intensity region decreases with increase in the tip clearance.

Considering the flow structure in the impeller with small tip clearance of 0.20 mm, it can be noted that the TLV structure is small, and the front part of the blade shows the high vortex intensity area. However, as the tip clearance increases, the TLV begins to develop in the runner rim area, and the high vortex intensity area of the TLV increases significantly. At the same time, the secondary leakage vortex structure begins to appear, affecting the middle of the runner rim and the lower-stage blade surface pressure at the outlet. When the tip clearance increases to 1.10 mm, the area of the high vortex intensity zone of the TLV is the largest, but the influence range of the secondary leakage vortex is significantly reduced only affecting the front part of the impeller rim-flow field and the front part of the PS of the lower stage blades. From the diagram of the TLV core in Fig. 7.13, it is obvious that for the large tip clearance of 0.8 and 1.10 mm, the line of the inlet leakage vortex

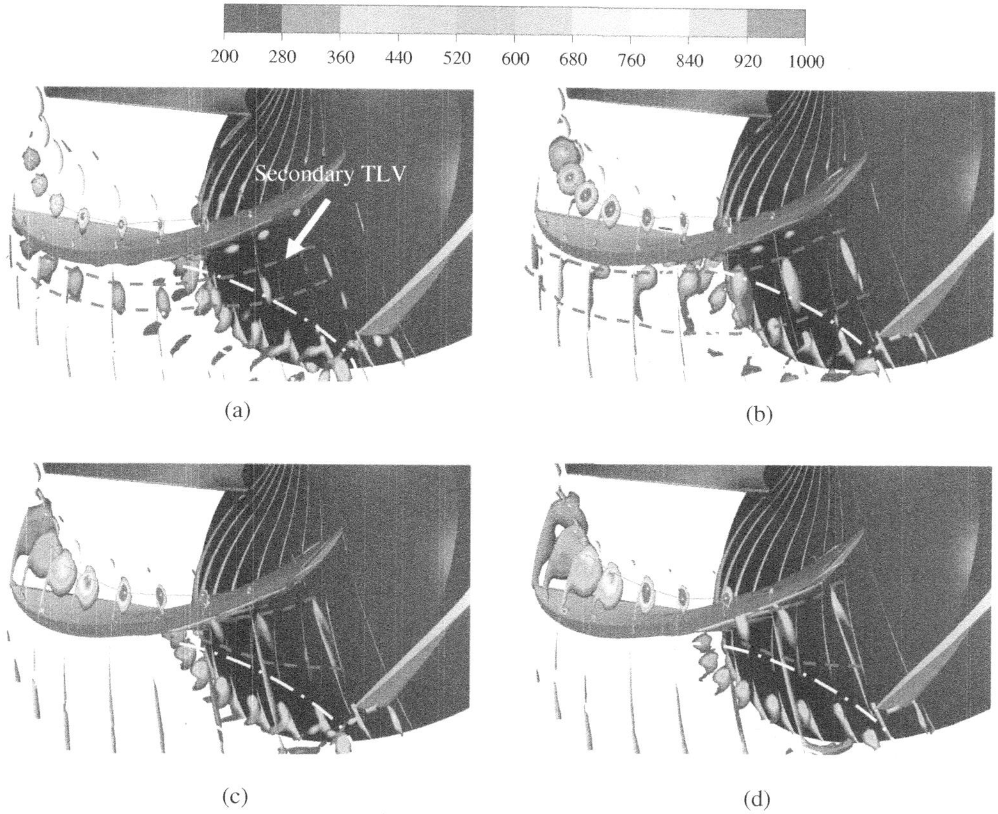


**Figure 7.12** Vortex intensity distribution of the tip leakage flow for various tip clearances. (a) $t = 0.20$ mm, (b) $t = 0.50$ mm, (c) $t = 0.80$ mm, (d) $t = 1.10$ mm.

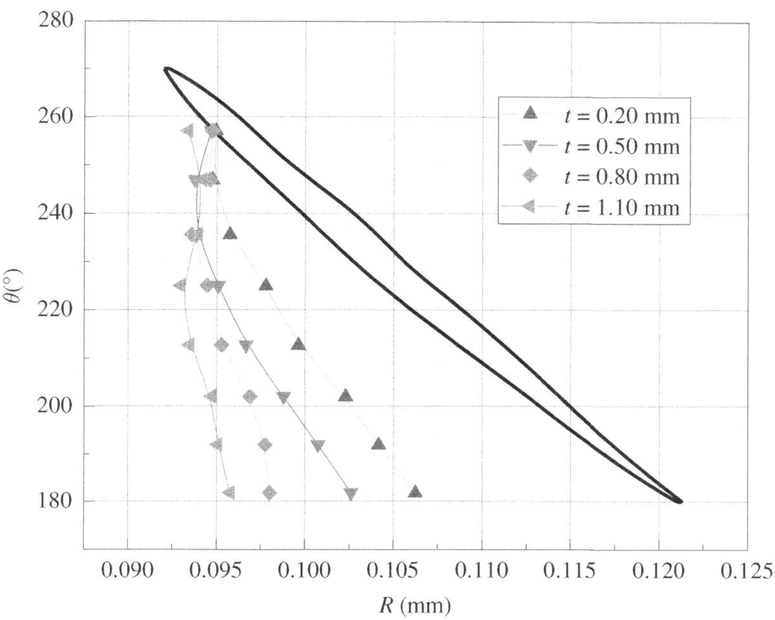


**Figure 7.13** Diagram of the core of the leakage vortex for various tip clearances.

core is different from that at smaller tip clearances, and the secondary fluctuations in the flow field appear.

In summary, it can be noted that the area of high leakage vortex intensity zone increases with increase in the tip clearance, and for the smaller tip clearance, the strengths of the outlet return vortex and the separation vortex are higher than those for the larger tip clearance impeller. Therefore, in order to reduce the impeller vortex loss and improve the efficiency of the pump, it is important to consider the impact of both the TLV and the return vortex.

## 7.3 Unsteady Flow Characteristics

### 7.3.1 Time-domain Analysis of the Pressure Pulsations

The time-domain diagrams of the pressure pulsation at the four monitoring points at the impeller inlet and outlet, in the middle of the impeller, and in the middle of the guide vane of the mixed-flow pump for various flow rates ($0.6Q_{des}$, $1.0Q_{des}$, and $1.4Q_{des}$) are given in Fig. 7.14. As can be seen from this figure at various flow rates, the pressure pulsation at each monitoring point shows the periodic behavior, and in one rotation of the impeller, it is influenced by the number of impeller blades. The pressure pulsation curve has four peaks and four troughs, which is consistent with the

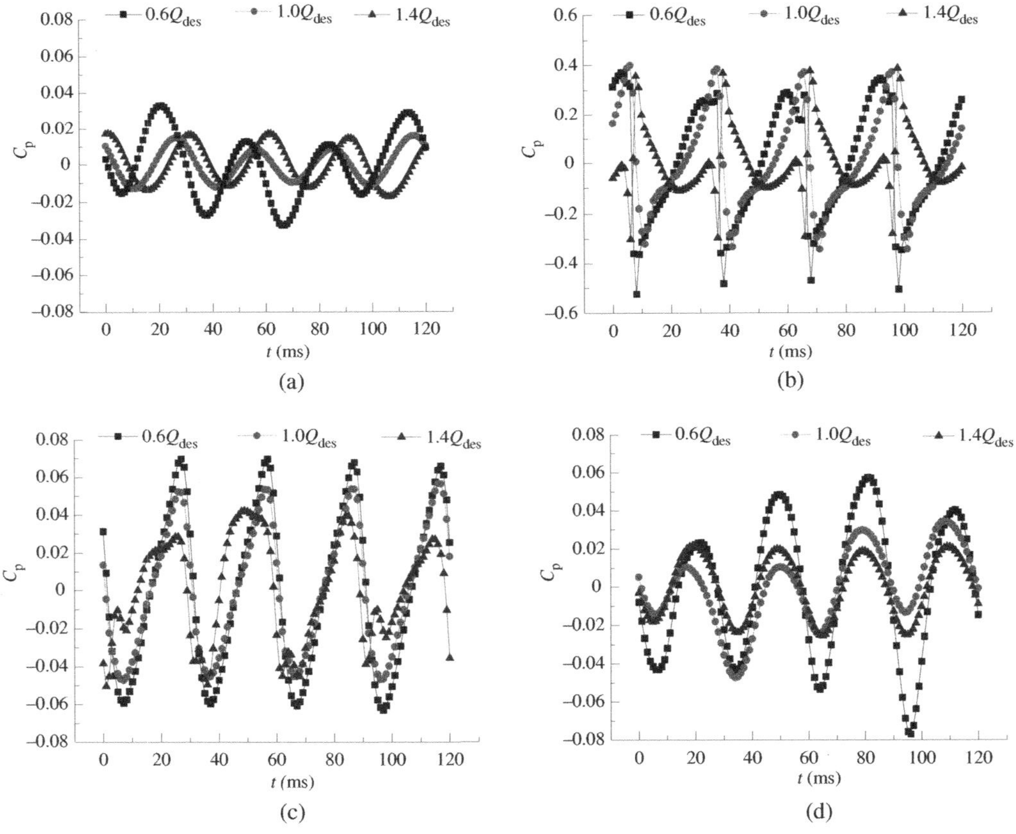


**Figure 7.14** Time-domain plots of the pressure fluctuation at various monitoring points. (a) Monitoring point A, (b) Monitoring point B, (c) Monitoring point C, (d) Monitoring point D.

blade passing frequency (BPF). At $0.6Q_{des}$ flow rate, the amplitude of the pressure pulsation at each monitoring point is high, and as the flow rate increases, the fluid flow gradually stabilizes, and the amplitude of the pressure pulsation gradually reduces. At the impeller inlet monitoring point, the amplitude of the pressure pulsation is not large, and the pressure pulsation coefficient $C_p$ varies in the range of 0.04, but the difference in the pressure coefficient at different flow rates is large. As the monitoring point moves to the inside of the impeller, at the middle point of the impeller, the pressure difference between the PS and suction surface of the blade is large, and there is a leaky flow at the top of the blade, which makes the pressure fluctuation appear to rise steeply and fall suddenly. However, the peak pressure coefficient at each flow rate is similar, and the influence of the flow rate on the amplitude of the pressure pulsation in the frequency domain is not obvious. In the region between the impeller outlet and the middle of the guide vane, the amplitude of the pressure pulsation coefficient is again larger since this is the region with the most serious dynamic and static interference between the impeller and the guide vane. It can also be noticed that the peak and the trough values of the pressure coefficient at each flow rate are different due to the dynamic and static interference, and the periodicity of the pressure pulsation is weakened at the $1.4Q_{des}$ flow rate, and a secondary wave crest structure appears. In the middle of the guide vane, the amplitude of the pressure pulsation coefficient decreases due to the rectification effect of the guide vane, but the amplitude of the pressure coefficient is higher at low flow rates.

In summary, at various flow rates, the influence of the flow rates on the location of each of the four monitoring points at the impeller inlet and outlet, in the middle of the impeller, and in the middle of the guide vane of the mixed-flow pump is significant on the pressure pulsation, especially in the dynamic and static interference zone of the impeller and the guide vane; at low flow rates, the amplitude of the pressure pulsation coefficient changes the most.

### 7.3.2 Frequency-domain Analysis of the Pressure Pulsation

The fast Fourier transform (FFT) time domain is applied to the time-domain numerical calculations of pressure pulsation at each monitoring point to obtain the frequency-domain plot of the pressure pulsation. It is found that the fundamental frequency of the pressure pulsation at each monitoring point is 24.17 Hz, and the main frequency is 96.67 Hz, which is four times the fundamental frequency. Figure 7.15 shows the frequency-domain plots of each monitoring point at various flow rates. It can be noted from this figure that at various flow rates, the main frequency in the impeller inlet, outlet, and middle part for pressure pulsation is 96 Hz, which is four times the impeller rotational frequency and is the same as the BPF, and the second harmonic frequency is 190 Hz, which is two times the blade passing frequency. There is no obvious high-frequency pulsation at each monitoring point, which indicates that the BPF of always appears to play a dominant role. At both the smaller flow rate and the higher flow rate relative to the design flow rate, the sub-frequency component increases significantly, while at the design flow rate, the presence of the sub-frequency component is not obvious, indicating that at the design flow rate, the fluid flow is more stable and the external characteristic curve for pump efficiency is higher in magnitude. Comparing the amplitude of the pressure coefficient, it can be seen that the pressure coefficient at the middle monitoring point of the impeller is the largest at different flow rates, and the pressure coefficient decreases at the monitoring point far away from the impeller position. The pressure pulsation coefficient is influenced by the leakage flow at the top of the impeller. Furthermore, with increase in the flow rate, the amplitude of the pressure coefficient at the middle monitoring point of the impeller gradually decreases, and this flow rate condition has the greatest influence on the frequency-domain amplitude at the middle position of the impeller.

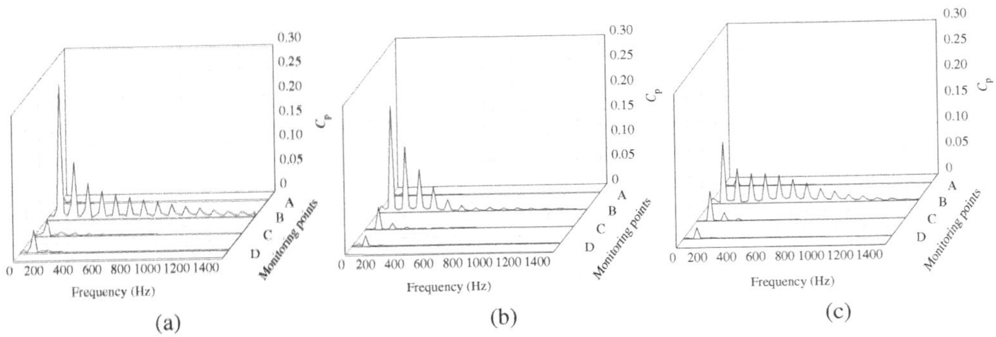


**Figure 7.15** Time-domain plots of pressure fluctuation at various monitoring points for various flow rates. (a) $0.6Q_{des}$, (b) $1.0Q_{des}$, (c) $1.4Q_{des}$.

## 7.4 Transient Flow Field Due to Rotor–Stator Interaction (RSI)

### 7.4.1 Flow Fields Before the RSI Zone

Figure 7.16 shows the flow fields near the impeller inlet before the RSI zone. It can be seen from this figure that under the three partial-loading conditions, the reduction in the flow rate results in the occurrence of the vortex structure in the blade passage because of the flow separation, although the flow is stable at the design flow rate. The vortex structure near the suction surface of blade changes more compared to the vortex near the hub in the partial-loading conditions. And the scale of the vortex structure near the blade suction surface and close to the end wall region becomes big enough to cover almost half of the cross-section of the blade passage when the flow rate is $0.2Q_{des}$. On the other hand, the vortex near the hub does not change much and it even has a shrinking trend for flow rates from $0.6Q_{des}$ to $0.2Q_{des}$ due to the squeeze effect on big vortex. Thus, the RSI zones are significantly affected by these vortices from the former blade passage.

### 7.4.2 Flow Fields in the RSI Zone

The RSI effect between the impeller and the guide vane in the mixed-flow pump is the main cause of the pressure fluctuation in this region. Therefore, in order to explore the flow field in the RSI region of the pump under the partial-loading conditions, the relative velocity distributions and streamlines on a series of sections with different phases through the rotor axis are analyzed for various flow rates. Since the impeller is rotating all the time, six phases in one guide vane passage (PA) are selected to describe the internal flow fields of RSI. These RSI zones under three partial-loading conditions are shown in Fig. 7.17.

It can be seen from Fig. 7.17, at flow rate of $0.2Q_{des}$, and $1/6\ T$ phase in the RSI zone, a sequence of circumferential vortex structures appears before the guide vane passage (PA), which are almost perpendicular to the axis of rotation. It can be noted that the vortices with smaller structures become bigger when the blade rotates from $4/6\ T$ phase to $1/6\ T$ phase. These vortex structures can be divided into two parts (A and B) influenced by the strong entrainment from the vortices in the guide vane passage. The part A gradually merges with the vortex structure inside the guide vane. Part B is also affected by internal vortices, and its vortex line direction is different from that of part A; however, it almost disappears since most of them are absorbed by the guide vane vortices. In the next phase, it is obvious that the two parts move away from the blade PS and shrink into part C and part D.

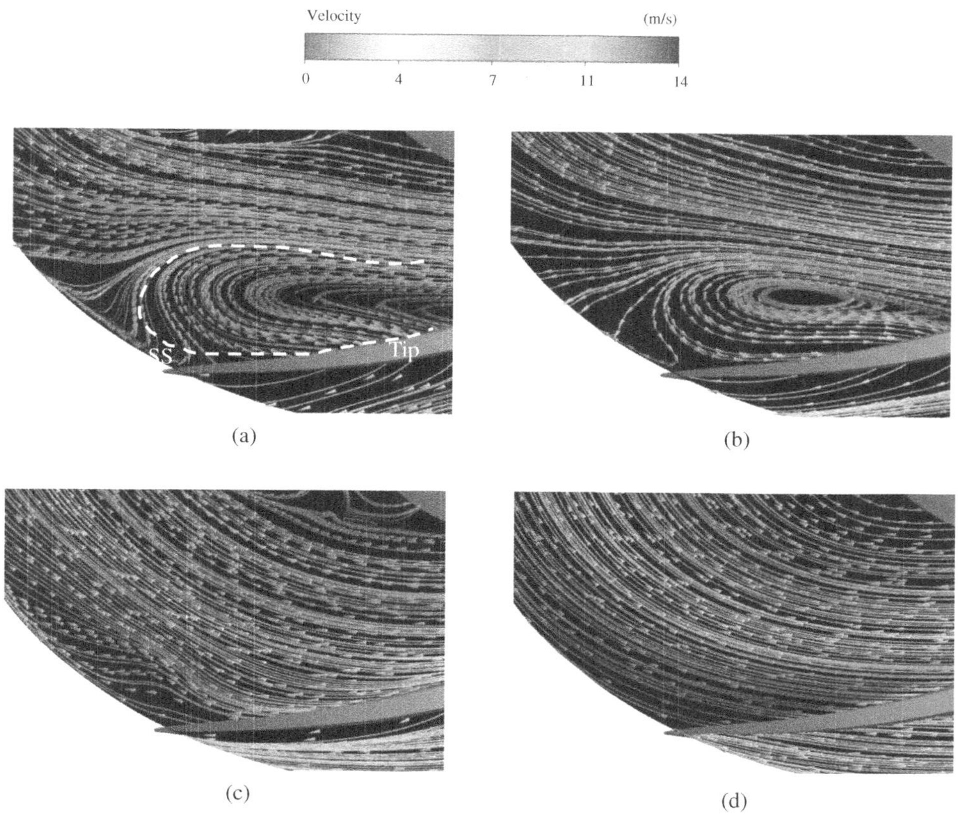


**Figure 7.16** Flow fields and streamlines in the impeller before the RSI zone. (a) $0.2Q_{des}$, (b) $0.4Q_{des}$, (c) $0.6Q_{des}$, (d) $1.0Q_{des}$.

Additionally, at flow rates of $0.4Q_{des}$ and $0.6Q_{des}$, even though the circumferential vortex structures still exist before the guide vane, they are closer to the stronger vortices in the guide vane passage. With increase in the flow rate, these vortices become closer to the inlet of the guide vane. For instance, at $0.4Q_{des}$ flow rate, when the blade inlet rotates across the middle section in 4/6 $T$ phase, a small vortex appears, which grows bigger in the guide vane passage (PA), and then it merges with the stronger vortex structures and disappears at 3/6 $T$ phase. Similarly, at $0.6Q_{des}$, these vortices are far away from the blade PS, and the center line of these vortices is nearly parallel to the inlet of the guide vane. Therefore, it can be concluded that the influence zone of RSI is closer to the guide vane when the flow rate is larger under the part-loading condition. At lower flow rates, the flow fields in the RSI zone become unstable and complex.

Figure 7.18 shows the streamlines on the blade-to-blade section of the impeller and the guide vane at the middle span in the three part-load conditions. It can be seen from this figure that the flow fields are relatively stable at $0.6Q_{des}$ compared to the other two partial-loading conditions. A separation vortex near the blade suction surface occurs and becomes strong when the flow rate decreases. Moreover, since the position of the vortex does not change with the rotation of the impeller, the core of the vortex could connect as a straight line, as shown in the orange dashed line in this figure. This vortex is not the result of the RSI effect but is due to the decrease in the flow rate. The flow fields in the RSI zone and the guide vane passage change a lot under the partial-loading

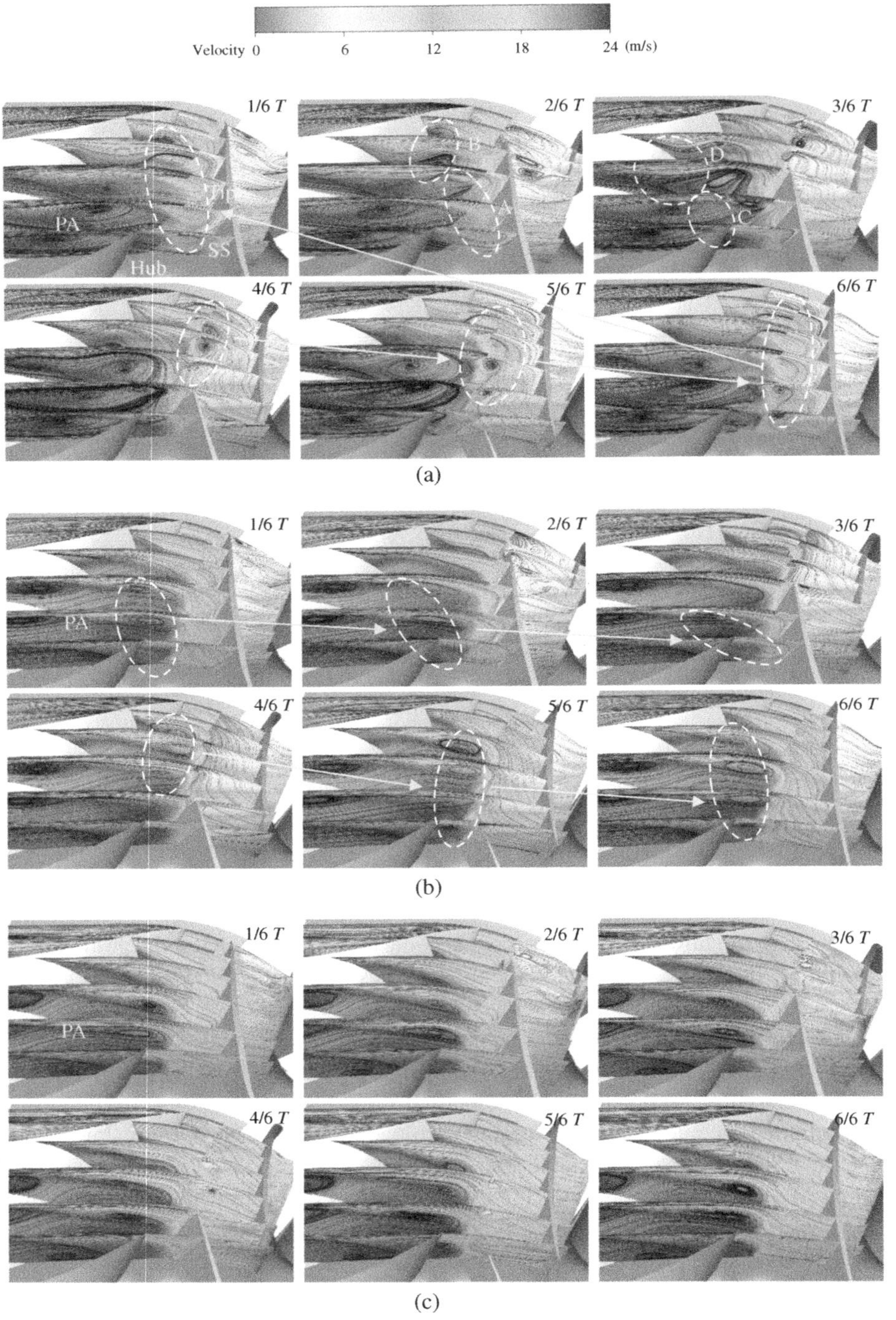


**Figure 7.17** Relative velocity distributions and streamlines in the RSI zones. (a) $0.2Q_{\mathrm{des}}$, (b) $0.4Q_{\mathrm{des}}$, (c) $0.6Q_{\mathrm{des}}$.

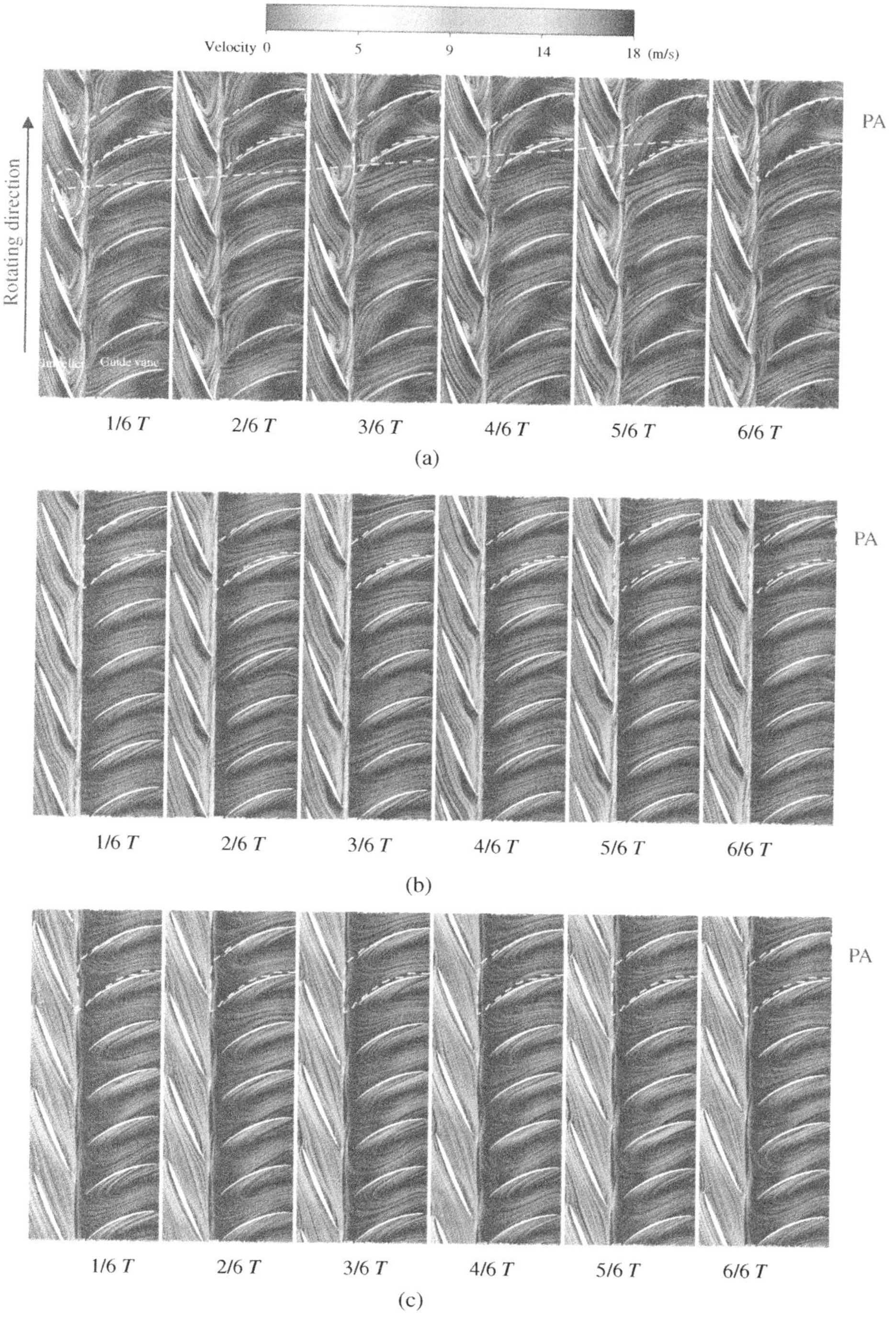


**Figure 7.18** Streamlines on the blade-to-blade sections at the middle span for the three partial-loading conditions. (a) $0.2Q_{des}$, (b) $0.4Q_{des}$, (c) $0.6Q_{des}$.

conditions. At $0.2Q_{des}$ flow rate, due to the rotation of the impeller, the monitoring point of the guide vane (PA) also changes. The fluid flowing into the guide vane passage (PA) could be decomposed into two parts at the same time. One is the fluid from the impeller blade passage that is rotating away from the guide vane passage (PA), and the other is the fluid from the impeller blade passage that is rotating across the guide vane passage (PA). At 1/6 $T$ phase, due to the outlet width of the single impeller channel being larger than that of the guide vane, most of the fluid in the impeller flows into the corresponding guide vane channel. However, after the impeller rotates some angle, especially from 1/6 $T$ phase to 3/6 $T$ phase, some of the fluid from the impeller channel starts to flow into the downstream guide vane channel, which makes the inflow angle of the local guide vane shrink. At $0.4Q_{des}$ and $0.6Q_{des}$ flow rates, the strong vortex in the guide vane due to backflow takes the dominant position and its size decreases with the increase in the flow rate.

Figure 7.19 shows the instantaneous iso-surfaces of the $Q$-criterion [3–6] in the three partial-loading conditions. It can be seen from this figure that at $0.2Q_{des}$ flow rate when the blade rotates across the guide vane passage (PA), the vortex in the impeller becomes erratic. In the RSI zone, vortices are generated between the impeller and the guide vane, and they move into the guide vane passage with the fluid. These vortices dissipate in the passage as they merge with the backflow vortices in the guide vane and eventually become a part of the backflow vortices. Furthermore, in the guide vane passage (PA), these vortices change with the impeller rotation and move toward the outlet of the guide vane. Since the function of the guide vane is to convert the momentum or kinetic energy of the fluid into potential energy, these vortices disappear near the end of the outlet of the guide vane passage. Likewise, in the other two partial-loading conditions,

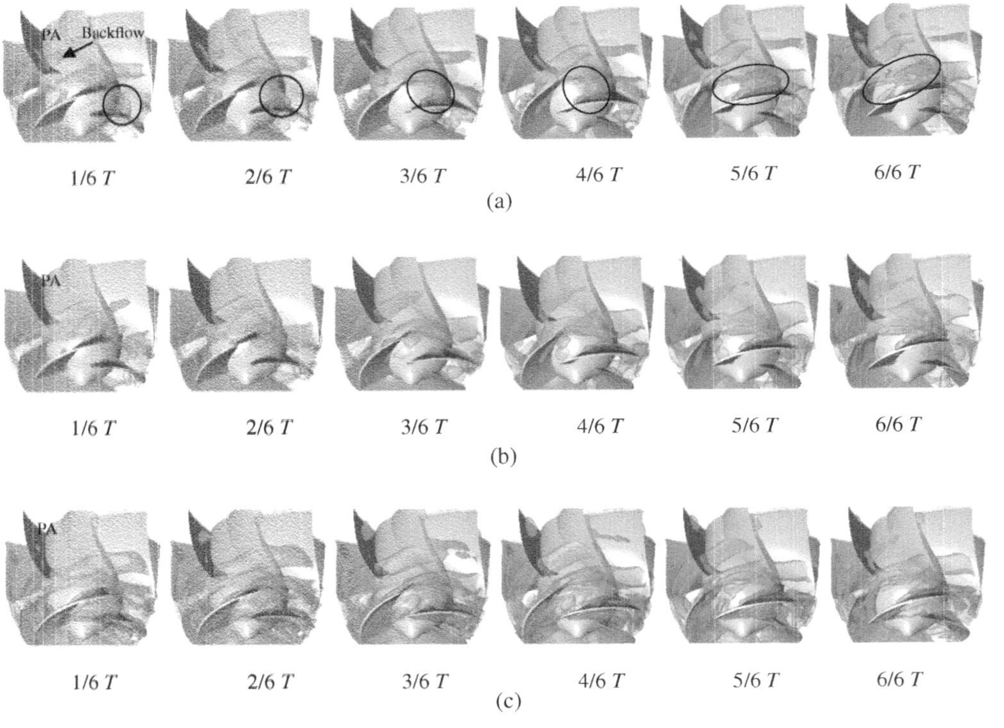


**Figure 7.19** Iso-surfaces at $Q$=37421.5 $s^{-2}$ in the impeller and the guide vane passages at three partial-loading conditions. (a) $0.2Q_{des}$, (b) $0.4Q_{des}$, (c) $0.6Q_{des}$.

the vortices are similarly generated, and they move into the guide vane, the only difference being that they are of different strength. Thus, the vortices in the guide vane consist of the backflow vortices and the vortices from the RSI zone.

## 7.4.3 Pressure Fluctuation in Partial-loading Conditions

### 7.4.3.1 Time-domain Analysis

In order to quantitatively analyze the unsteady flow characteristics of the mixed-flow pump at small flow rates, six monitoring points are used from the inlet to the outlet of the pump along the main flow direction of the pump to record the data of the pressure fluctuation. The region between the impeller outlet and the guide vane inlet is the RSI zone; the fluid flow is relatively unstable in this region. Therefore, to analyze the influence of the RSI on the internal flow field in the mixed-flow pump, two additional monitoring points are employed in the RSI region. The locations of the monitoring points are shown in Fig. 7.20.

Figure 7.21 shows the pressure pulsation plots of different monitoring points in the time domain along the main flow in various partial-loading conditions. It can be seen from these plots that the amplitude of the pressure pulsation at various monitoring points along the mainstream flow shows a similar trend in all three partial-loading conditions, wherein the peak amplitude of the pressure pulsation grows to the maximum value close to the RSI zone (monitoring point D1) and then decreases at the monitoring point F. In one impeller rotation cycle ($t = 0.042$ second), most of the monitoring points show a periodic feature with four peaks and troughs in the time-domain plots due to the BPF. The rotational shaft frequency is about 24.17 Hz, and the BPF, which is four times the rotational frequency, is 96.68 Hz, which corresponds to the rated speed of the mixed-flow pump (1450 r/min) with four impeller blades. This shows that the unsteady flow in the pump is greatly influenced by the number of blades in the partial-loading conditions, and the periodic rotation of the blade is the main cause of the oscillation in the pressure fluctuation curve.

By comparing the amplitude of the pressure pulsation at the same monitoring point in various flow rate conditions, it can be noted that the amplitude of the pressure pulsation is affected by the partial-loading condition. At the monitoring point A, which is at the inlet of the pump, the pressure pulsation amplitude at $0.6Q_{des}$ flow rate approach 0, and it is lower than at the $0.2Q_{des}$ and the $0.4Q_{des}$ flow rate, which indicates that the flow is unsteady at smaller flow rates. Interestingly, at $0.4Q_{des}$ flow rate, the pressure pulsation curve does not show the periodic feature like for the

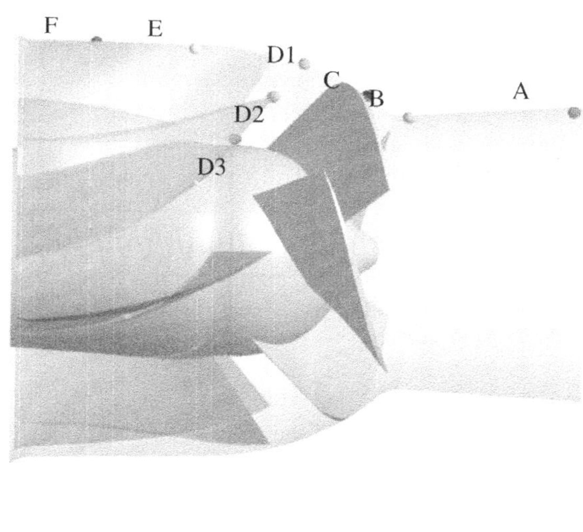


**Figure 7.20** The spatial distribution of monitoring points.

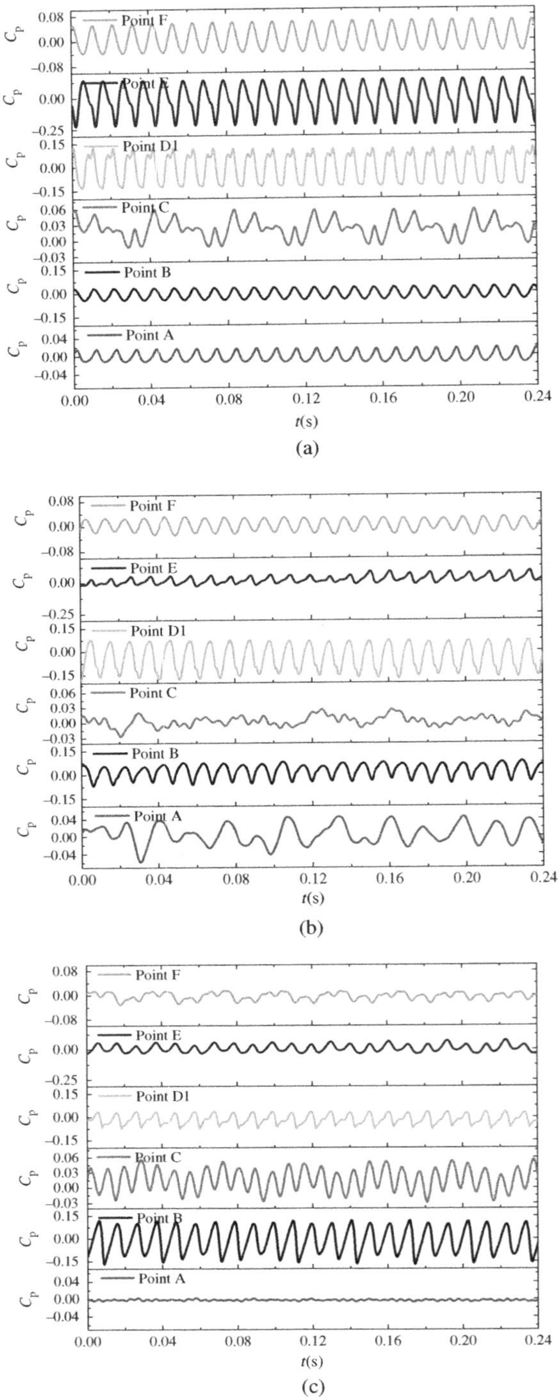


**Figure 7.21** Pressure fluctuation in the time domain along the main flow for different flow rates. (a) $0.2Q_{des}$, (b) $0.4Q_{des}$, (c) $0.6Q_{des}$.

other two flow rates. It is most likely that the rotating stall occurs near this flow rate of $0.4Q_{des}$. At the monitoring point B, which is at the inlet of the impeller, the amplitude of the peak and the trough value of the pressure pulsation at $0.6Q_{des}$ flow rate is higher than that at $0.2Q_{des}$ and $0.4Q_{des}$ flow rates. Obviously, at this monitoring point, the larger partial-loading condition shows that the flow is much more unstable. In the blade tip region (Point C), the pressure pulsation changes a lot with increase in the flow rate. The cycling time of the pressure pulsation curve in different partial-loading conditions is different. At $0.2Q_{des}$ flow rate, the peak value of the pressure coefficient $C_p$ changes approximately in one impeller rotation cycle, while it changes in a quarter of the impeller rotation cycle at $0.6Q_{des}$ flow rate. At $0.4Q_{des}$ flow rate, due to the rotating stall, the pressure pulsation does not show a periodic characteristic. In the RSI zone, the pressure pulsation amplitude becomes the largest among all the monitoring points at various partial-loading conditions, and they all show unsteady flow characteristics. And, the maximum value of the pressure coefficient $C_p$ almost reaches 0.15.

To further understand the nature of the pressure pulsation in the RSI zone, pressure pulsation is compared at the two additional monitoring points D2 and D3, as shown in Fig. 7.22. It can be noted that with increase in the flow rate, the pressure fluctuation coefficient decreases, and all three monitoring point positions (D1, D2, and D3) show four peaks and troughs, which means that they are all affected by the BPF of the impeller. At $0.2Q_{des}$ flow rate, at point D1, a secondary peak appears in the curve, indicating a strong secondary flow in these zones. It also implies much more energy loss, which is in accordance with the energy performance loss at smaller flow rates. In the guide vane, the two monitoring points (point E and point F) show similar changes; at the smaller flow rate, the amplitude of pressure pulsation becomes bigger. In addition, when the monitoring point D1 deviates from the rotor–stator region, the amplitude of pressure pulsation decreases.

Thus, the periodic rotation of the impeller has great influence on the pressure pulsation in the mixed-flow pump in partial-loading conditions. The periodicity of the pressure fluctuation curve at each monitoring point is obvious, and the number of peaks and troughs is the same as the number of impeller blades, which is four in the present case. The amplitude of the pressure fluctuation is the largest in the RSI region. In the RSI region, the pressure coefficient gradually decreases with the increase in the flow rate. Overall, the amplitude of the pressure coefficient of each monitoring point decreases as the monitoring point position deviates from the interference zone. Especially in the guide vane, the amplitude of the pressure coefficient decreases gradually along the direction of the main flow.

#### 7.4.3.2 Frequency-domain Analysis

The pressure pulsation data at each monitoring point are analyzed by the FFT, and the frequency-domain plot of each monitoring point is obtained, as shown in Fig. 7.23. Since the rated speed of the mixed-flow pump is 1450 r/min, the rotational frequency $F$ is about 24.17 Hz. Moreover, since the number of impeller blade $Z$ is 4, the BPF $F_b$ is four times the rotational frequency and is around 96.68 Hz.

It can be seen from Fig. 7.23 that in the various partial-loading conditions, the main frequency of the pressure fluctuation is the BPF at most of the monitoring points, and there are higher harmonics with two and three times the BPF. However, the maximum value of the amplitude of the pressure fluctuation occurs at different monitoring points for different flow rates. At $0.2Q_{des}$ flow rate, the maximum value of the pressure coefficient $C_p$ appears at 96.3 Hz at monitor point E, which is close to the guide vane inlet, and the second peak value also occurs at 96.3 Hz in the RSI zone. At other monitoring points for this low flow rate of $0.2Q_{des}$, the peak value of pressure fluctuation is low, but the frequency is 96.3 Hz, which is close to the BPF. At $0.4Q_{des}$ flow rate, the maximum value

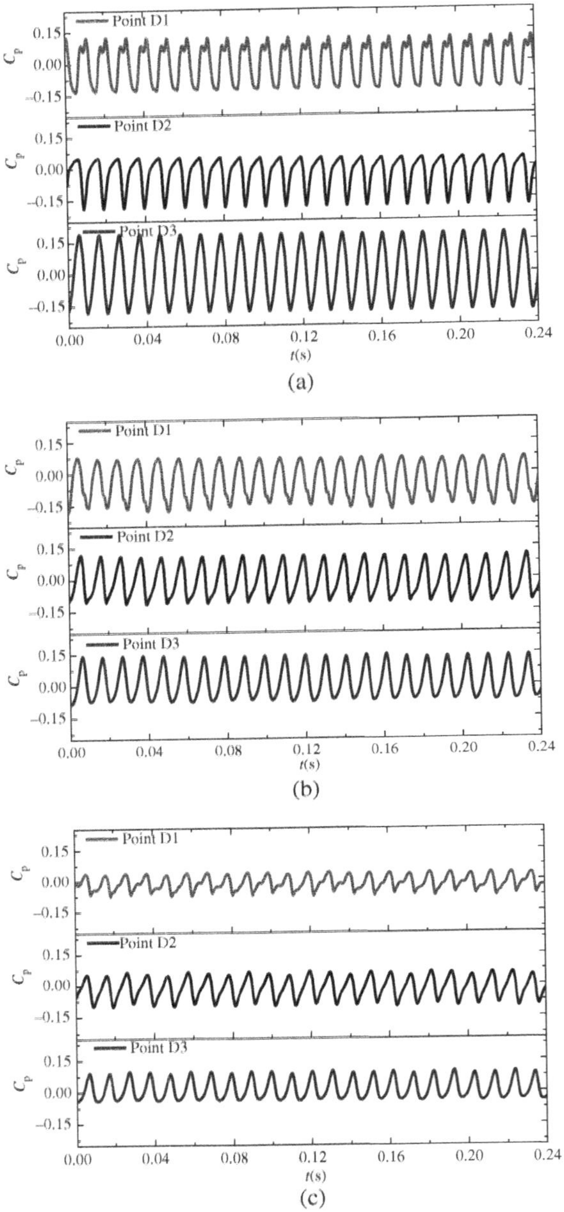


**Figure 7.22** Pressure fluctuation in the time domain in the RSI zone at different flow rates. (a) $0.2Q_{des}$, (b) $0.4Q_{des}$, (c) $0.6Q_{des}$.

in pressure pulsation occurs in the RSI zone, and the pressure coefficient $C_p$ is 0.11221. At $0.6Q_{des}$ flow rate, the maximum value in the pressure pulsation occurs at the monitoring point B, which is different from the other two lower flow rates. It should be noted that the main frequency changes at flow rate rates of $0.2Q_{des}$ and $0.4Q_{des}$. At $0.4Q_{des}$ flow rate, the main frequency decreases to 56.6 and 53.8 Hz, respectively, at point A and point C. However, at $0.6Q_{des}$ flow rate, the main frequency increases to 229.4 and 113.3 Hz at point A and point C, respectively. Overall, with the change in the flow rate, the change in the pressure coefficient $C_p$ is more obvious in the RSI region than in other locations.

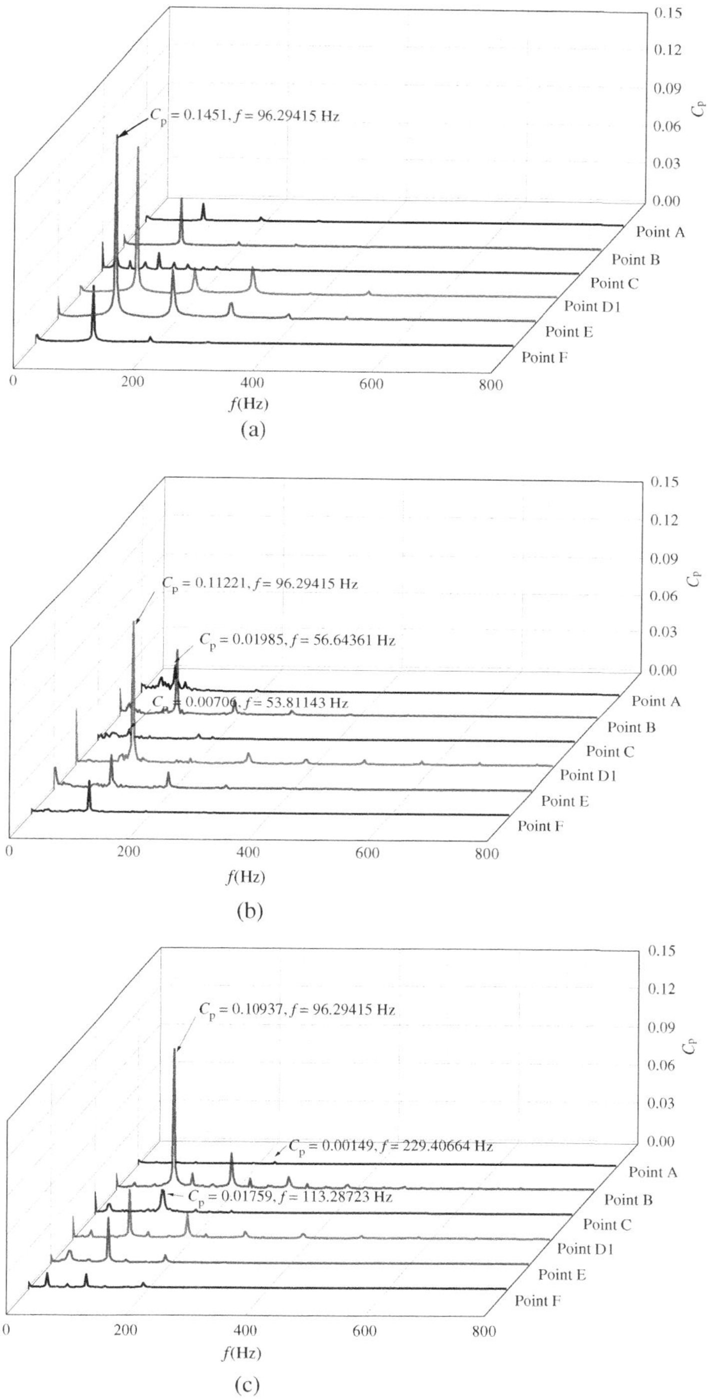


**Figure 7.23** Pressure fluctuation in the frequency domain along the main flow at different flow rates. (a) $0.2Q_{des}$, (b) $0.4Q_{des}$, (c) $0.6Q_{des}$.

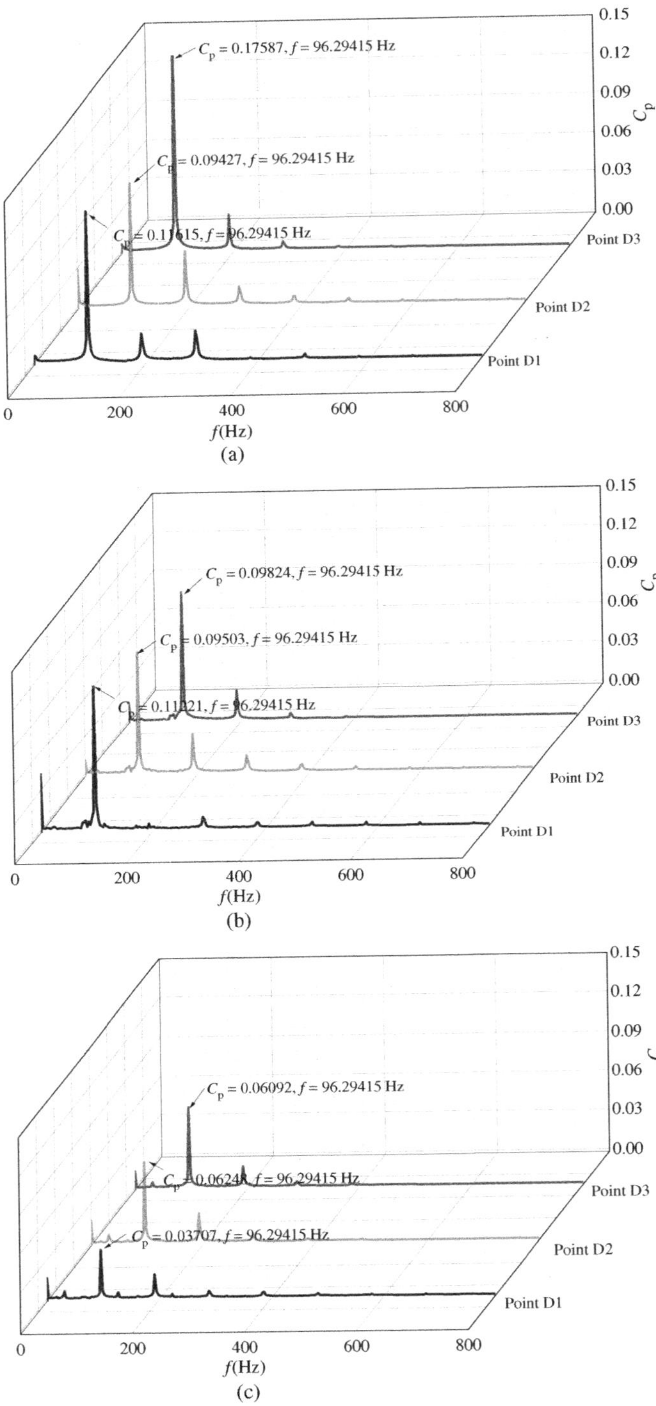


**Figure 7.24** Pressure fluctuation in the frequency domain in the RSI zone at different flow rates. (a) $0.2Q_{des}$, (b) $0.4Q_{des}$, (c) $0.6Q_{des}$.

Figure 7.24 shows the frequency-domain plot of point D1, D2, and D3. It can be noted from this figure that the range of the frequency domain is from 0 to 300 Hz at various monitoring points. In the three partial-loading conditions, the main frequency is the BPF, which indicates the strong effect of the number of impeller blades on the main frequency in the RSI zone. In the frequency domain plot, the doubling and tripling of the BPF occur in the second and third place. When the pump operates at the smallest flow rate of $0.2Q_{des}$, the amplitude of the pressure fluctuation at different monitoring points is higher than at the other two relatively higher flow rates of $0.4Q_{des}$ and $0.6Q_{des}$. At flow rates of $0.2Q_{des}$ and $0.4Q_{des}$, when the monitoring points are close to the end wall, the amplitude of pressure coefficient $C_p$ at the BPF is relatively large. When the monitoring points are close to the hub, the amplitude of the pressure coefficient $C_p$ is greatly influenced by the flow rate. At $0.2Q_{des}$ flow rate, the amplitude of the pressure coefficient $C_p$ at point D3 has the highest value, while it decreases to the lowest value at flow rate of $0.6Q_{des}$. The higher harmonics also have a similar trend.

From the above analysis, it can be concluded that the BPF, which is caused by the rotation of the blade and other two higher harmonics, are the dominant frequencies in the partial-loading conditions. In the RSI zone, the amplitude of the pressure pulsation in the frequency domain changes with change in the flow rate, and the double and triple frequencies appear in the frequency domain. With increase in the flow rate, the amplitude of pressure pulsation in the frequency domain decreases.

## 7.5 Summary

The summary of the contents of this chapter is given below:

(1) At the design flow rate, the head loss coefficient in the impeller of the mixed-flow pump is impacted by the size of the tip clearance. As the tip clearance increases from 0.2 to 1.1 mm, the head-drop loss coefficient in the impeller increases 1.62 times. Thus, the energy losses in the impeller are closely related to the size of the tip clearance. At the design flow rate, the total entropy production in the mixed-flow pump increases with increase in the tip clearance but is not linear with the size of the clearance. The increase in the tip clearance increases the TLF, which increases the energy losses in the impeller; however, the hydraulic losses in the guide vane are reduced. It is found that in the impeller, the high entropy production zone is located in the middle and upper parts of the blade rim region in the direction of the TLF. However, in the guide vane, the hydraulic losses are negatively correlated with the size of the tip clearance, since the high-speed fluid has more kinetic energy to dissipate and convert it into potential energy.

(2) The increase in the tip clearance results in increase in the leakage volume and allows for increase in the growth and strength the TLV in the rim area flow field. At the same time, the increase in the size of the TLV affects the leakage flow on the edge of the lower blade. Increase in the tip clearance also increases the blade leading-edge (LE) leakage flow, which results in high hydraulic loss in this region. When the tip leakage gap is greater than 0.2 mm, the high vortex intensity area of the TLV increases with the gap, while the high vortex loss area is affected by the return flow, and the secondary leakage flow is significantly reduced. Therefore, to improve the efficiency of the mixed-flow pump, it is important to consider the influence of TLV and the return vortex.

(3) At most flow rates, the influence of the BPF dominates at the impeller inlet and in the middle of the guide vane, pressure pulsation is the dominant effect. At small flow rate, the pressure pulsation coefficient experiences the largest change in the amplitude. As the flow rate increases, the pressure pulsation amplitude in the frequency domain decreases. When the mixed-flow pump operates at partial-loading conditions, the flow fields in the RSI zone changes rapidly. With decrease in the flow rate, the flow fields become more chaotic, and more vortices are generated in the RSI zone, which results in more energy loss. With the rotation of the impeller, the vortices generated in the RSI zone move into the guide vane and dissipate in the guide vane passage. The BPF and its higher harmonics frequencies of the pressure pulsation are the dominant frequencies, and the frequency distribution range is small. In the RSI zone, the pressure pulsation amplitude in the frequency domain at different monitoring points is higher, the distribution range of the frequency domain becomes larger, and the frequency component increases. Meanwhile, with the increase in the flow rate, the amplitude of the pressure pulsations in the frequency domain at different monitoring points decreases.

## References


1 Shi, W.D., Shao, P.P., Zhang, D.S. et al. (2014). Influence of operational condition and tip clearance size on trajectory of tip leakage vortex. *Journal of Drainage and Irrigation Machinery Engineering* 32 (5): 373–377.

2 Yaojun, L., Jinfeng, S., and Yiping, H. (2013). Effect of tip clearance on unsteady flow in axial-flow pump. *Journal of Drainage and Irrigation Machinery Engineering* 31 (8): 667–673.

3 Lucius, A. and Brenner, G. (2010). Unsteady CFD simulations of a pump in part load conditions using scale-adaptive simulation. *International Journal of Heat and Fluid Flow* 31 (6): 1113–1118.

4 Zhan, J., Li, Y., Wai, W., and Hu, W. (2019). Comparison between the Q criterion and Rortex in the application of an in-stream structure. *Physics of Fluids* 31 (12): 121701.

5 Zhang, Y., Wang, X., Zhang, Y., and Liu, C. (2019). Comparisons and analyses of vortex identification between Omega method and Q criterion. *Journal of Hydrodynamics* 31: 224–230.

6 Zhang, Z., Che, X., Nie, W. et al. (2017). Study of vortex in flow fields induced by surface dielectric barrier discharge actuator at low pressure based on Q criterion. *Plasma Science and Technology* 20 (1): 014006.

# 8

# Rotational Stall in a Mixed-flow Pump

## 8.1 Energy Characteristics

### 8.1.1 Energy Performance of the Pump Along the Axial Flow Direction

Since the impeller of the mixed-flow pump is its main working component, the determination of the average static pressure difference at different cross-sections before and after the impeller of the pump along the axial direction is very important. Then, the values of the pressure difference can be converted in terms of the energy head $H$ and recorded as $\Delta H$, which represents the energy acquired by the pump at different axial positions in the process of doing the work. Figure 8.1 shows the energy head difference $\Delta H$ between different cross-sections along the axial direction at the design condition and two stall conditions (critical stall and deep stall). It can be seen from this figure that the effective head of the two stall conditions at the impeller inlet ($-0.35 < z < -0.15$) is almost 1m lower than that at the design condition. This indicates that there is a large hydraulic loss at the impeller inlet under stall condition. In the impeller domain, the trend in the variation of the effective head at different axial cross-sections is consistent with the energy performance characteristics of the entire mixed-flow pump.

The $\Delta H$ and $H$ are highest at the critical stall condition but lowest at the design condition. However, at the deep stall condition, the energy head $\Delta H$ is less than that at the critical stall condition when $z = -0.1$ m which is in the middle of the impeller. It indicates that the head under deep stall condition is at the bottom of the saddle region, and the energy loss is much more in the deep stall condition than in the critical stall condition. At the guide vane, the energy head $\Delta H$ decreases more at both the stall conditions compared to the design flow rate condition, which, of course, is also influenced by the vortex in the guide vane. In conclusion, due to the swirl flow in the inlet part and the vortex in the impeller and guide vane in both stall conditions, there is significant energy loss, which causes drop in the head.

### 8.1.2 Energy Loss Mechanism Under Stall Condition

#### 8.1.2.1 Theoretical Analysis of Head Drop in Saddle Area

In the mixed-flow pump, the theoretical head of the pump is normally defined as

$$gH_t = u_2 v_{u2} - u_1 v_{u1} \tag{8.1}$$

$$u_2 = \pi D_{2s} n / 60 \tag{8.2}$$

In Eqs. (8.1) and (8.2), $u_2$ is the peripheral speed of impeller outlet in m/s, $v_{u2}$ is the absolute peripheral speed component of impeller outlet in m/s, $u_1$ is the peripheral speed component of impeller

*Mixed-flow Pumps: Modeling, Simulation, and Measurements*, First Edition.
Wei Li, Leilei Ji, Ramesh Agarwal, Weidong Shi, and Ling Zhou.

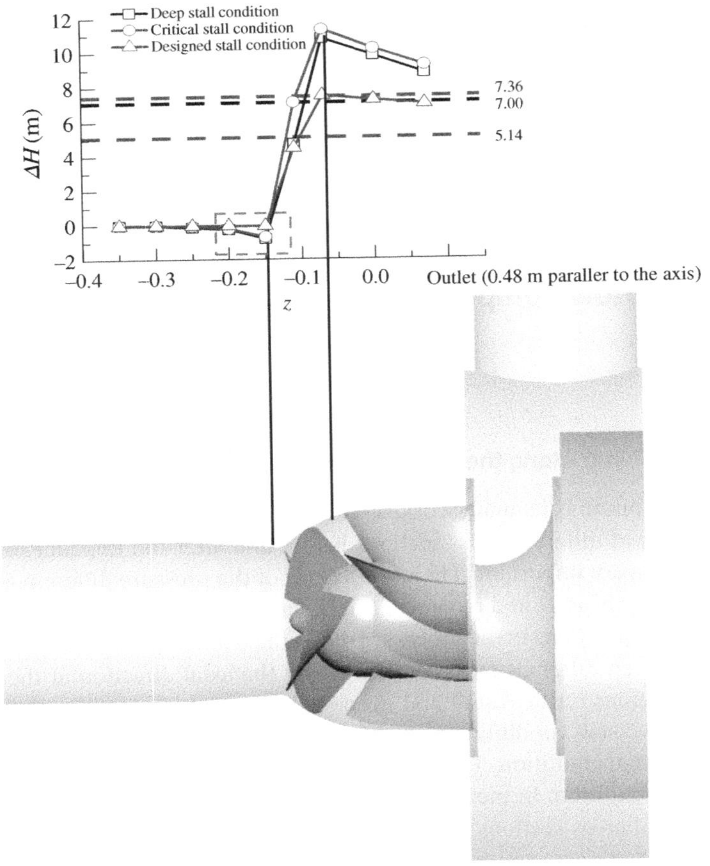


**Figure 8.1** Variation in the energy head difference $\Delta H$ along the direction of the pump axis.

inlet in m/s, $v_{u1}$ is the absolute peripheral speed component of the impeller inlet in m/s, $g$ is the acceleration due to gravity in m/sec$^2$, $D_{2s}$ is the flange diameter of the blade outlet in m, and $n$ is the impeller rotating speed in rad/s.

If the inlet has no pre-rotation, i.e., $v_{u1} = 0$, then selecting the rim surface as the analysis object and expanding the theoretical head formula in Eq. (8.1), one obtains:

$$gH_t = v_{u2}u_2 = (\pi n D_{2s})^2 - \pi n D_{2s}\frac{Q_t}{\psi_{2s}A_2 \tan\beta_{2s}} \tag{8.3}$$

where $Q_t$ is the theoretical flow rate in m$^3$/s, $\psi_{2s}$ is the outlet expelling coefficient of the rim surface t, $\beta_{2s}$ is the outlet flow angle at the rim of the blade, $A_2$ is the impeller outlet area in m$^2$, which can be obtained from the formula for the platform side area.

The hydraulic loss $\eta_H$ and leakage loss $\eta_v$ are now substituted in Eq. (8.3) without considering the velocity slip, and then the pump head can be calculated as:

$$H = \frac{\eta_H}{g}\left[(\pi n D_{2s})^2 - \pi n D_{2s}\frac{Q}{\psi_{2s}A_2 \tan\beta_{2s}\eta_V}\right] \tag{8.4}$$

According to Eq. (8.4), there is a low-speed vortex area near the rim region and the blade outlet for both stall conditions, and the stall vortex structure in the passage is smaller for the critical stall condition. In the deep stall condition, the stall vortex and the vortex structure at the blade outlet are

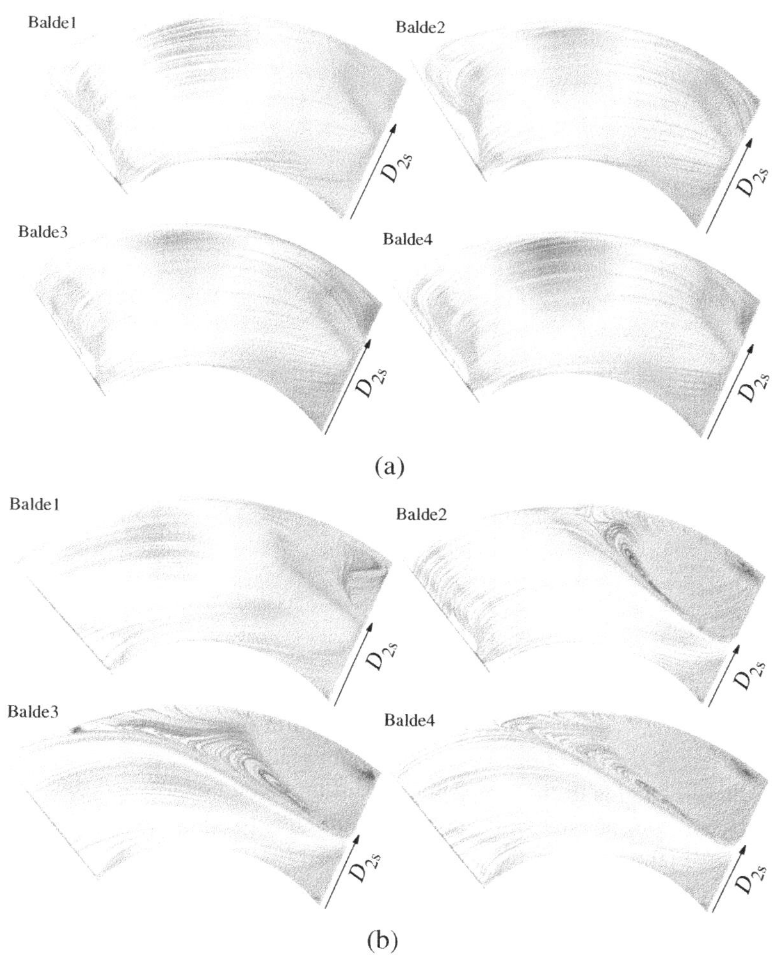


**Figure 8.2** Streamlines distribution on the blade surface in stall condition: (a) critical stall condition and (b) deep stall condition.

larger than in the critical stall condition, and the effective outlet diameter $D_{2s}$ at the blade outlet is reduced, which reduces the effective flow area $A_2$ in Eq. (8.4). Therefore, in the deep stall condition, there is more drop in the head of the mixed-flow pump. By observing the distribution of streamlines on the suction surface in stall condition, as shown in Fig. 8.2, it can be observed that the difference in the streamlines structure on the surface of the four blades in deep stall condition is significantly different than that in the critical stall condition due to the influence of the tip leakage flow, stall vortex, and backflow. Additionally, the effective flow area at the blade outlet is further reduced, which indicates that under deep stall condition, not only is there more drop in the head of the mixed-flow pump, but also more hydraulic imbalance occurs in the flow passages, which further increases the vibration and noise of the pump and causes more hydraulic losses.

#### 8.1.2.2 Flow Distribution in Each Channel

To compare the flow rates through a single impeller passage under different operating conditions, Fig. 8.3 shows the flow rates in different passages of this outlet section. The flow rate of a single

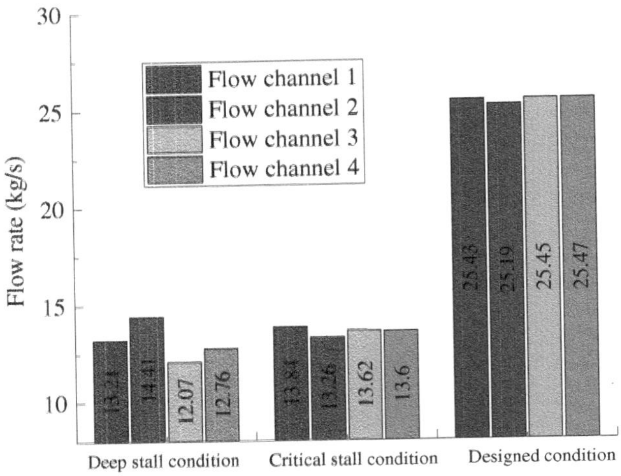


**Figure 8.3** Flow rates distribution in each passage at the impeller outlet at various flow rates.

flow channel can be written as:

$$Q = \bar{V}_S A \tag{8.5}$$

where $\bar{V}_S$ is the area-averaged streamwise velocity in the cross-section of the single flow channel in m/s, and $A$ is the area of the cross-section of the single flow channel in m$^2$.

Examining the nature of the flow fields in the deep stall condition, it can be observed that the differences in the flow fields in the four passages are very significant and are related to the leakage vortex, stall vortex, and backflow in the four impeller passages. In order to compare the flow rates in each passage of the impeller under the design, critical, and deep stall conditions, the flow function of a single passage is defined as $x_{\text{passagei}\ (i=1,2,3,4)}$, where $Q = x_{\text{passage1}} + x_{\text{passage2}} + x_{\text{passage3}} + x_{\text{passage4}}$. In data processing, the standard deviation can reflect the degree of dispersion in the value relative to the average value. In the ideal situation, the flow rate distribution in each passage of the impeller should be the average value. Therefore, the standard deviation $s$ is used to calculate the standard deviation of the three working conditions (design, critical stall, and deep stall) as given in Eq. (8.6):

$$S = \sqrt{\frac{n\sum x^2 - \left(\sum x\right)^2}{n(n-1)}} \tag{8.6}$$

After calculation, in the 2680th calculation time step ($t = 0.58$ seconds), the standard deviations of the design condition, critical stall condition, and deep stall condition are 11.35%, 20.74%, and 85.21%, respectively. It implies that the circumferential flow asymmetry in the four passages is most prominent in the deep stall condition, which is consistent with the previously mentioned observation that the surface plots of streamlines on the four blades are significantly different in the three operating conditions (design, critical stall, and deep stall) and the effective flow area at the outlet is reduced in deep stall condition. This situation also leads to the circumferential propagation of stall vortex in the impeller passage of the mixed-flow pump. Additionally, the asymmetric moment caused by the uneven circumferential hydraulic force distribution can also intensify the vibration of the mixed-flow pump.

#### 8.1.2.3 Region of the Maximum Energy Loss in the Impeller

Figure 8.4 shows the turbulent kinetic energy (TKE) distribution on the 50% span section of the blade for the design condition, critical stall condition, and deep stall condition. At the same scale, high-value TKE areas A and B in the four impeller channels at the critical and deep stall conditions can be observed, and the high-value areas of TKE are located at the impeller outlet and behind the suction surface of the blade outlet. Combined with the velocity distribution in Section 8.1.2.2, it can be seen that the high-value region of TKE basically corresponds to the backflow region of the

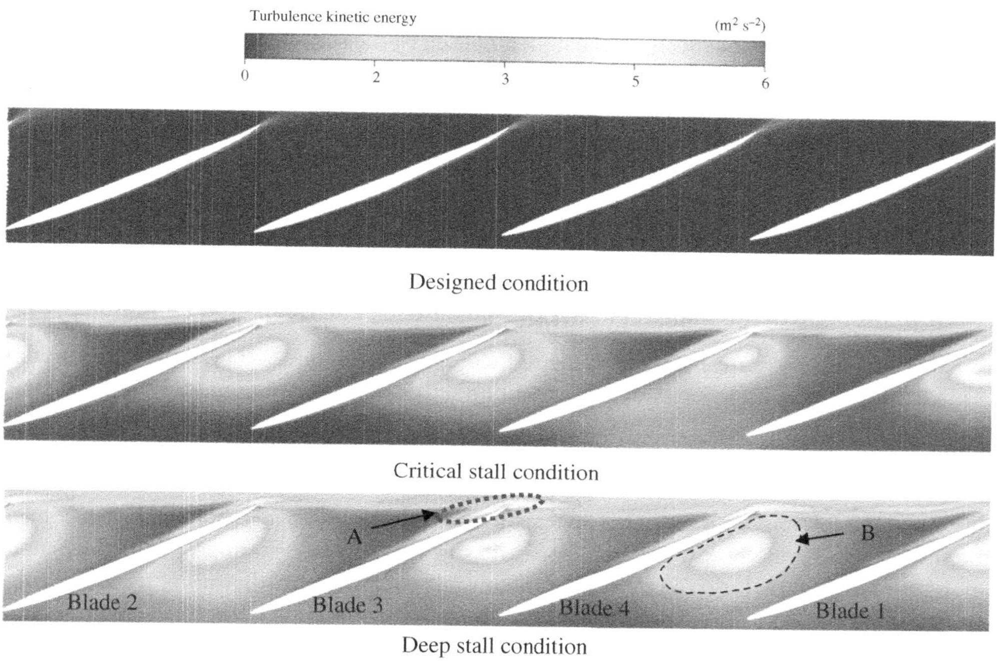


**Figure 8.4** Turbulent kinetic energy distribution at 50% span location at various flow rates.

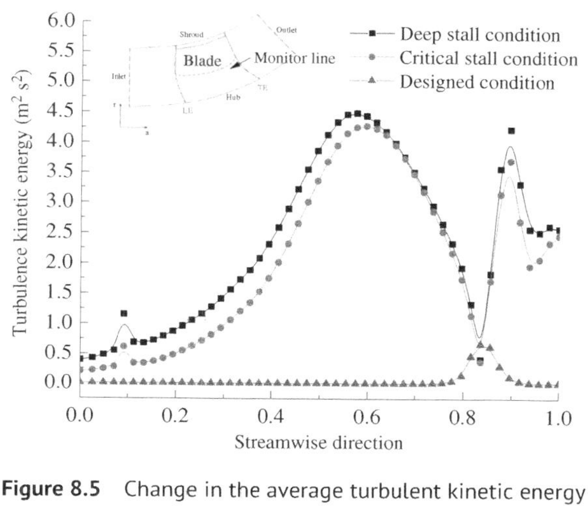


**Figure 8.5** Change in the average turbulent kinetic energy in the four passages at various flow rates.

impeller outlet and the separation region of the boundary layer near the blade suction surface. This indicates that the separation vortex or stall vortex in the flow passages and the backflow at the impeller outlet in stall condition leads to a greater increase in energy loss in the impeller.

Figure 8.5 shows the average TKE distribution in each flow section from the impeller inlet to the outlet in the design condition, critical stall condition, and deep stall condition. At the design operating condition, the TKE of the fluid is basically stable (no large variations and fluctuations) along the streamwise direction, and the level of the TKE is low, which only increases in the 80–90% chord length region, but the increase is small. Under both the stall conditions, the TKE in the impeller increases significantly and its maximum value appears near the impeller outlet. It can be noted that the maximum TKE at 60% chord length is caused by the stall vortex in the passage, while the maximum TKE at 90% chord length is caused by the backflow at the outlet. Furthermore, the average TKE in the passages is the highest under the deep stall condition, which implies that the energy loss in the impeller is the highest in the deep stall condition.

## 8.2 Flow Structure in the Critical and Deep Stall Conditions

### 8.2.1 Flow Structure in the Impeller in the Design, Critical, and Deep Stall Conditions

Figure 8.6 shows the pressure field and velocity field distribution on the blade-to-blade surface of the mixed-flow pump impeller in design condition, critical stall condition, and the deep stall condition.

It can be seen from Fig. 8.6 that the difference in the pressure and velocity field distributions in the four flow channels of the impeller is most obvious in the deep stall condition. Using the same pressure scale, the low-pressure core area appears in the impeller passage in both stall conditions, and the number and scale of low-pressure core areas in the impeller channels in the deep stall condition are significantly larger than in the critical stall condition. Furthermore, by comparing the velocity distribution in the cross-section, it can be seen that the fluid in the flow channels has boundary layer separation at the trailing edge (TE) of the blade suction surface, which results in low-speed zones in both stall conditions. In the deep stall condition, the low-speed zone of the fluid caused by the boundary layer separation is slightly larger than that in the critical stall condition, and the distribution of the high-speed zone of fluid at the impeller inlet channel is uneven. In summary, in both types of stall condition, not only is the non-uniformity of flow field more prevalent in each impeller channel of the pump, but also the boundary layer separation occurs, which makes the state of flow in the impeller channels quite complex.

Since the flow in the impeller of a mixed-flow pump in stall condition is very complex, in order to accurately analyze the structure and position of the vortices, the $Q$-criterion method is employed to identify the vortices. This quantity ($Q > 0$) can be used to identify the vortex regions where the vorticity magnitude prevails over the strain-rate magnitude, excluding the part caused by the pure shearing motion due to the vorticity magnitude. Here, the typical blade # 3 and blade # 4 and the flow channel between them are selected for further study. Figure 8.7 shows the distribution of the vortex core in the impeller in the design condition, critical stall condition, and deep stall condition. Figure 8.7(a) shows the enlarged view of the flow fields in the impeller. It can be seen from this figure that the tip leakage vortex (TLV) and the small-scale backflow vortices at the impeller outlet are the main vortex structures in the impeller in the design condition. Due to the influence of the mainstream, the tip leakage flow (TLF) is entrained at the leading edge of the blade and forms a leakage vortex strip. Compared to the mainstream flow, the TLF is weak; therefore, the leakage

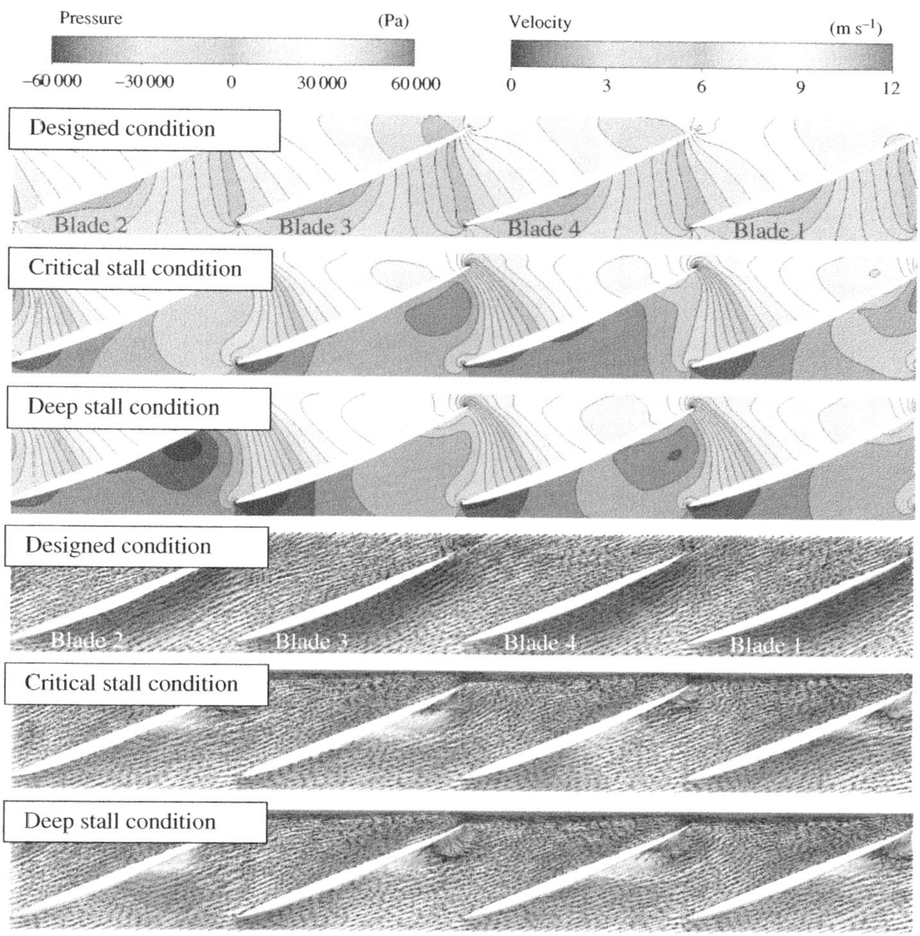


**Figure 8.6** Pressure and velocity distribution at 50% blade height at various flow rates.

vortex strip passes through the impeller passage rapidly with the mainstream, and thus, there is no accumulation and blockage in the passage. Furthermore, the flow lines in the middle of the flow channel and at the hub are smooth (without fluctuations), and no large-scale vortex is formed to block the channel. However, in both the critical and deep stall conditions, the vortex structure in the flow channel is significantly larger and stronger and occurs in different shapes. In the rim zone, the rolled-up position of the leakage vortex is offset toward the middle of the flow channel. Influenced by the stall vortex in the channel, the downstream leakage vortex strip is almost at an angle of 90° to the mainstream. Although the leakage vortex strip appears to be smaller than that in the design condition, it nevertheless has a blocking effect. From the end wall to the middle of the channel, the flow is affected by the inlet swirl, and the inflow angle of blade # 3 increases, resulting in flow separation of the fluid in the middle of the flow passage near the suction surface of the blade outlet forming a large-scale stall vortex, which not only affects the development of the leakage flow in the end wall area but also causes the blockage of most of the flow passage. Furthermore, as the flow rate decreases, a backflow vortex develops from blade #3 hub to blade #4 rim at the outlet of the impeller. Comparing the two stall conditions, the size of the stall vortex in the deep

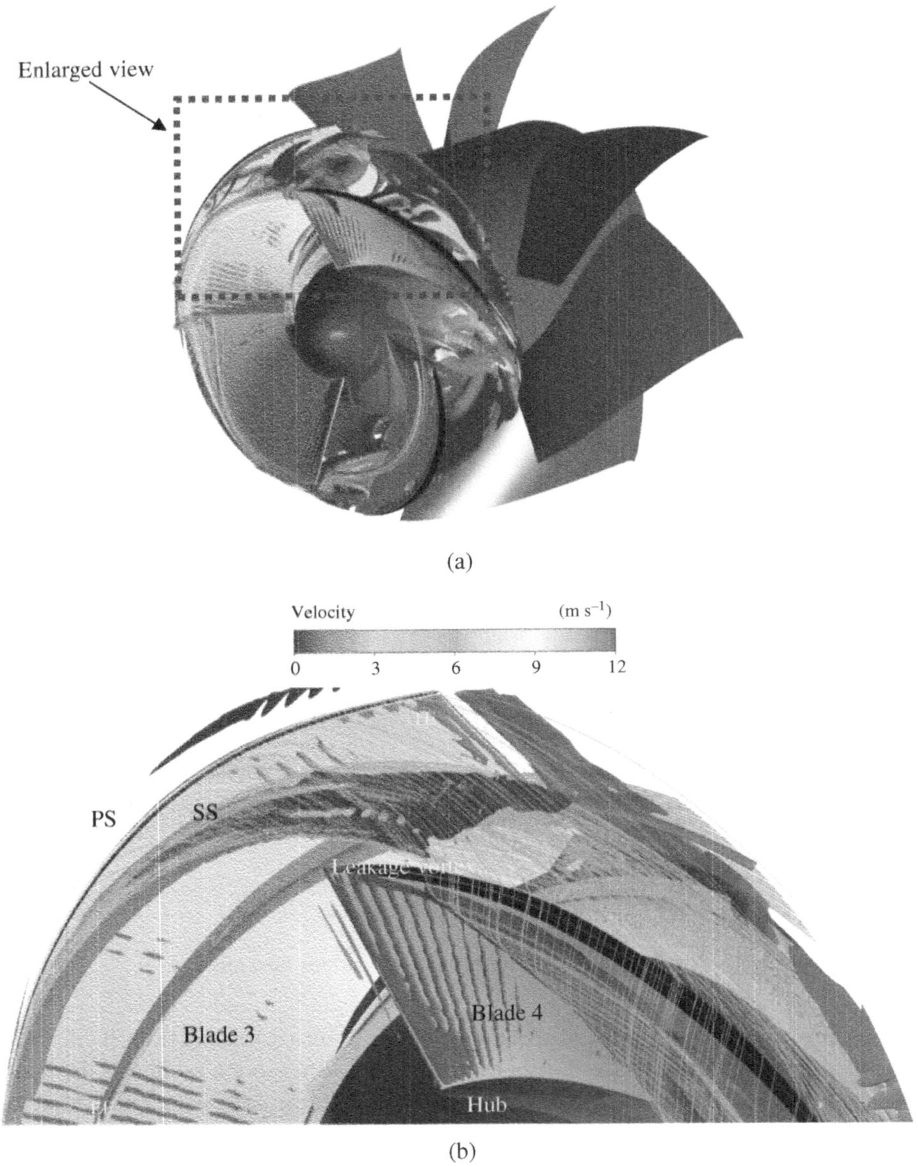


**Figure 8.7** Distribution of vortex core in the impeller. (a) Enlarged view of the flow field in the impeller, (b) Design condition, (c) Critical stall condition, (d) Deep stall condition.

stall condition is slightly larger, and the core trajectories of the stall vortex and the leakage vortex are slightly different. In summary, the formation of the stall vortex, the increase in the backflow, and the generation of the leakage vortex are the main reasons for blocking of the impeller passage.

## 8.2.2 Key Factors Causing the Energy Loss

### 8.2.2.1 Influence of the Inlet Swirl on Energy Loss

Three-dimensional streamline distribution in the inlet part of the mixed-flow pump and the axial velocity component of the flow fields on the cross-section at $z = -0.2$ m in three flow conditions

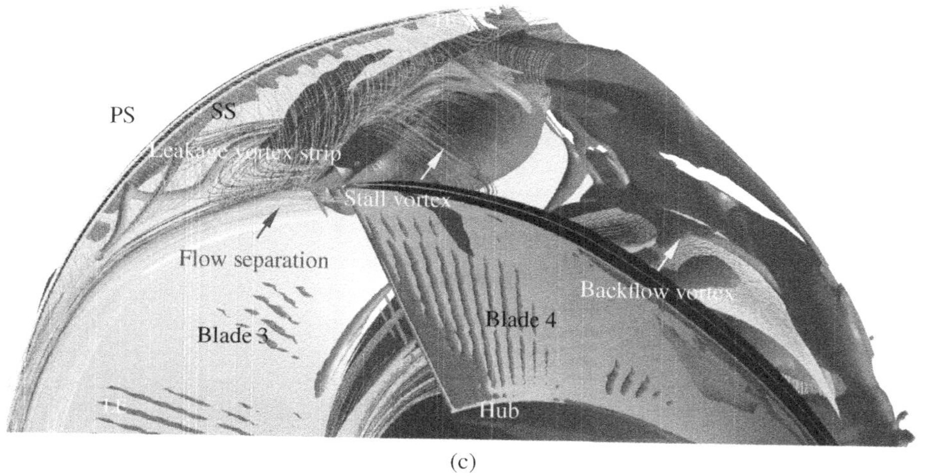


(c)

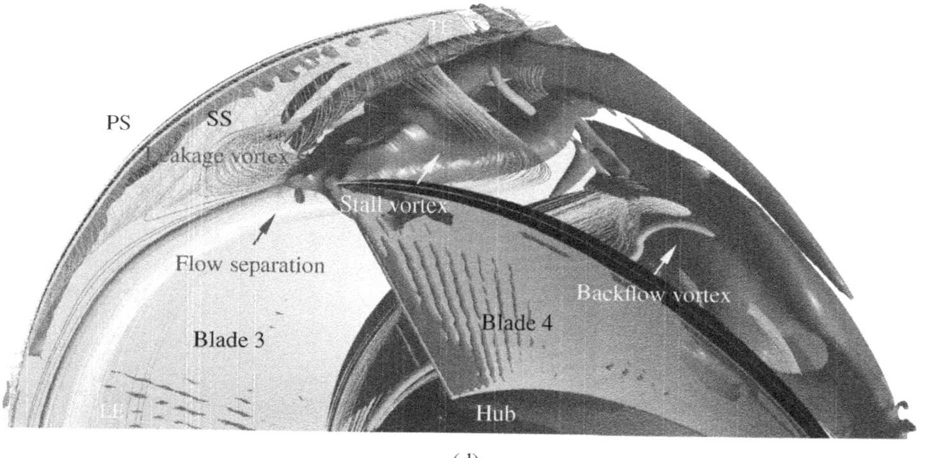


(d)

**Figure 8.7** (*Continued*)

(design, critical stall, and deep stall) are shown in Fig. 8.8. For the design condition, the impeller inlet streamline is smooth and no vortex occurs. In both stall conditions, as the flow rate decreases, the circumferential velocity around the impeller inlet increases due to the inertia of the liquid; thus, the energy near the pipe wall increases so that the energy gradient necessary to maintain the flow of liquid along the streamline no longer exists resulting in the backflow near the impeller inlet.

Due to the influence of the inlet swirl, the inflow angle of the blade also changes significantly, as shown in Fig. 8.9. For the design flow condition, the inflow angle of the fluid is the smallest at the rim, and the inflow angle from the rim to the hub gradually increases, and the maximum inflow angle is 34°. In addition, the inflow velocity at the leading edge of the blade increases gradually from the hub to the rim, and the flow angle with the maximum velocity shows the smallest difference from the actual blade angle. For both the critical and deep stall conditions, the inflow angle and the inlet velocity distribution on the blade change significantly; the inflow angle in the rim area increases, and the high-speed zone occurs in the middle of the flow channel. For the critical stall condition, the inflow angle deviates from the suction side of the blade, the maximum inflow angle in the rim area is 56°, and the zone with the highest inflow velocity moves from the rim to 0.75 times the blade height. In the deep stall condition, the trend of the inflow angle distribution is

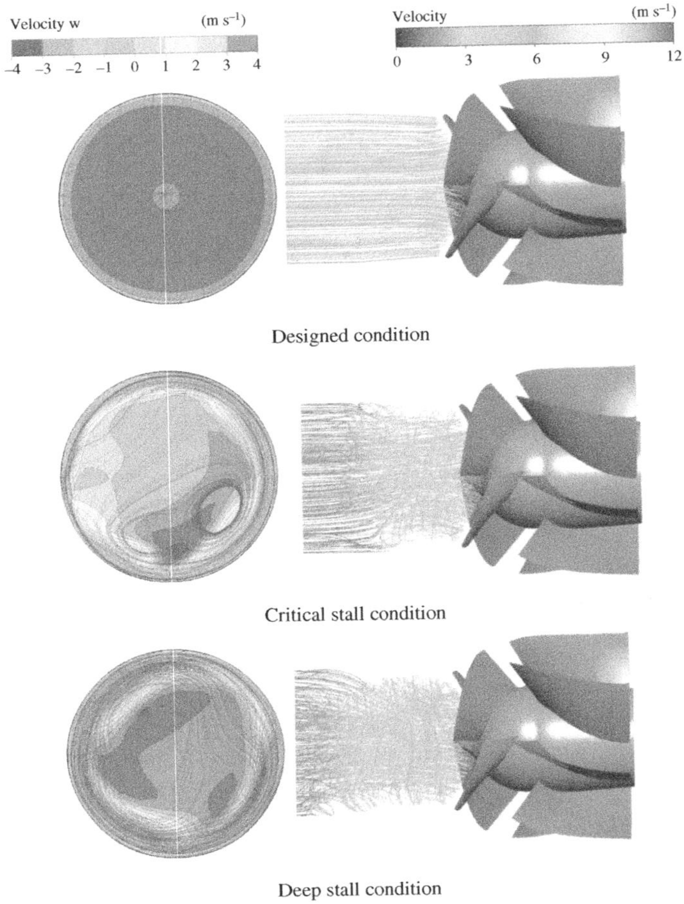


**Figure 8.8** 3D streamlines at the impeller inlet and 2D streamlines at the axis section at inlet $z = -0.2$ m.

consistent with that for the critical stall condition. The maximum inflow angle in the rim area is 72°, and the high-speed zone moves to 0.6 times the blade height. In addition to the influence of inlet pre-swirl, the inflow angle may also be affected by the blockage due to the TLV. For both the stall conditions, the flow field and the inflow angle of the blade in the impeller passage affected by the inlet pre-swirl are mainly concentrated in the upper half of the blade, which is more than 0.5 times the blade height. This is consistent with the amount of the swirl in the inlet section. It indicates that the inlet swirl has a great influence on the fluid entering the impeller, which not only makes the flow inception angle increase by 38° but also induces the flow separation at the blade inlet. It results in a large energy loss but also makes the inflow high-speed zone along the blade profile move from the rim to the middle of the blade, leading to a continuous decrease in the work done by the blade.

#### 8.2.2.2 Influence of the Tip Leakage Vortex on the Energy Loss

TKE is a commonly used variable to estimate the turbulence intensity, which is directly related to the momentum transport process in the boundary layer region. Generally, the TKE is defined in terms of the velocity fluctuations in the velocity components as given below; the superscript (′)

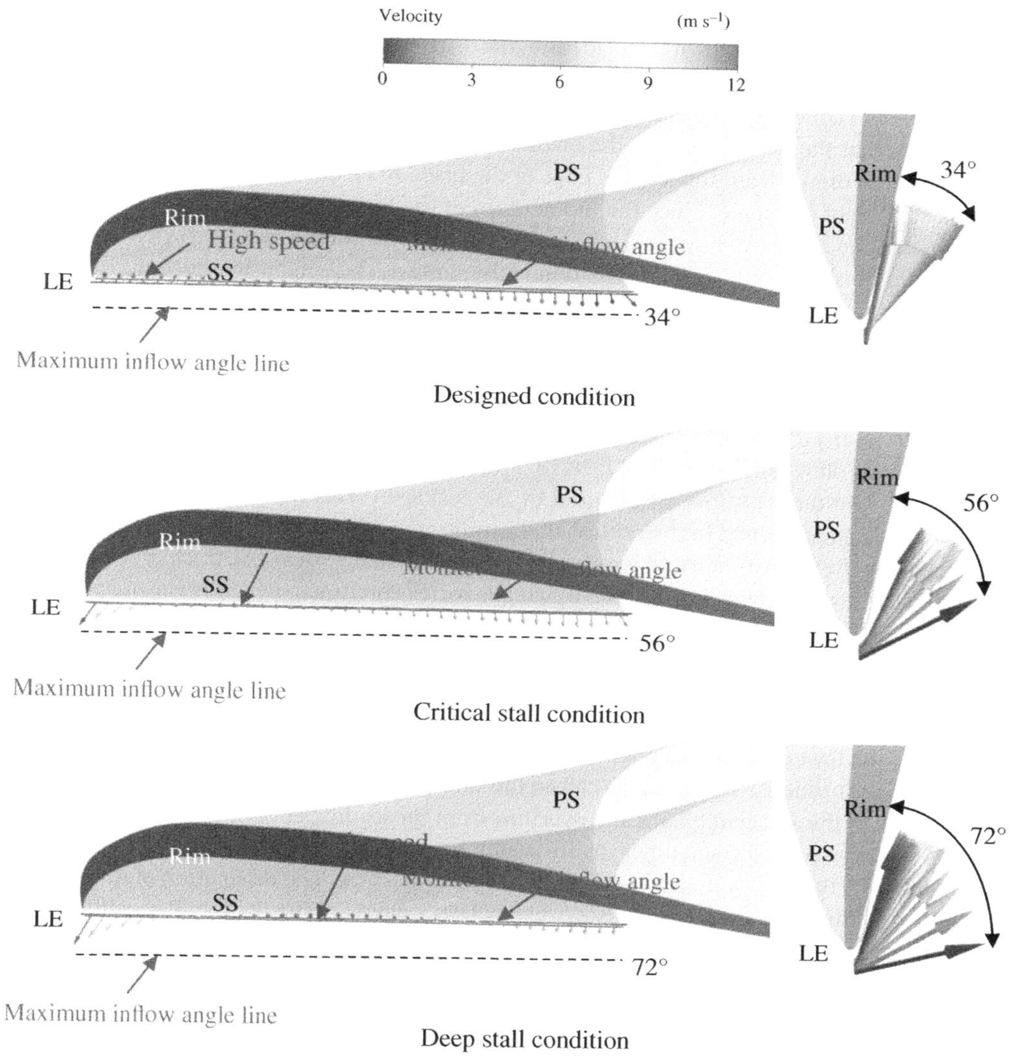


**Figure 8.9** Inflow angle of blade number # 3 for various operating conditions.

denotes the fluctuations about the mean, and the bar denotes the time-averaged value:

$$k = \frac{1}{2}\left(\overline{(u')^2} + \overline{(v')^2} + \overline{(w')^2}\right) \tag{8.7}$$

The dissipation rate of the TKE has some advantages in studying the mechanical energy loss at a small scale. In isotropic turbulence, the dissipation rate can be simplified to be expressed in terms of the length scale and the root mean square fluctuation velocity $v'rms$ as:

$$\varepsilon = 15\nu(\boldsymbol{v}rms'/\lambda)^2 \tag{8.8}$$

where $\lambda$ represents the Taylor microscale.

Thus, the TKE can be used to intuitively analyze the high loss zone caused by the stall vortex core and other secondary vortex structures in the impeller, while the dissipation rate can reflect the turbulence kinetic energy dissipation in the end wall region. Figure 8.10 shows the distribution of the three-dimensional streamlines at the blade rim, the distribution of the TKE in the cross-section of the blade passage, and the nephogram of the turbulent energy dissipation rate at section A of the pump in the design condition, critical stall condition, and the deep stall condition. It can be seen from this figure that the leakage flow at the blade rim is stable in the design flow condition, and the leakage flow is drawn from the leading edge of the blade to form a leakage vortex away from the suction surface of the blade. Since the momentum of the leakage flow is much less than the momentum of the mainstream flow, the direction of the trajectory of the leakage vortex is close to the mainstream flow direction. This observation is the same as the conclusion by Goltz et al. [1] achieved by experiments on the leakage flow using high-speed photography. In their experiment, the included angle between the downstream trajectory of the vortex core and the chord length of the blade was about 14°. The leakage flow flows out of the tip clearance at a relatively high speed, resulting in a sudden increase in the velocity of the flow with the mainstream mixing zone at the rim. The large velocity difference causes the dissipation of the TKE, thus forming a high loss zone at the blade rim as shown in zone B of Fig. 8.10. Furthermore, the energy loss in the rim area is relatively small, and the high loss area caused by the leakage vortex does not coincide with the core area of the leakage vortex. In the critical stall condition, with a decrease in the flow rate, the TLF is affected by the mainstream and is also squeezed by the stall vortex. The track of the leakage vortex can be divided into two sections: the upstream vortex core track, which is shorter, and the longer leakage vortex track, whose angle with the blade chord increases. Starting from the leading edge of the blade, the direction of the vortex core is approximately parallel to the circumferential direction, presenting a completely different morphological structure. Not only is the secondary vortex structure formed in the impeller passage, but also part of the leakage flow overflows the leading edge of the lower blade. This phenomenon is consistent with the conclusion by Vo et al. [2] in studying the gas turbine, they found that when the stall occurred, the TLV track would reach the pressure surface of the adjacent blades and overflow from the leading edge of the adjacent blades. The leakage vortex strip is formed downstream which has a sharp angle of 30° with the TE of the blade, which spans the end wall area of the entire impeller passage. The dissipation of the TKE in the rim region of the cross-section increases, as shown in zone C. In the deep stall condition, the leakage vortex strip formed by the leakage flow is enhanced; the angle between the downstream leakage vortex trajectory and the TE of the blade is further reduced; the loss of the TKE caused by the leakage flow in section A is further increased; and a larger zone with high energy loss is formed in the rim area as shown in zone D. However, the leakage flow that overflows the leading edge of the lower blade decreases or even disappears, which was not indicated by Vo et al. [3].

Figure 8.11 shows the distribution of the velocity field on the pressure equivalent surface of the mixed-flow pump in the design condition, critical stall condition, and deep stall condition. The static pressure value of the iso-surface is −8 kPa. It can be noted from this figure that the pressure iso-surface is basically perpendicular to the flow cross-section in the design condition; however, the TLF not only affects the pressure gradient in the rim area in both stall conditions, which makes the pressure equivalent surface extend toward the inlet direction but also causes the flow velocity in the rim area to be significantly reduced and forms a high-speed zone in the middle of the blade outlet; however, the high-speed zone in the deep stall condition is significantly larger than that in the critical stall condition. In fact, in the stall condition, the high-speed zone might be affected by the complex flow from the rotor-stator interaction region between the impeller and the guide vane. It can be concluded that there are a large number of vortices in the flow channel from the rim to the high-speed zone blocking part of the passage, and the leakage flow is one of the important factors.

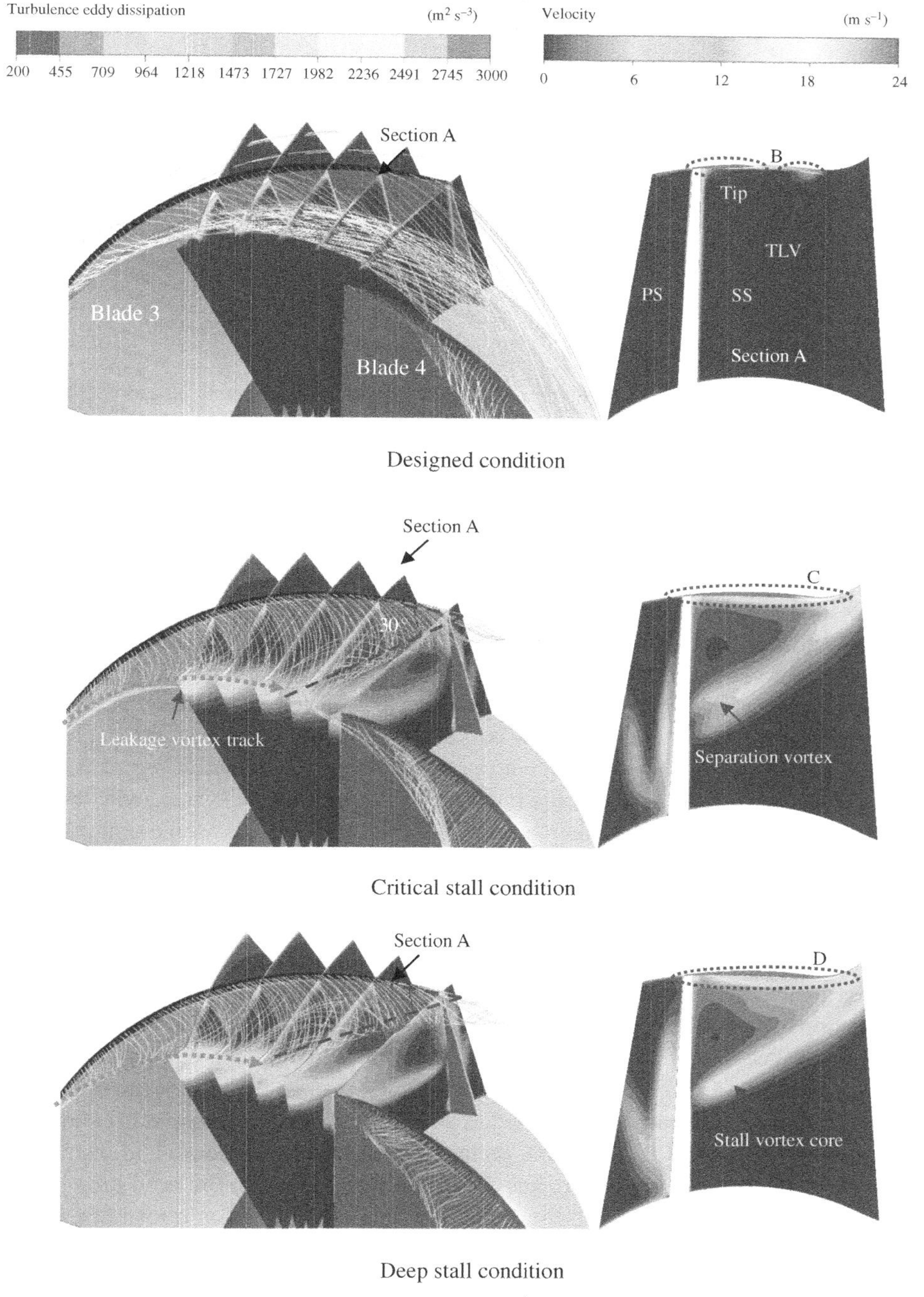


**Figure 8.10** Turbulent kinetic energy dissipation in the impeller at different cross-sections for various operating conditions.

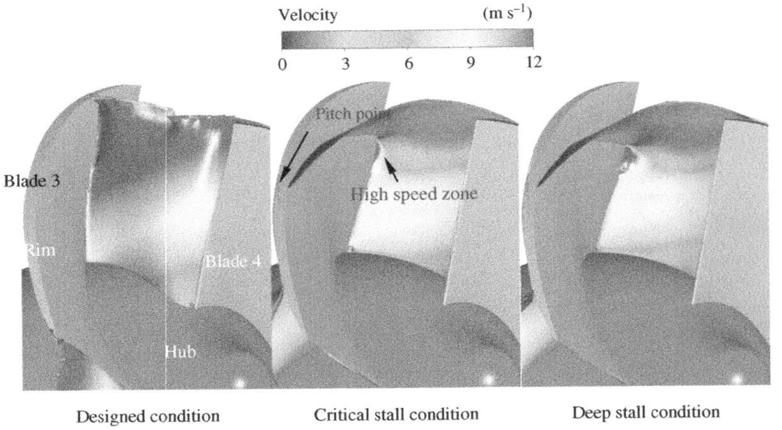


**Figure 8.11** Distribution of velocity field on the pressure iso-surface of the impeller at various operating conditions.

#### 8.2.2.3 Influence of the Stall Vortex on the Energy Loss

Figure 8.12 shows the inducement for the generation of stall vortex and the distribution of the TKE on the surface of the stall vortex. It can be seen from this figure that in the design condition, although there TLV exists since the rim region is affected by the TLF, in the rim region, the overall trend of the flow streamlines is consistent with a good flow pattern as shown in region A of the figure. As the flow rate decreases to the critical stall condition, the attack angle increases at the blade inlet, and the flow separation occurs near the suction surface at the outlet of the blade, forming a large separation vortex. Part of the fluid is blocked by the separation vortex, and the flow direction deflects and flows into the next stage of the flow passage, thus inducing the flow separation and generation of the stall vortex in the next stage of the flow passage, as shown in region B of the figure. It can be noted from this figure that the impeller rotates counterclockwise, while the stall vortex formed by the flow separation rotates clockwise. At this time, the scale of the stall vortex is relatively small, and only about one-third of the flow passage is blocked. According to the previous literature also [4], the stall vortex propagates in the direction opposite to the impeller rotation, which is consistent with the classical interpretation of the rotating stall [5]; it results in blockage in more passages with large amount of energy losses. In the deep stall condition, the area of the flow separation is greater, the scale of the stall vortex is larger; the blockage area is larger; and the effective work section of the impeller is further reduced. As can be seen from region C in this figure, the stall vortex has blocked nearly half of the impeller flow passage in this deep stall condition, and the streamlines at the leading edge of the stall vortex core have reached the inlet of the next stage blade and have overflowed. This can be considered as a typical feature of the deep stall condition, which echoes the performance of the TLF, and seems to have mutual influence and restriction between them. By comparison, the stall vortex is the most critical factor in the head drop in the mixed-flow pump. From the distribution of the TKE with the same $Q$-criterion iso-surface, it can be noted that in the design condition, the flow pattern in the impeller is better, and the overall TKE level is lower, which indicates that the flow in the passage is relatively stable, and the energy loss caused by the vortices is smaller. As the flow rate decreases to the critical stall condition, the stall vortex forms a large blocking area in the passage near the outlet, and the TKE in this region increases significantly. It can also be observed that the stall vortex is close to the inlet rim region of the next stage blade. In the deep stall condition, the overall distribution of the vortices in the passage

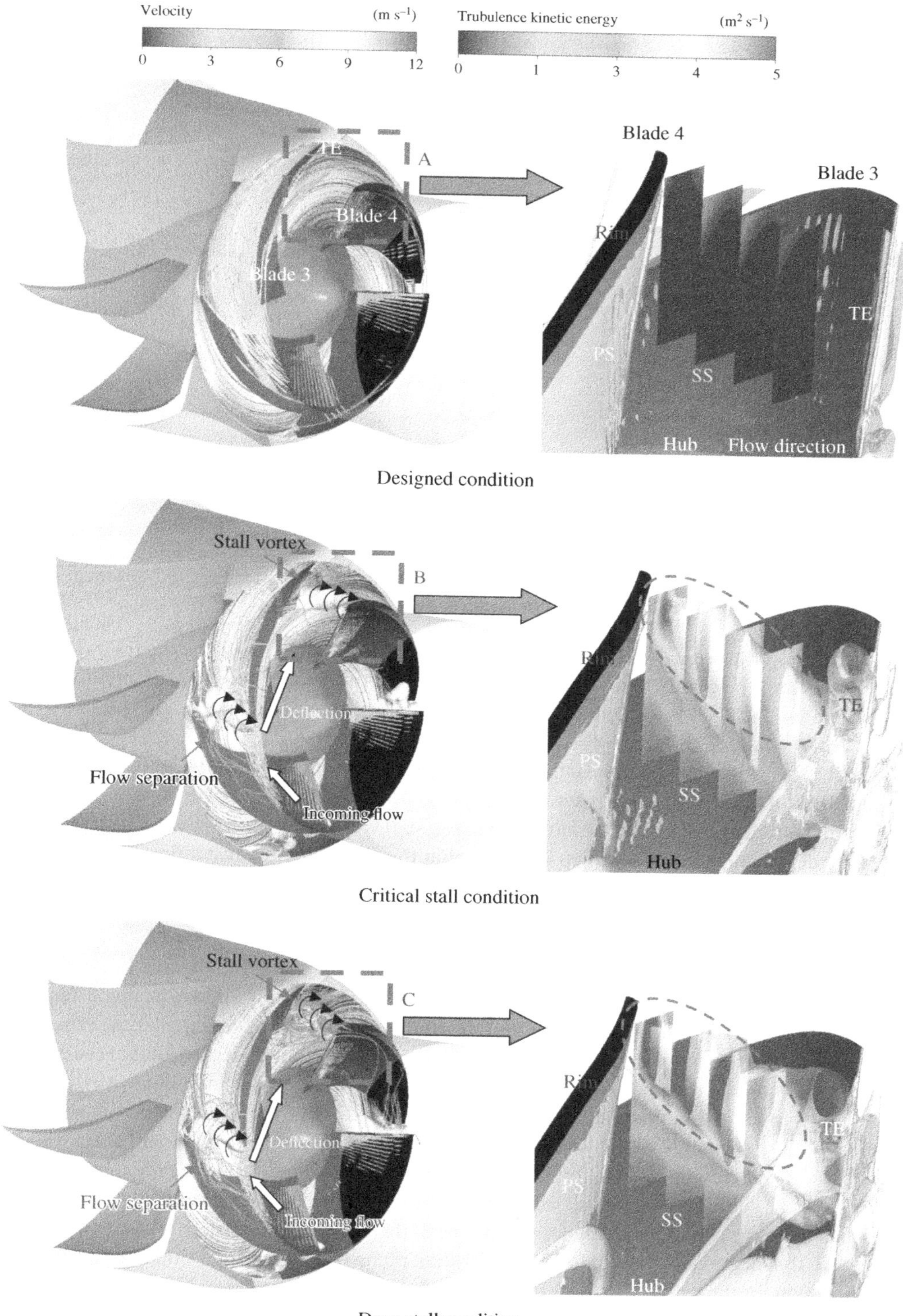


**Figure 8.12** Inducement of stall vortex generation and turbulent kinetic energy distribution on stall vortex surface.

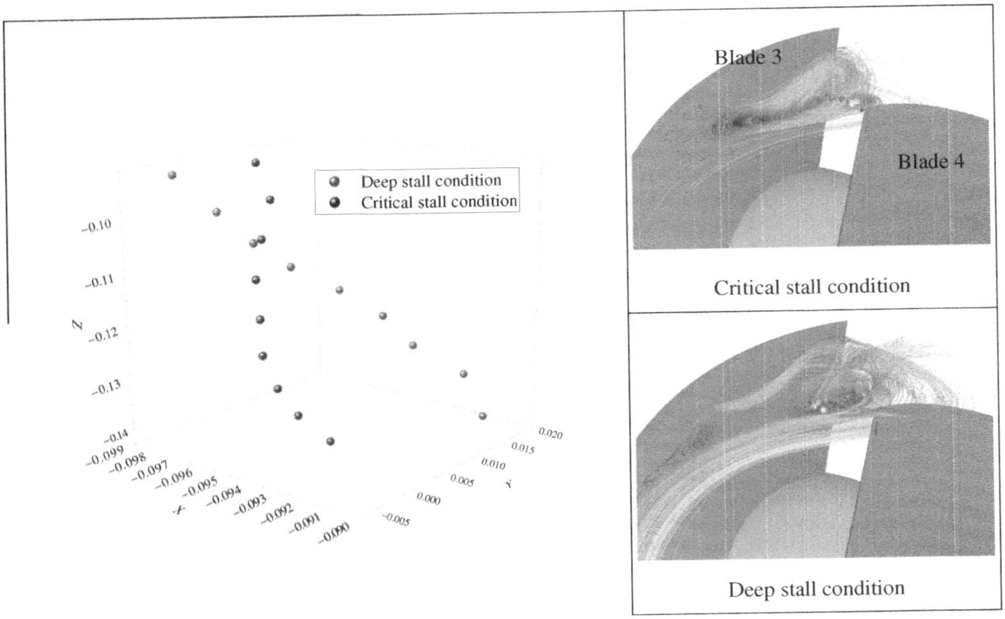


**Figure 8.13** Distribution of stall vortex core in the impeller passage at stall condition (*X*: horizontal direction, *Y*: vertical direction, *Z*: axial direction).

is similar to that in the critical stall condition, but the blockage region caused by the interaction of the stall vortex and the mainstream is larger and extends to the flow field in the rim region of the next stage blade. The TKE value on the surface of the stall vortex is also larger, and therefore the energy loss in the passage is larger.

Figure 8.13 shows the core of the stall vortex (line of maximum TKE) in the stall condition of the pump. The direction of each axis represents the coordinate value of the vortex core. It can be seen from this figure that the distribution range of the stall vortex core in the *X* and *Y* directions is larger in the deep stall condition, which indicates that the region of flow instability caused by the stall vortex in deep stall condition is larger.

#### 8.2.2.4 Influence of the Backflow Vortex on the Energy Loss

The flow cross-section diagram of the impeller outlet is shown in Fig. 8.14. The axial velocity distribution in the circumferential direction of the four outlet channels is shown in Fig. 8.15. The positive value of the axial velocity indicates that the fluid flows out from the impeller, and the negative value indicates the backflow into the impeller. It can be seen from this figure that in the design condition, the axial velocity is positive, and most of the fluid flows out of the impeller, with the maximum velocity appearing at the blade outlet. In the stall conditions, the axial velocity in the four flow passages has a negative value, which indicates that there is backflow at the impeller outlet. These backflows are close to the suction surface. In the deep stall condition, the region of negative axial velocity is larger than that in the critical stall condition, which indicates that the backflow in the impeller increases, and the energy loss increases in the deep stall condition.

Figure 8.16 shows the velocity vectors in the meridian plane B at the impeller outlet. In the design condition, except for the small-scale backflow in area C near the hub region and the guide vane, the overall quality of the flow field at the impeller outlet is very good. In the critical stall condition, the recirculation area in the guide vane increases, and the stall vortex core A appears in the impeller

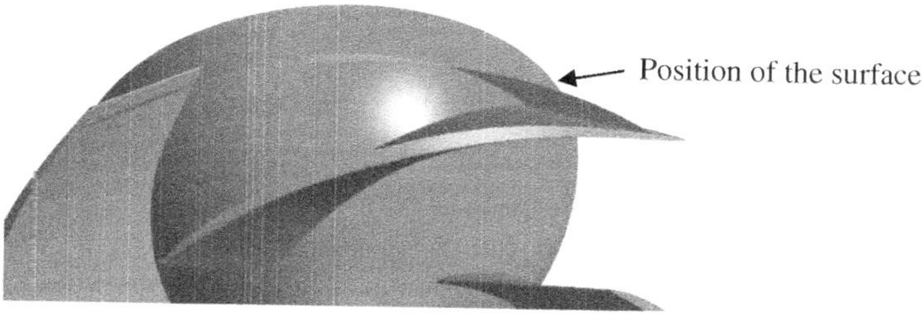


**Figure 8.14** Flow cross-section diagram at the impeller outlet.

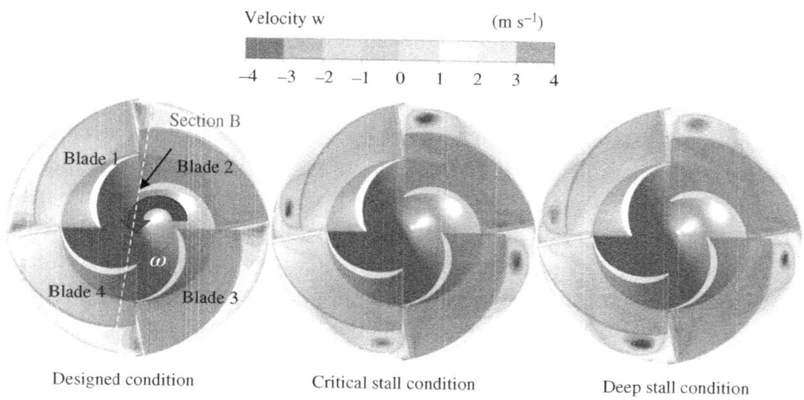


**Figure 8.15** Axial velocity distribution at the impeller outlet for various operating conditions.

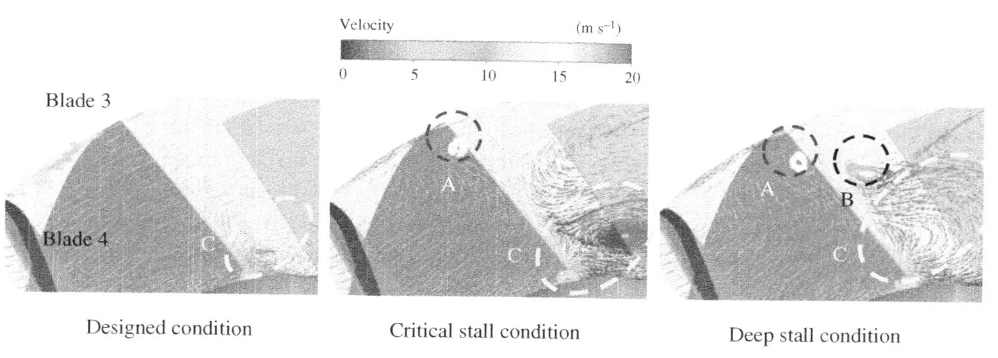


**Figure 8.16** Velocity vector distribution in the meridian plane B of the impeller outlet for various operating conditions.

passage. In the deep stall condition, there are not only the two kinds of vortices mentioned above, but also the influence area of the backflow in the guide vane is further enlarged, which affects the flow field structure in the rotor and stator interference region, and the interference backflow in area B is formed in the middle position of the impeller. It can be noted that in the deep stall condition, the backflow at the impeller outlet has a significant impact on the main flow, resulting in a large hydraulic loss.

## 8.3 Effect of the Tip Clearance on the Rotating Stall

### 8.3.1 TLF Patterns and TLV Core Trajectories for Different Tip Clearances

Due to the open impeller structure of the mixed-flow pump, a throttling effect is produced in the small blade tip clearance region due to the strong pressure gradient between the pressure surface and the suction surface of the blade during the operation of the pump, where the velocity of the fluid increases sharply. The fluid ejected from the small gap forms a long and spiral vortex structure in the impeller flow channel due to the entrainment effect of the main flow; it is called the TLV. The existing literature on the subject has shown that the existence of the TLV not only affects the flow passage in the impeller's main flow channel but also consumes some of the power, which results in a decrease in the efficiency of the mixed-flow pump.

Since the TLF of the mixed-flow pump in the design condition is stable, his section presents the influence of the various tip clearances on the generation of TLF, and the trajectory of TLV core in the mixed-flow pump at the design flow rate is presented. Figure 8.17 shows the TLF fields of the mixed-flow pump and the span distance of the vortex core from the blade suction surface for different tip clearances in the design condition, where the TLV core is the entrainment center of the vortex in the blade rim region. The inverse rainbow colormap is used in the contours to better show the TLF patterns. It can be noted from this figure that the TLF of the mixed-flow pump with different tip clearances is rolled into the TLV structure in the impeller flow passage in the design condition. The TLV flows to the impeller outlet in a "snail shell" shape and entrains the fluid from the blade rim leading edge and then further develops in the impeller flow channel. However, as the tip clearance increases, the TLF increases, which not only increases the blockage area influenced by the TLV but also changes the trajectory of the TLV core along the blade profile. If the intersection point of the tangent line to the TLV core curve from the *Point 1* and the blade surface is defined as the starting point of the TLV core, it can be noted that as the tip clearances increase, not only the starting point of the TLV moves away from the blade inlet side but also the TLV begins to keep away from the blade surface.

Figures 8.18 and 8.19 show the TLF fields of the mixed-flow pump for the three tip clearances in the critical stall condition and the deep stall condition. Since the TLF is affected by the mainstream flow, the TLV does not form in some of the sections in the blade rim region (there are eight sections shown in Fig. 8.18). Thus, the core of the vortex structure closest to the tip clearance in each section is defined as the TLV core. It can be seen from these figures that in the stall condition, the TLF intensity at the blade rim is significantly weakened; therefore, it is closer to the blade suction surface under the influence of the mainstream flow. At the maximum head condition, except for the TLV tail, the distance between the TLV core and the blade suction surface in each section is basically the same. At the blade outlet of this stage, where it is close to the inlet of the lower blade, the distance from the TLV core to the blade suction surface increases with the increase in tip clearance. At the minimum head point, the effect of the TLV is more prominent. In Section 8.2, it is affected by the unstable flow in the inlet pipe, and the vortex core tends to be far away from the blade surface. As the section moves toward the middle of the blade (Sections 8.4 and 8.5), the distance from the TLV core to the blade suction surface begins to decrease again, indicating that the influence of the unstable flow on the TLF begins to decrease. In Sections 8.7–8.9 of these figures which are near the blade outlet, although the vortices in these sections are loosely defined as the TLVs, the formation of these vortex structures in these sections is more affected by the mainstream flow.

Figure 8.20 shows the comparison of the distance from the TLV core to the blade suction surface or the end wall for the three tip clearances. It can be seen from Fig. 8.20(a) that the distance between

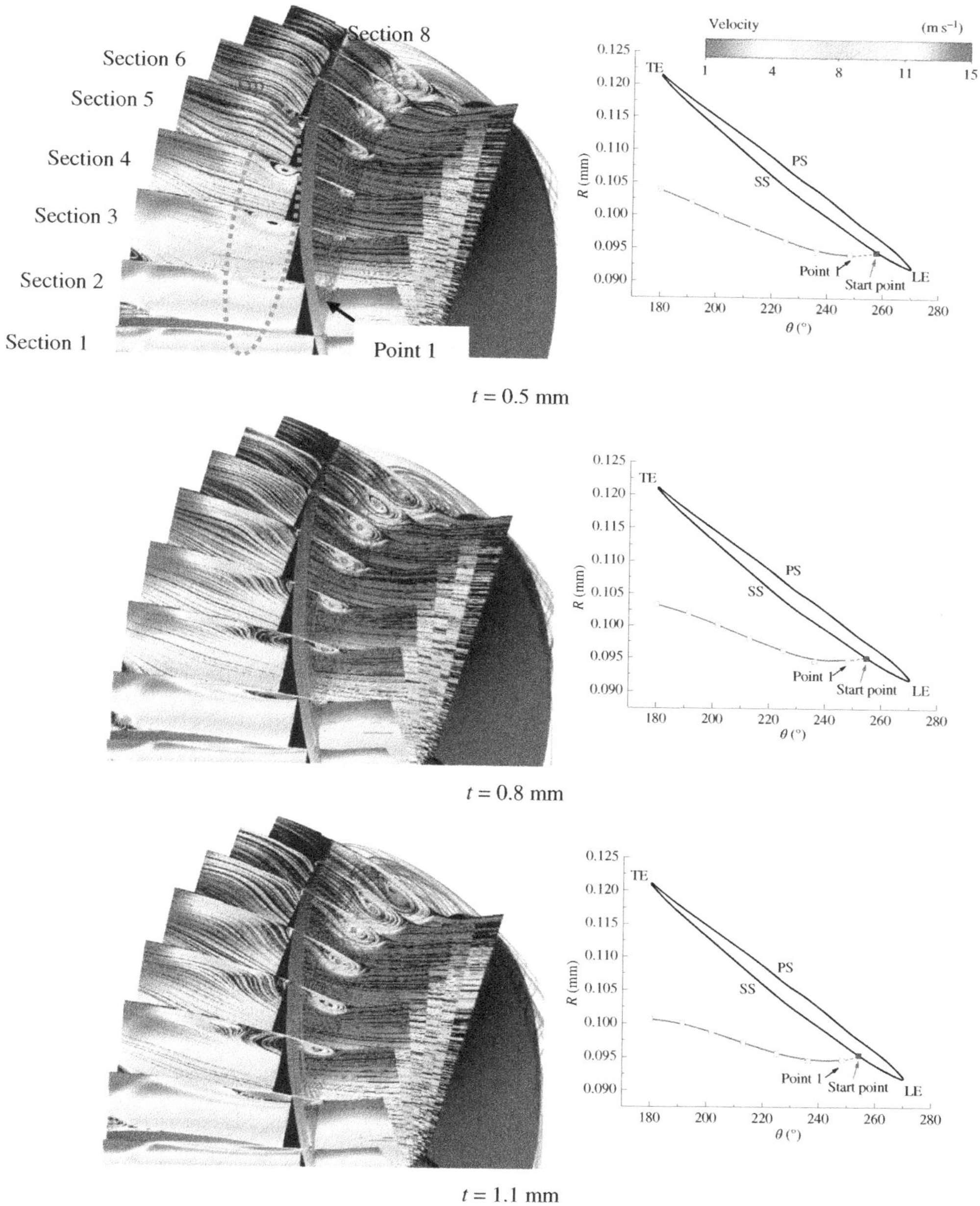


**Figure 8.17** TLF fields for various blade tip clearances at design flow rate.

the TLV core and the chamber end wall gradually increases with the increase in the tip clearance. Furthermore, the TLV core gradually moves toward the hub from the blade inlet to the blade outlet. This is a result of the combined effect of the increased blade tip clearance and the oblique outflow feature of the mixed-flow pump. It can be seen from Fig. 8.20(b) that in the front part of the blade rim, the core area of the TLV is farthest from the blade suction surface when the tip clearance is 0.5 mm, but the distance decreases with increase in the tip clearance. However, in the vicinity of the

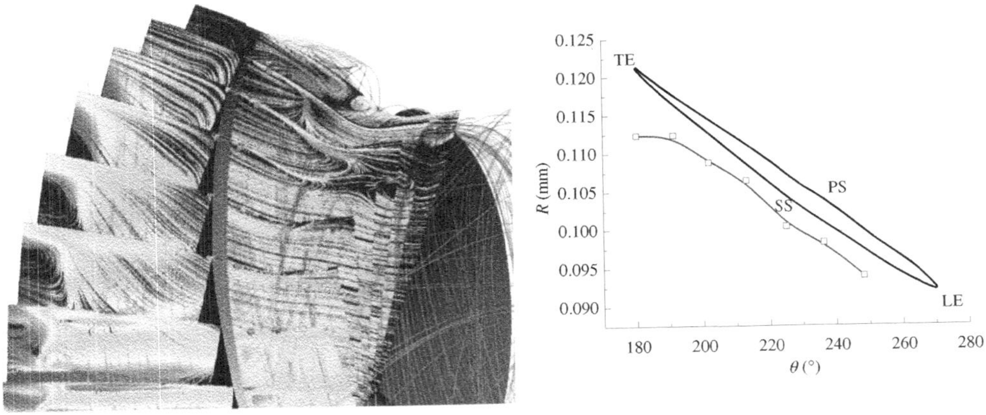


$t = 0.5$ mm

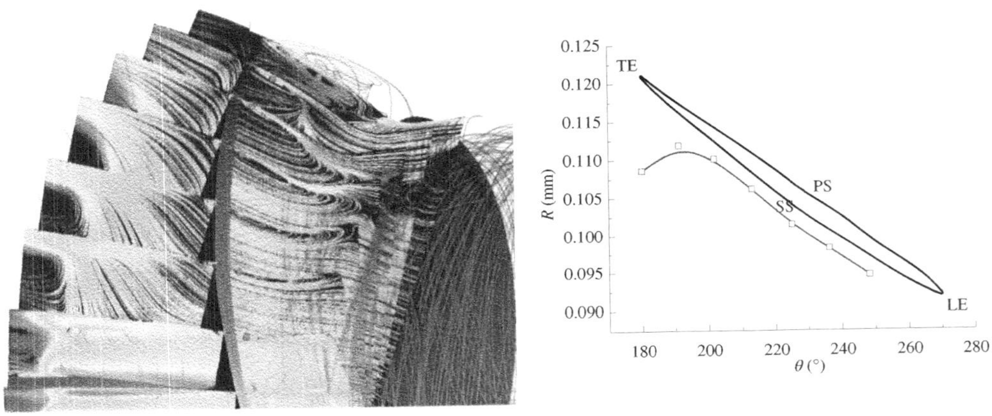


$t = 0.8$ mm

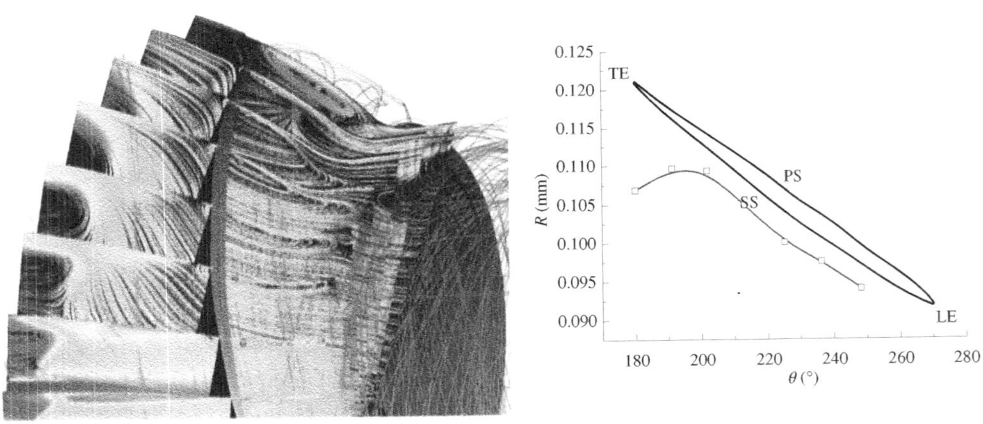


$t = 1.1$ mm

**Figure 8.18** TLF fields for various blade tip clearances at the largest head in the rotating stall condition.

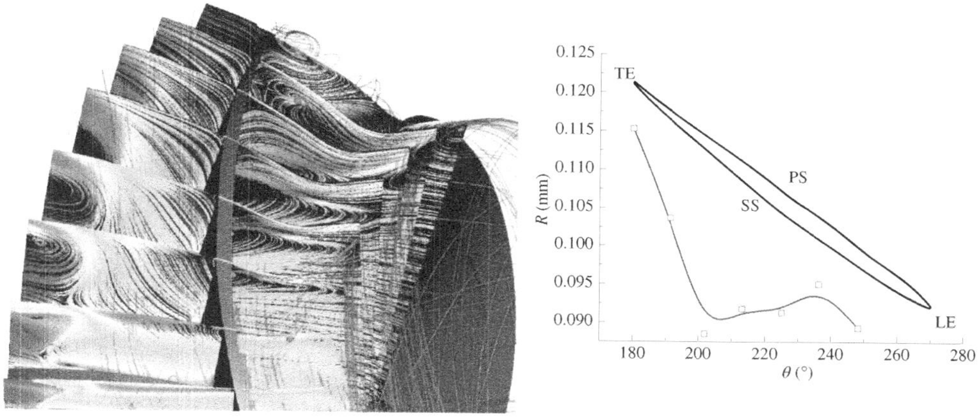


$t = 0.5$ mm

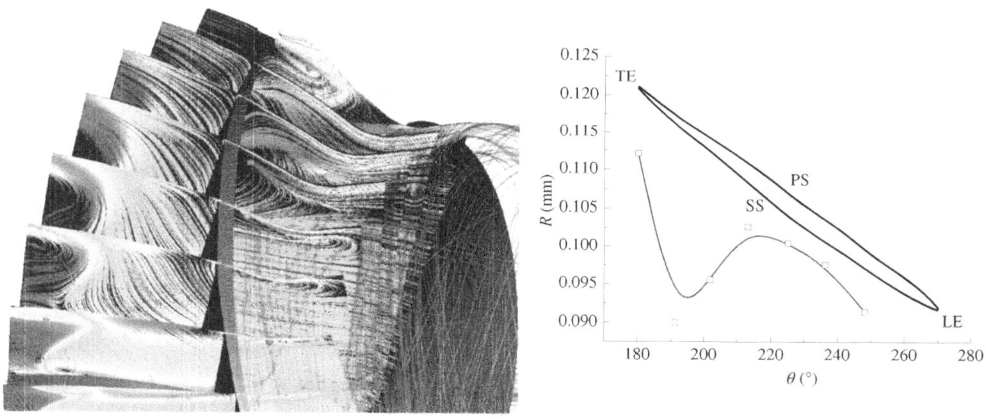


$t = 0.8$ mm

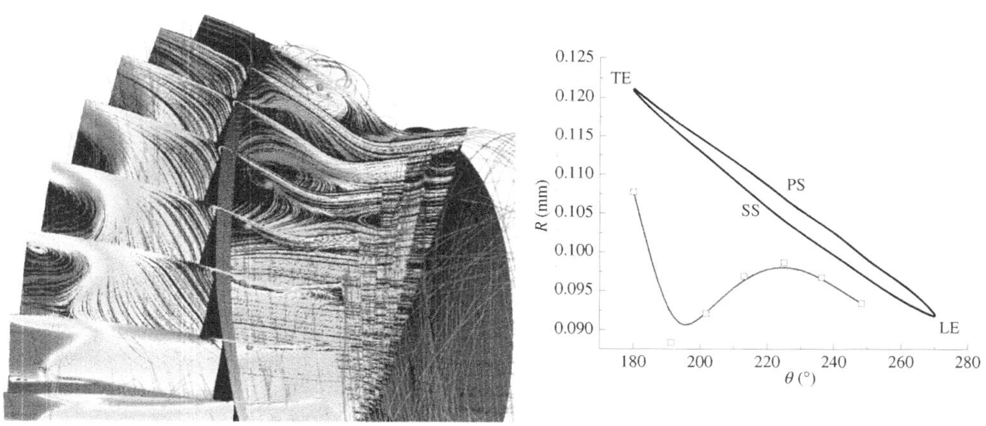


$t = 1.1$ mm

**Figure 8.19** TLF fields for various blade tip clearances at the lowest head in the rotating stall condition.

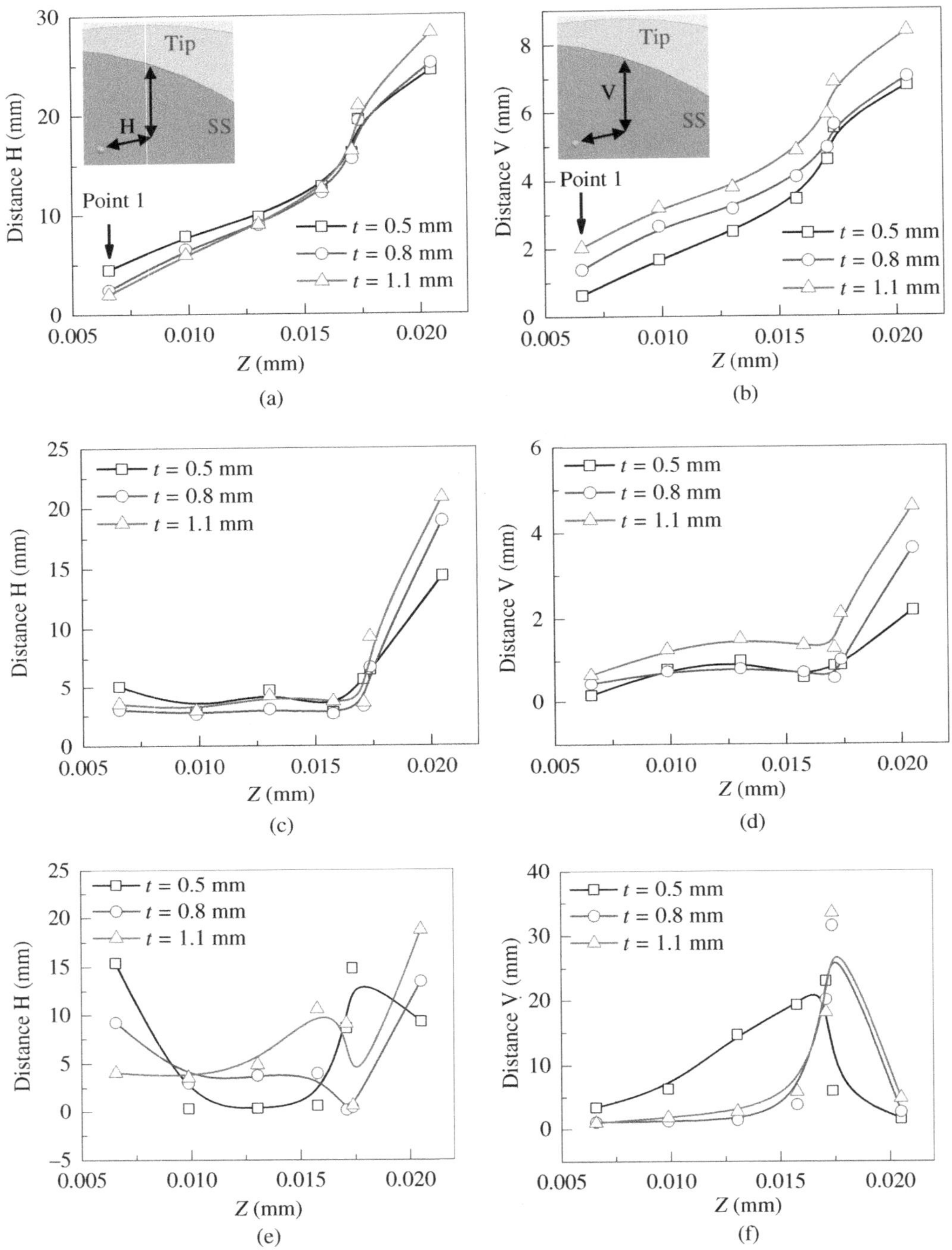


**Figure 8.20** Distance of TLV core for various tip clearances (a) to suction surface at design flow rate, (b) to end wall at design flow rate, (c) to suction surface at largest head in the rotating stall condition, (d) to end wall at largest head in the rotating stall condition, (e) to suction surface at lowest head in the rotating stall condition, and (f) to end wall at lowest head in the rotating stall condition.

blade outlet rim, the distance between the TLV core and the blade suction surface changes with an opposite trend. It is because the area affected by the upper blade leakage vortex is small in smaller tip clearance while the energy in the core area is higher. Hence, it requires more time for the TLV to vanish in the flow channel, and the distance of TLV is longer, which affects the blade rim inflow at this stage of the blade. It can be further noted by the results of Fig. 8.20(a) and (b) that the TLF affects the fifth section when the tip clearance is 0.5 mm, while the distance between the TLV core and the blade suction side increases from this section to the impeller outlet with increase in the tip clearances.

### 8.3.2 Influence of the Inlet Swirl Flow and the TLV Characteristics

It was shown in the previous section, that there is a close relationship between the TLF and the inlet flow fields of the mixed-flow pump. Figure 8.21 shows the average axial velocity for different Span parameters at the pump inlet. It quantitatively gives the exact impacted region caused by the swirling flow. From the 0.5 mm to the 1.1 mm tip clearance, the areas affected by the flow instability in the inlet are approximately 35%, 52%, and 64%, respectively. Thus, the reason for the high energy losses at the pump inlet in stall conditions is the increase in the strength of the inlet swirl flow. The increase in the tip clearance further enhances the swirling flow at the pump inlet, which can be attributed to the backflow of the TLF or the backflow in the impeller. In fact, since the vortex is attached near the pipe's end wall, this in turn affects the leakage volume and the intensity of the TLF. Thus, the effects of TLF and the inlet swirl complement each other. Overall, the increase in the tip clearance influences the inflow instability at the pump inlet and causes high energy losses induced by the enhanced swirling flow and the vortex. Furthermore, the increase in the swirling flow with an increase in the tip clearance also affects the flow in the impeller. Therefore, it is important that the tip clearance is not increased beyond a certain size.

### 8.3.3 Distorted Flow in the Blade Rim Region Below the TLV

Figure 8.22 shows the flow fields near the blade rim of the mixed-flow pump for various tip clearances when the span factor of the blade-to-blade surface is 0.85. Since the blade-to-blade surface is situated below the TLV, the flow fields in this region are less affected by the TLV. Since this region is close to the blade suction surface and is below the TLV, the fluid velocity in this region is relatively low. Furthermore, the flow in each flow channel of the impeller for the three tip clearances is stable and the velocity vectors in the mainstream flow are consistent with the blade placement angle. However, in the rotating stall condition, a low-velocity zone and a vortex structure appear

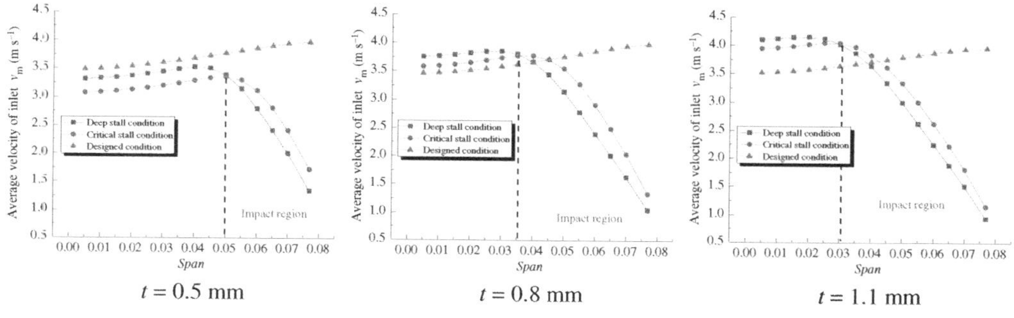


**Figure 8.21** Average axial velocity for various span parameters in the inlet axial section.

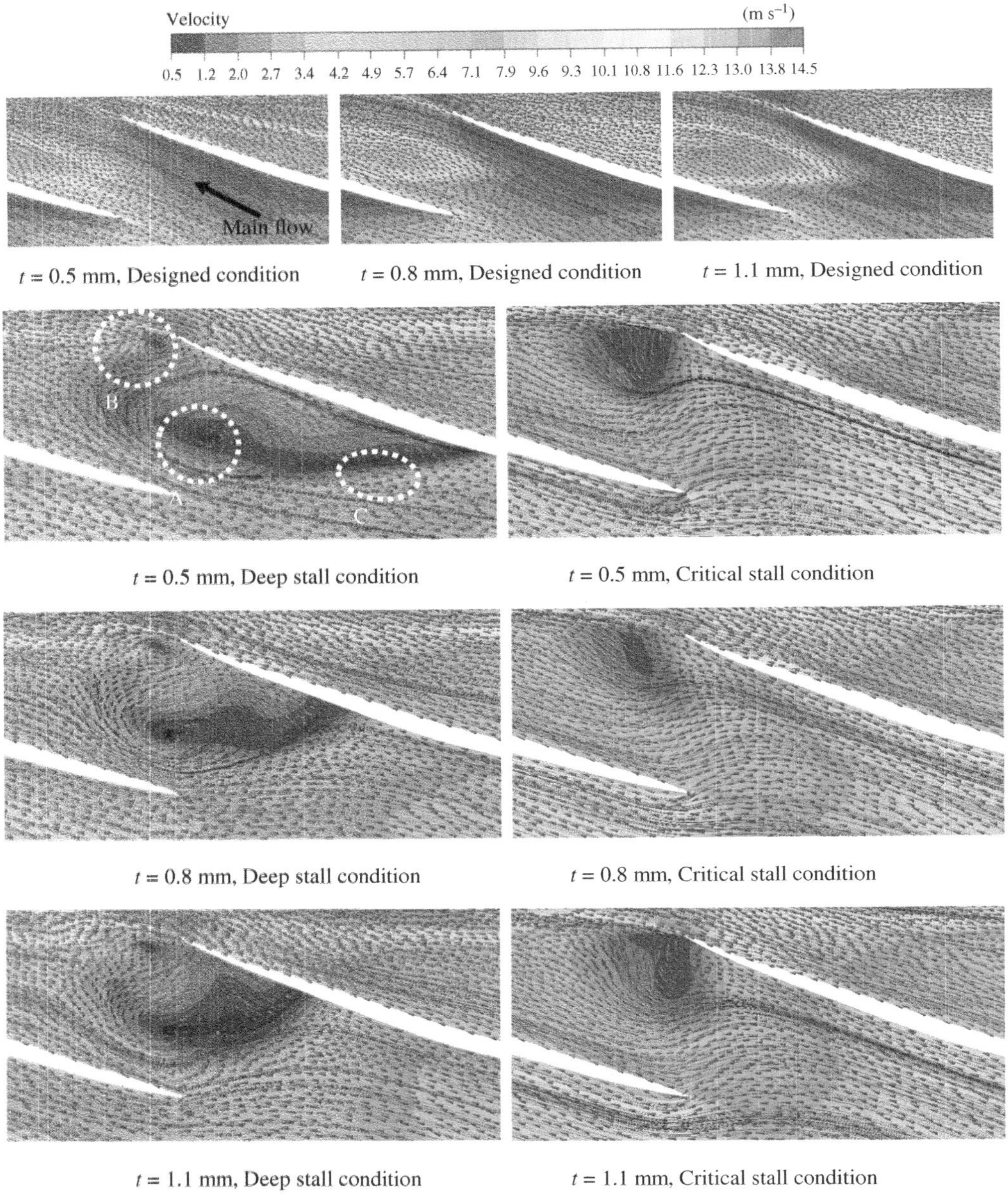


**Figure 8.22** Velocity on the blade-to-blade section for various span parameters.

on the monitoring surface, which are close to the blade pressure surface and is near the outlet of the impeller channel. The area and number of low-velocity zones and vortex structures at the minimum head operating conditions are significantly greater than that at the maximum head operating conditions in the rotating stall condition. When the mixed-flow pump operates in the most severe stall condition, vortex A and vortex B appear in the flow fields for all three tip clearances due to the

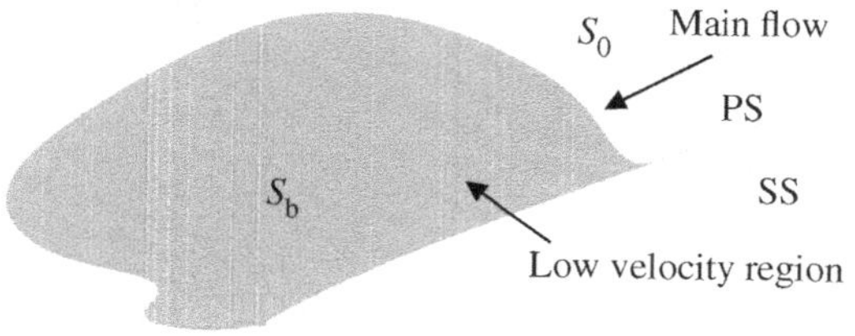


**Figure 8.23** Schematic diagram of the disordered factor.

flow separation and the instability in the TLF. Especially at the smallest tip clearance of 0.5 mm, not only does the low-velocity zone caused by the flow separation have a large area, but also a small vortex C appears near the starting position of the flow separation, which increases the unstable disturbance and aggravates the unstable flow inside the impeller.

In order to conduct the quantitative analysis of the degree of distortion in the flow field in the rim region below the TLV, the flow distortion factor $B_b$ of the flow field in the rim region is defined in Eq. (8.9), whose schematic diagram is shown in Fig. 8.23. The flow distortion factor $B_b$ is defined as follows:

$$B_b = \frac{S_b}{S_0} \tag{8.9}$$

where $S_0$ represents the total area of the impeller flow channel in the blade rim region, and $S_b$ represents the area of the low-velocity region caused by the vortex structures in the rim region. The threshold value (TV) of the relative velocity is approximately defined as 4.8, since the minimum average velocity in the section is about 4.8 m/s, as shown in Table 8.1. Five sets of TVs less than 4.8 are selected. It can be seen from Eq. (8.9) that larger $B_b$ value indicates that the rim flow fields caused by the vortex structure are more distorted, which implies that there is significant secondary flow and more energy losses in these flow fields. When $B_b = 1$, it implies that the vortex structures completely cover the flow field in the tip region.

The flow distortion factor $B_b$ caused by the vortex structures at Span = 0.85 for blade-to-blade surface interaction in the blade rim region of the mixed-flow pump for different tip clearances can be quantitatively obtained. The results of computation are shown in Table 8.1. (The flow distortion factor in the design flow rate condition is not listed in Table 8.1 since its value is zero.) From Table 8.1, it can be noted that the flow distortion factor for the minimum value of the head is greater than that for the maximum value of the head for the three tip clearances in the rotating stall condition; therefore, the flow fields in the impeller rim region are more distorted under these conditions of minimum head value, which indicates the high kinetic energy loss in the rim region as well as the head drop in the mixed-flow pump. In this section, the flow distortion factor has been calculated by setting the threshold of the vortex structure region, and the smaller this TV, the closer the center of the vortex structure. According to Table 8.1, when the relative speed threshold is less than 3.5 m/s, the flow distortion factor at the two extreme conditions of minimum and maximum head value (deep and critical stall conditions) increases with the increase in tip clearances. But, when the relative speed threshold is greater than 4 m/s, this trend is broken, and the difference in the distortion factors in the three tip clearances reduces with the increase in tip clearance.

**Table 8.1** The flow distortion factor $B_b$ and the average velocity at the monitoring section near the blade tip region.

| Case parameters | $B_b$(TV = 2.0 m/s) | $B_b$(TV = 3.0 m/s) | $B_b$(TV = 3.5 m/s) | $B_b$(TV = 4.0 m/s) | $B_b$(TV = 4.5 m/s) | Average velocity (m/s) |
|---|---|---|---|---|---|---|
| $t = 0.5$ mm, Deep stall condition | 0.1418 | 0.2074 | 0.2751 | 0.3288 | 0.3652 | 4.815 |
| $t = 0.5$ mm, Critical stall condition | 0.0108 | 0.0448 | 0.0770 | 0.1091 | 0.1423 | 7.007 |
| $t = 0.8$ mm, Deep stall condition | 0.1558 | 0.2274 | 0.2806 | 0.3128 | 0.3482 | 5.037 |
| $t = 0.8$ mm, Critical stall condition | 0.0492 | 0.0835 | 0.1168 | 0.1521 | 0.1881 | 6.502 |
| $t = 1.1$ mm, Deep stall condition | 0.1938 | 0.2624 | 0.2943 | 0.3293 | 0.3717 | 4.843 |
| $t = 1.1$ mm, Critical stall condition | 0.0821 | 0.1146 | 0.1471 | 0.1815 | 0.2194 | 5.982 |

## 8.4 Propagation Characteristics of Rotating Stall

### 8.4.1 Pressure Fluctuation Characteristics of the Flow in the Stall Condition

When the rotating stall occurs in a mixed-flow pump, there is a large pressure drop in the flow path. The periodic generation and disappearance of stall vortex (SV) often induce unstable pressure pulsation. Therefore, it becomes important to analyze the pressure fluctuation characteristics in the stall condition of the pump to understand the mechanism of the rotating stall. For this purpose, the four monitoring points are created in the four flow paths of the impeller in the circumferential direction with the extended wingspan. The location of the monitoring points is shown in Fig. 8.24, where L1, L2, L3, and L4 are located at the impeller inlet, Y1, Y2, Y3, and Y4 are in the middle of the flow path, and T1, T2, T3, and T4 are located at the impeller outlet, and all the monitoring points are located at 50% blade span.

It is important to note that the stall cell propagates in the circumferential direction opposite to the direction of the impeller rotation. Figure 8.25 shows the time domain plots of various flow paths in

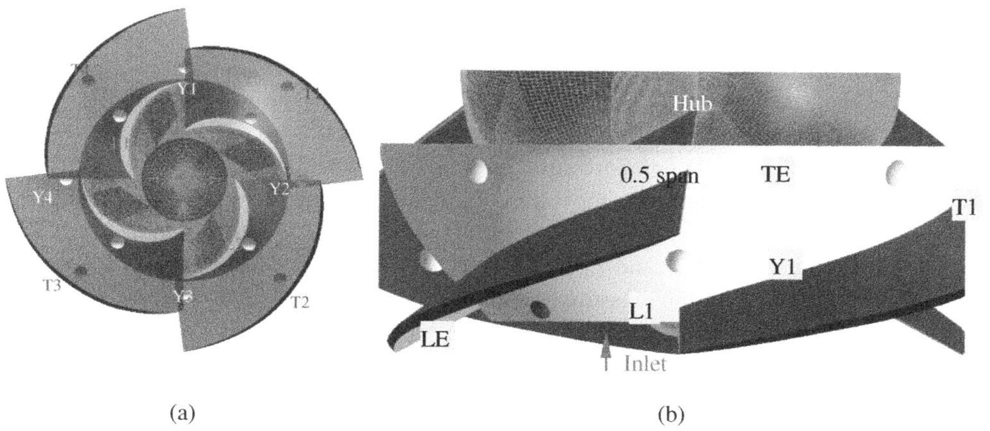


**Figure 8.24** Positions of the impeller monitoring points. (a) Front view and (b) Side view.

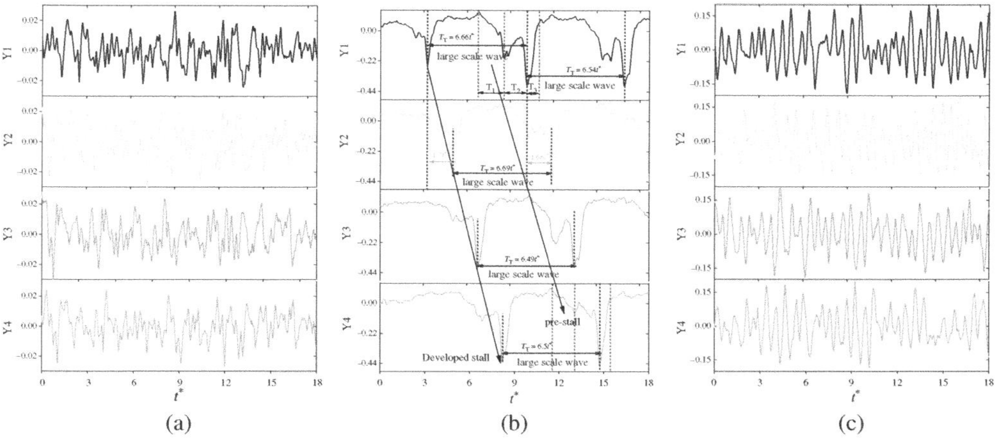


**Figure 8.25** Time domain characteristics of the pressure coefficient in the impeller for various flow rates at (a) design condition, (b) critical stall condition, and (c) deep stall condition.

the design, critical, and deep stall conditions. Since the speed of rotation of the model mixed-flow pump is 1450 r/min, the time required for it to rotate one revolution is 0.041 seconds, which is represented in this figure as $1t^*$, which is also defined as the period of the mixed-flow pump. The time range shown in Fig. 8.25 is from $0t^*$ to $18t^*$, which is the time required for the impeller to rotate 18 circles.

It can be noted from Fig. 8.25 that at the design flow rate, pressure in various flow paths fluctuates periodically due to the rotor-stator interaction. Furthermore, due to the inducement of the TLF, the pressure fluctuation waveform is serrated, its amplitude is much smaller than in the stall conditions, and the pressure distribution is relatively stable. At the critical stall point (CSP), the pressure drop is about 10 times higher than that in the design condition due to the influence of the rotating stall, and the wave crest and trough representing the impeller rotation can be neglected. The pressure fluctuation curves at the four monitoring points Y1, Y2, Y3, and Y4 show that the waveforms of the pressure coefficient $Cp$ in different flow paths are almost the same; however, there is a large phase difference in the waveforms between the adjacent monitoring points, which is caused by the circumferential propagation of the stall cell in the flow paths. The large period of the waveform is the propagation period of the stall cell in the impeller. The period of the large waveforms in the four flow paths is calculated as shown in Fig. 8.25(b), and the average value of the time period is obtained. The approximate propagation period of the stall cell is $6.59t^*$, which is equivalent to 6.59 times the shaft period since $t^*$ denotes one shaft period. Hence, the stall propagation speed is much lower than the rotational speed of the impeller. The examination of the large waveform shows that after a period of steady fluctuations, the pressure coefficient $Cp$ first shows a downward trend and then a small increase, finally dropping to the lowest point. The pressure waveform has two distinct troughs in each large wave. Since the stall is often accompanied by a sudden drop in pressure, the corresponding time intervals $T_1$, $T_2$, and $T_3$ are divided into the pre-stall stage, the stall stage, and the stall collapse stage based on the pressure fluctuation. When the pressure coefficient $Cp$ at Y3 has the lowest value, Y4 is in the first trough, and both Y4 and Y1 show consistent value in the pressure pulsation. It can therefore be concluded that when a certain flow path is in the stall stage, the clockwise adjacent path is about to enter the pre-stall stage with weak flow distortion, which is consistent with the flow field of the impeller outlet section at CSP.

From Fig. 8.25(b), it can be concluded that the phase difference is about $1.7t^*$ when the pre-stall occurs at Y1 and Y2; the exact value is $1.66t^*$ when the fully developed stall cell is formed, and the interval angle between Y1 and Y2 is 90°. Since the large fluctuation periods of the four flow paths are almost the same, Y1 and Y2 are selected as the representative points for determining the phase difference between the pre-stall stage and the developed stall cell. Therefore, the two important speeds can be calculated as follows:

$$\omega_{\mathrm{s}} = \frac{P_{\mathrm{m}}}{P_{\mathrm{s}}}\omega_{\mathrm{r}} = \frac{90}{1.7 \times 360}\omega_{\mathrm{r}} = 0.147\omega_{\mathrm{r}} \tag{8.10}$$

$$\omega_{\mathrm{T}} = \frac{P_{\mathrm{m}}}{P_{\mathrm{T}}}\omega_{\mathrm{r}} = \frac{90}{1.66 \times 360}\omega_{\mathrm{r}} = 0.151\omega_{\mathrm{r}} \tag{8.11}$$

In Eq. (8.10), $\omega_{\mathrm{s}}$ represents the propagation speed of the pre-stall stage, and in Eq. (8.11), $\omega_{\mathrm{T}}$ represents the stall cell propagation speed. $\omega_{\mathrm{r}}$ is the impeller rotational speed, and $P_{\mathrm{S}}$ and $P_{\mathrm{T}}$ represent the phase difference between the two monitoring points in the pre-stall state and the stall state, respectively.

The stall cell propagation period $T_T$ is approximately $6.59t^*$, then the number of stall cells $N_c$ can be calculated as:

$$N_c = \frac{P_T}{P_m \times T_T} = \frac{1.66t^*}{\frac{90}{360} \times 6.59t^*} \approx 1 \tag{8.12}$$

At the deep stall point (DSP), the flow field in the pump becomes very complex. Different number of vortices occur in the four flow paths; several stall cells circumferentially propagate in the impeller, and the pressure coefficient *Cp* exhibits highly irregular behavior. The pressure waveform shows a sharp cone shape with straight ascending and straight descending lines. When a flow path gets rid of the disturbance due to stall cell by the increased pressure, the next stall cell spreads rapidly into this path to induce the new distortion in the flow. Although the amplitude of the pressure fluctuation is not as large as in case of CSP, it is nevertheless much higher than the pressure fluctuation caused by the rotor–stator interaction (RSI). The period of sudden drop in the pressure due to the flow distortion is quite short; it is only 1/4 times the period of the impeller shaft. It further illustrates the high stall cell propagation speed and the adverse flow condition in DSP.

To analyze the frequency domain characteristics of the pressure fluctuation, pressure data in the time domain is transferred to the frequency domain using the Fast Fourier Transform (FFT). Figure 8.26 shows the frequency spectrum of the pressure fluctuation in the impeller. Since the pressure fluctuation spectra of the monitoring points in different flow paths are nearly the same, the spectra of the same monitoring point in the design, critical stall, and deep stall conditions can be compared. At the design flow rate, the impeller rotation frequency is 24.17 Hz, corresponding to the impeller rotational speed of 1450rpm. For the impeller, the blade passing frequency (BPF) is $f_{b1} = 96.68$ Hz, and for the guide vane, BPF is $f_{b2} = 169.19$ Hz.

From Fig. 8.26, it can be noted that at the design flow rate, the main frequency is the shaft frequency $f_s = 24.17$ Hz, and there exist higher harmonics since the RSI plays a dominant role and compared to the amplitude of the impeller BPF $f_{b1} = 96.68$ Hz, it is relatively small. In both the critical and deep stall conditions, the frequency spectra changes greatly compared to the design condition, and the main frequency changes from $f_s$ to the stall propagation frequency $f_{stall}$. In the critical stall condition, the main frequency is $f_{stall}$ and its harmonics $2f_{stall}$ and $3f_{stall}$, etc. The amplitudes of the $f_{stall}$ and its higher harmonics are much higher than that of $f_s$ and $f_{b1}$. The peak frequency is $f_{stall} = 3.82$ Hz, and its amplitude is 0.092, which is about 12 times that of $f_s$. Low-frequency signal of high amplitude confirms that a stall has occurred. According to the stall cell numbers based on the frequency domain diagram of Miyabe given in the article [6], the number of stall cells at CSP is 1, and Miyabe's result is consistent with that given in Eq. (8.12). While at the DSP, due to the more complex flow field and the more severe degree of stall in the flow paths, stall cell propagation frequency changes a great deal. Furthermore, since the amplitude of pressure pulsation caused by the rotating stall is much higher than that due to the RSI, the amplitude of $f_{stall}$ in the frequency domain diagram is higher than that of $f_s$. Therefore it can be concluded that there exist at least two different frequencies for circumferential propagation of the stall cells in the deep stall condition, namely $f_{stall1} = 35.65$ Hz and $f_{stall2} = 42.73$ Hz. Hence, it can be concluded that multiple stall cells appear simultaneously in the impeller and propagate circumferentially in the flow paths.

The propagation speed of the stall cells ($\omega_s$) is determined by the angle of rotation ($\theta$) of the pressure field and its duration in time. Therefore $\omega_s$ can be obtained as:

$$\omega_s = \frac{\Delta\theta}{\Delta t} = \frac{2\pi f_{stall}}{N} \tag{8.13}$$

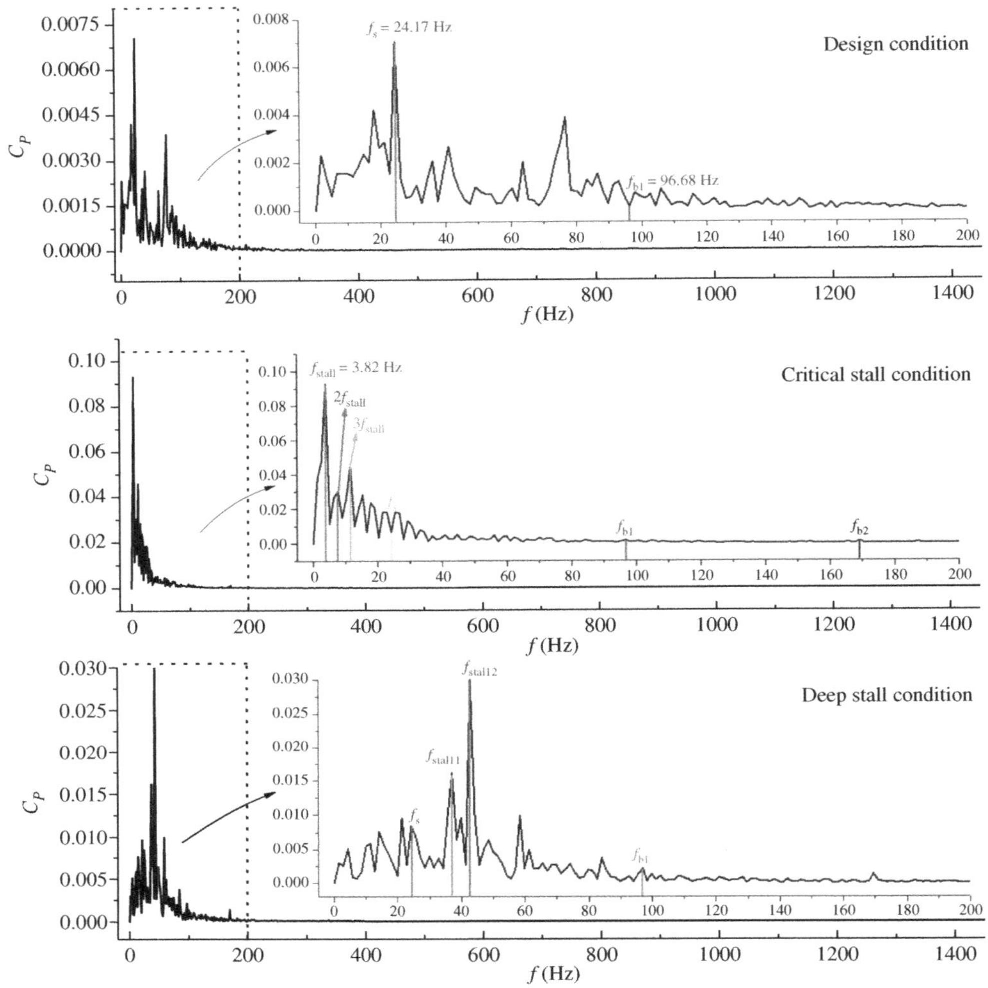


**Figure 8.26** Frequency domain distribution of pressure coefficients in the impeller at various flow rates.

From Eq. (8.13), at the critical stall condition, the stall propagation speed is 0.993 rad/s, which is 16% of the rotor speed. In the deep stall condition, again from Eq. (8.12), the propagation speed of the stall in mode1 is 9.264 rad/s, and in mode2 is 11.1 rad/s, which is much higher than the propagation speed in the critical stall condition. Because of the influence of several stall cells and the rapid circumferential propagation of stall cells, the four flow paths of the mixed-flow pump get blocked to varying degrees, resulting in a great dissipation of energy leading to a sharp drop in the performance of the energy head curve.

### 8.4.2 Correlation Between the Transient Flow Field and the Energy Characteristics of the Pump

Since there exist multiple stall cells and the stall propagation velocity in the circumferential direction of the impeller is very high in the deep stall condition, the flow characteristics in each flow path of the impeller become very complex and difficult to properly identify and characterize. Therefore, this section focuses on the analysis of the stall mechanism in critical stall condition.

In the critical stall condition, due to the periodic propagation of single stall cells, the head fluctuates alternately with sudden drop and rise. Figure 8.27 shows the transient head pulsation and the transient pressure pulsation in the pump at the monitoring points L1, Y1, and T1 in the same flow path, where L1 is located at the impeller inlet, Y1 is in the middle of the flow path, and T1 is located at impeller outlet. It can be seen from this figure that the pressure fluctuation at the impeller inlet monitoring point L1 is like the pressure pulsation at the monitoring point Y1 due to the influence of the stall cell described in Section 8.4.1. After a period of stabilization, pressure drops sharply, and the pressure drop reaches more than 40 kPa. Pressure fluctuation at the monitoring point Y1 in the middle of the flow path is highly affected by the stall, and the pressure drop is over 70 kPa. In addition, the pressure drop at Y1 lags behind the pressure drop at L1. When the pressure fluctuation at the monitoring point L1 is at its peak, the pressure fluctuation at the monitoring point Y1 (located downstream of the same flow path) reaches its lowest level, and when the pressure fluctuation at the monitoring point L1 is the lowest, the opposite is true. The pressure at the monitoring point T1 at the TE fluctuates irregularly in a zigzag fashion due to the RSI, and the pressure drop is relatively much less. Considering the transient change in the pump head with time, it can be noted that after five cycles of unstable fluctuations, the pump head shows periodic fluctuation. It shows alternating sudden drop and rise in the pressure with a period of about $1.6t^*$, which is the same as the time required for the stall cell to propagate to the next flow path of the impeller. Furthermore, considering the fluctuation in the transient head of the pump in the design condition, it can be noted that the fluctuation in the head in the design condition also has periodicity due to the RSI; however, its amplitude is much smaller than that in the critical stall condition. Therefore, it can be concluded that the sudden change in the flow characteristics in the impeller flow paths affected by the stall is related to the fluctuations in the pump head. It is shown in this figure that when there is sudden pressure at the monitoring point L1, the head approaches the peak value, and when the pressure drops sharply at the monitoring point Y1, the head also drops sharply. Therefore, it can be concluded that the initial stall cell originates from the LE (with very small energy dissipation)

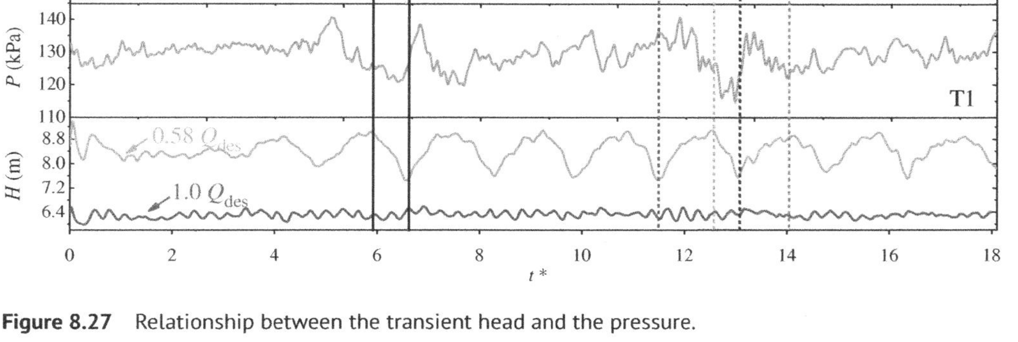


**Figure 8.27** Relationship between the transient head and the pressure.

and gradually develops downstream into a fully developed stall cell in the middle of the flow path resulting in a sharp drop in the pump head. Finally, the fully developed stall cell gradually decays toward the TE and propagates to the next flow path. It is similar to the conclusion by Yin [7] in the investigation of compressor stalls.

In order to describe more intuitively the influence of the rotating stall on the internal flow characteristics of the pump, four moments in time $t_1$, $t_2$, $t_3$ and $t_4$ are selected, and the internal flow fields of the pump at these four moments in time are investigated as shown in Fig. 8.28. Figure 8.28(a) shows the velocity streamlines diagram at the impeller outlet, and the flow path

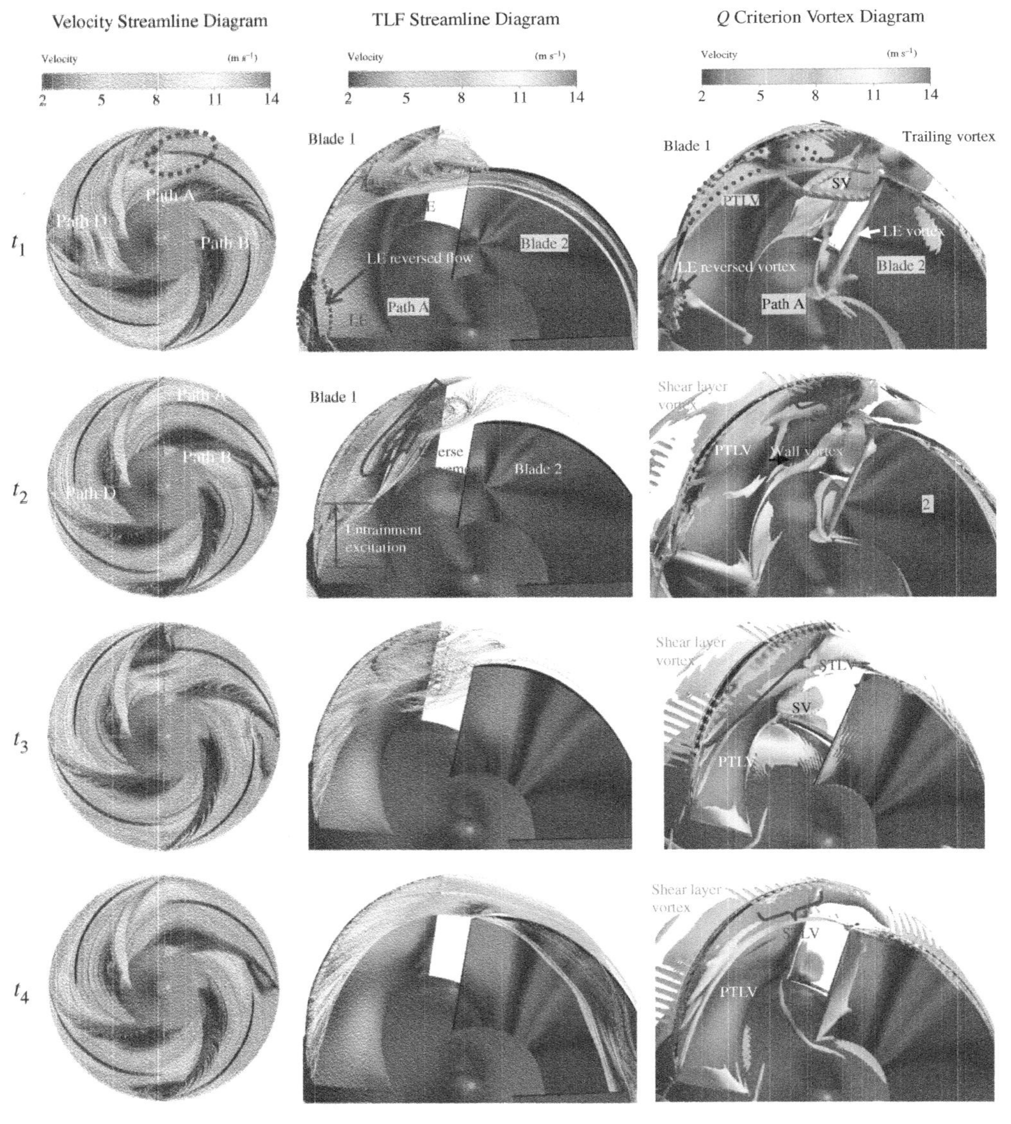


**Figure 8.28** Charts of flow field in the impeller at different time.

marked by the circle is consistent with that of the monitoring points L1, Y1, and T1; Fig. 8.28(b) shows the three-dimensional TLF diagram for this flow path; and Fig. 8.28(c) shows the vortex diagram based on the $Q$ criterion, where $Q = 300\,000\ \mathrm{s}^{-2}$ in the flow path which is consistent with the flow fields in Fig. 8.28(a) and (b).

At time $t_1$, the pressure at L1 is the highest, while pressure at Y1 in the middle of the flow path shows a downward trend; thus there is no vortex generated due to the flow separation at the LE. The vortex appears near the suction surface at the TE due to the flow distortion, which leads to the drop in pressure at Y1. It can be noted however that large number of vortices exist in the flow path D at the outlet, but the speed of convection of the vortices is lower than that in the flow path A and the stall is more prominent. The flow path A is in the pre-stall stage; there is a large pressure drop in the middle part of the adjacent flow path D located counterclockwise to flow path A, and the flow in path D is in the stall stage. Fig. 8.28(b) shows the TLF in path A as the pressure at the LE reaches its peak value at this moment in time $t_1$, and the pressure in the middle part of the flow path D is very low due to the stall. The LE leakage flow has large recirculation along the pressure gradient, which further interferes with the flow field in path D, resulting in energy dissipation. At the same time, due to the drop in pressure in the middle part of path A, the TLF converges and eventually mixes with the stall vortex (SV), further generating a large area of vortices in the middle and rear of the flow path D. The vortex structure and its position in flow path D can be captured more intuitively by the vortex diagram shown in Fig. 8.28(c). At time $t_1$, there exists a large area of TLV in the rim region. TLV can be divided into two types: PTLV and the secondary tip leakage vortex (STLV) [7]. The PTLV is mainly distributed in the main flow path D and is initially generated at the LE, and then develops into the flow path gradually approaching the PS of the lower blade. The STLV is mainly distributed near the tip clearance area in the form of scattered bars. Since there exists SV perpendicular to the wall and in the middle and rear of the whole flow path D, its blockage effect is significant. The TLF is disturbed by the SV, resulting in large quantities of low-energy fluids accumulating at TE, forming trailing vortices.

At time $t_2$, the pressure at the LE of the flow path A drops sharply to the lowest point, and the large-scale vortices appear, which are significantly disturbed by the initial stall cell at the LE. However, the rest of the flow path is stable, and there is no flow separation. Compared to the flow field characteristic at $t_1$, the four outlets of the impeller are not significantly affected by the SV; therefore, the transient head is at the peak due to less energy dissipation. Because of the entrainment effect of the initial stall cell at the LE, the momentum of the TLF at the LE is very low and is squeezed to SS. The pressure at the LE drops sharply with the formation of the initial stall cell, while the flow in the middle and rear section of the flow channel remains stable, and its pressure is higher than that at the LE. Part of the fluid has the characteristics of the recirculation flow along the pressure gradient, which is reflected in the middle section of the TLF. After the TLF flows back to the LE, it is impacted by the high-speed mainstream flow and flows downstream along the mainstream direction. Since the TLF is close to SS, the mixing effect with the mainstream flow becomes weaker, and no large-scale vortices are formed in the flow path. The vortex diagram further confirms this conclusion that the TLV is weaker at time $t_2$ compared to at time $t_1$, and the partial TLF adheres to the SS, forming the wall vortex, while there are no large-scale vortices in this flow path. Furthermore, the clearance height considered here is only 0.2 mm, which is much less than the large tip clearance height studied by Tan et al. [8]. Due to the extremely short distance between the tip and the shroud wall, strong shear flow occurs in the rim region. Therefore, the shear layer vortex appears in the rim region.

At the moment of time $t_3$, the transient pressure at the monitoring point Y1 is at the lowest point of the large pressure wave, and the stall affects the middle and rear sections of the flow path A.

The flow field in the impeller is like that at time $t_1$, except that flow path B replaces flow path A in the pre-stall stage, while flow path A replaces flow path D in the critical stall stage. During the time $(t_3 - t_1)$, the stall cell completes the clockwise propagation in a single flow path. At this moment, in addition to the existence of SV in the flow path, there also exists banded PTLV and many STLVs in the rim area. The PTLV initiated from the LE appears in the form of the reverse flow due to the blocking of the flow path by the SV and cannot further develop in this flow path. Therefore, the PTLV is present only as a thin band, while STLVs are distributed as strips occupying the rim area.

At moment $t_4$, the flow field is similar to that at $t_2$; however, since the inlet pressure is relatively high and the scale of vortices is relatively small, the TLF has no reflux characteristics. The TLF is stable and does not interfere with the mainstream flow, resulting in flow separation, and the energy dissipation is low. The vortex diagram at $t_4$ shows that there are no vortices in the flow path except the TLV and the trailing vortices, and the flow in the flow path is stable. Since the flow path is not affected by the rotating stall, the TLF is not affected by the entrainment and blockage effects of SV, and the PTLV is fully developed in the flow path.

In conclusion, in the critical stall condition, the large-scale vortices at the LE are caused by the flow distortion induced by the increase in the blade angle of attack and the formation of the initial stall cell. However, it is not the main cause of the turbulent dissipation, and the energy dissipation is relatively low, resulting in a transient pump head with higher amplitude at this moment in time. On the other hand, with the development of stall cells, large-scale vortices come to exist in the rear of the flow path due to the inducement of the developed stall cells, leading to high turbulent dissipation, which affects the flow patterns in the flow path; it is the main inducement for a steep drop in the transient head from the high amplitude. Different from the small amplitude periodic wave motion in the design condition due to RSI, the periodic generation, development, and collapse of the stall cell in the critical stall condition leads to a large periodic fluctuation in the transient head.

### 8.4.3 Morphology and Propagation Mechanism of Rotating Stall

In order to visually demonstrate the circumferential propagation mechanism of the rotating stall, Fig. 8.29 shows the transient internal flow characteristics in the impeller exit section. To illustrate the stall propagation cycle, these images are displayed in a rotating reference frame, and the initial position is set at $0t^*$.

At the $0t^*$ position, there exists the SV close to the wall in the flow path A. Since the SV firmly occupies the rear of this path and blocks it, the upstream fluid near the SS is squeezed. The streamlines are deflected, a large amount of fluid moves away upstream of the SV, and a high-speed fluid region appears in the middle of the flow path around the SV. The clockwise adjacent path B (adjacent to A) enters the pre-stall stage with weak distortions in the flow. It should be noted that when a flow path enters the stall stage, it first undergoes a pre-stall stage with weak flow distortion. After the 3/8th revolution of the impeller to position $3/8t^*$, the area occupied by the vortices in the flow path A decreases, the velocity of the vortices increases, and the strength of SV in the flow path A decreases. At the impeller position $3/4t^*$, the stall cell in the flow path A has completely disappeared and has propagated clockwise to the flow path B. The flow in path A is relatively stable, and only the outlet streamlines are slightly distorted. Further development of the stall cells leads to an increase in the blockage effect in the flow path B. At the impeller position $3/2t^*$, the area of the SV in the flow path B increases significantly the convection speed of the vortices decreases further, and the strength of the SV increases. At the same time, there exists a high-speed fluid region around the SV. The stall cell further grows into a bigger stall cell, while the flow path A no longer has any stall effect at this time. Subsequently, the SV in the flow path B gradually decays and disappears,

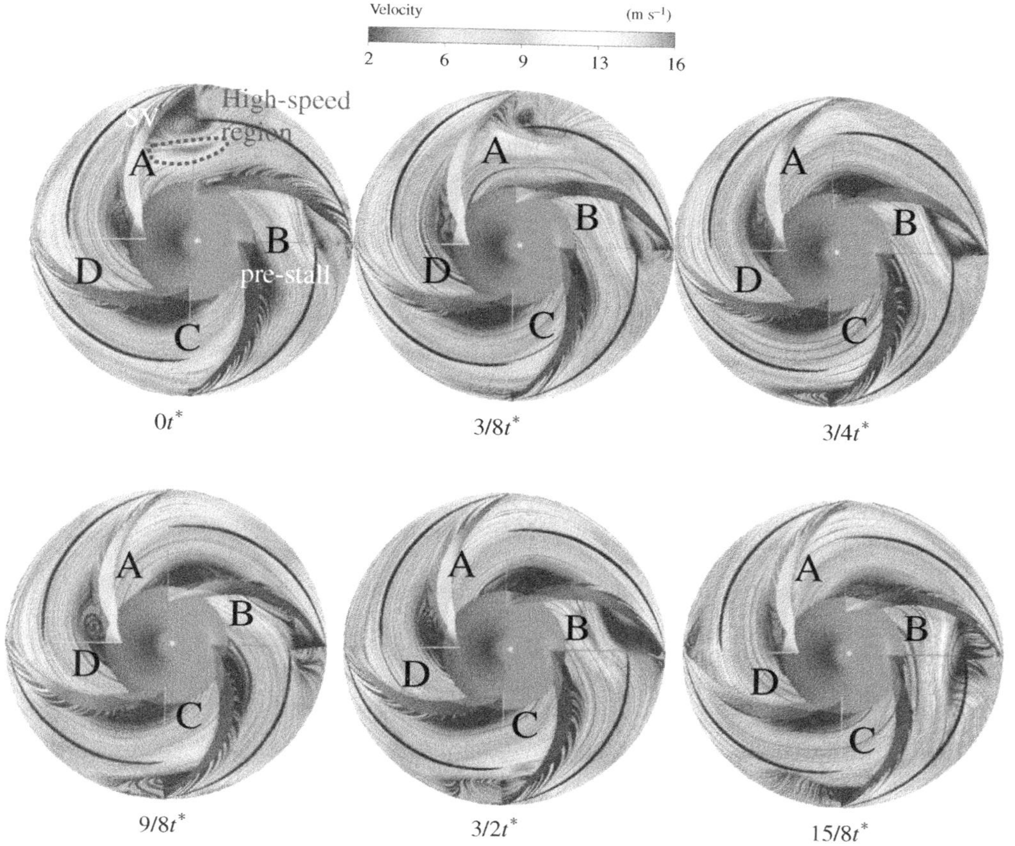


**Figure 8.29** Stall propagation process under CSP.

while the flow path C gradually enters the stall stage. As can be noted from Fig. 8.30, the period required for stall cell to propagate from path A to path B is about $1.5t^*$, which is approximately the same as that obtained from the diagram of pressure fluctuation.

In the above discussion, the process of circumferential propagation of stall cells has been described intuitively; however, the process of the generation, development, and dissipation of the stall cells in one flow path has not been fully captured. In this part, the evolution process of the stall cell propagation in a single path is described considering the pressure fluctuation. Figure 8.30 shows 27 monitoring points at the shroud, 0.5 span, and the hub of impeller suction surface. Among them, nine monitoring points S1–S9 are at the shroud from LE to TE, nine monitoring

**Figure 8.30** Distribution of monitoring points on the suction surface.

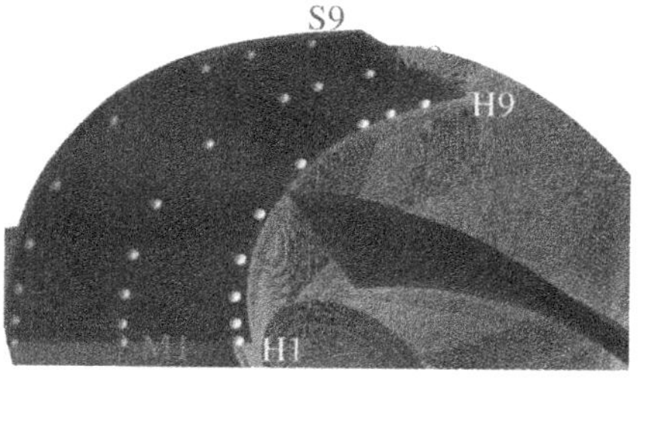

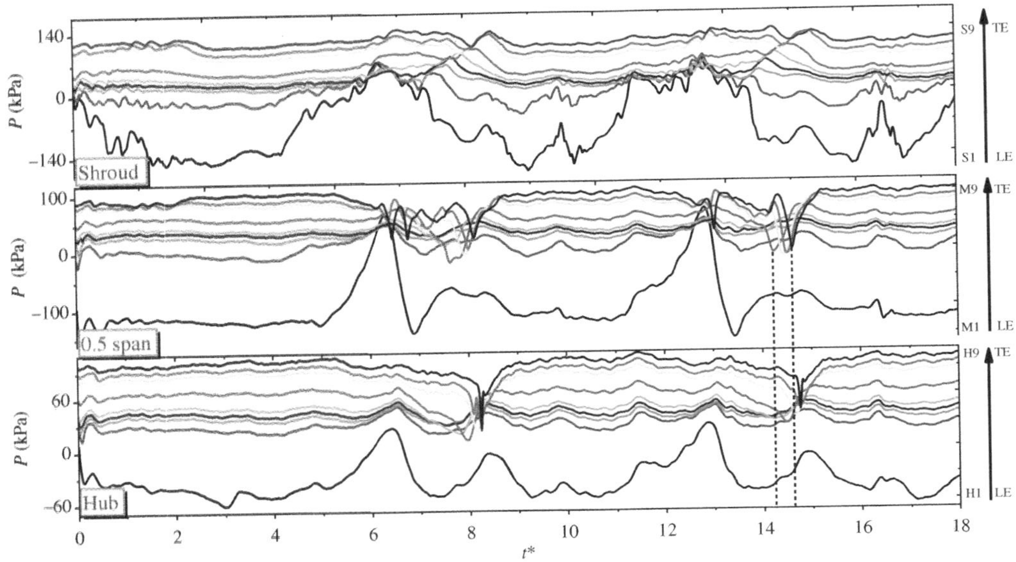


**Figure 8.31** Pressure fluctuation characteristics of the monitoring points at various spans.

points M1–M9 ate at mid-span, and nine monitoring points H1–H9 are at the hub. Figure 8.31 shows the pressure fluctuations at these 27 monitoring points.

As shown in Fig. 8.31, the pressure fluctuations on the shroud, the hub, and 50% span increase gradually from LE to TE along the wingspan, and all monitoring points show pressure drop at different times $t^*$. The time delay in the pressure drop at the same span shows the hysteresis effect along the wingspan, which indicates that the stall cell propagates from the LE to the TE. In the rim area, there is no significant pressure drop except at S1 and at the same chord length; the pressure drop at H1 in the hub area is less than that at the 50% span and the rim area. Comparing the pressure pulsation characteristics at the monitoring points M6–M9 and H6–H9 with the same chord length, it can be noted that the hysteresis in the pressure drop always occurs in the hub area. Thus, it can be concluded that the stall cell is initiated at the LE near the shroud and extends gradually toward the hub side of the TE. The pressure drop at H9 in the hub area is the most severe, and the pressure waveform shows a sharp cone shape that descends in a straight line and then rises in a straight line. It shows that the stall cell decays and collapses instantaneously when it approaches the TE.

In the above paragraph, the development process of the rotating stall is presented and analyzed, considering the pressure fluctuations in various parts of the mixed-flow pump. Now, the propagation and evolution process of the stall cells in a single path is further explored considering the transient flow field. Since the pressure drop always occurs in the path of the stall propagation, the area of the low-pressure region generally represents the position of the SV. Figure 8.32 shows the transient evolution of the low-pressure region around the SV in the flow path. As shown in this figure, along with the flow separation in the rim region at the blade inlet, the low-pressure region is initiated at LE, showing the downstream vortex shedding along the chord length. At the edge of the distorted flow region, the low-speed region appears at the tip of the rotor, which can be considered as the initial position of the stall cell. After $1/4t^*$cycle, the vortices shed from the low-pressure region at the LE develop into a cylindrical shape, which is perpendicular to the PS. Due to the influence of the stall cells, the low-speed vortex region appears near the low-pressure region. During the next $1/2t^*$ cycle, the low-pressure region gradually propagates downstream and further develops

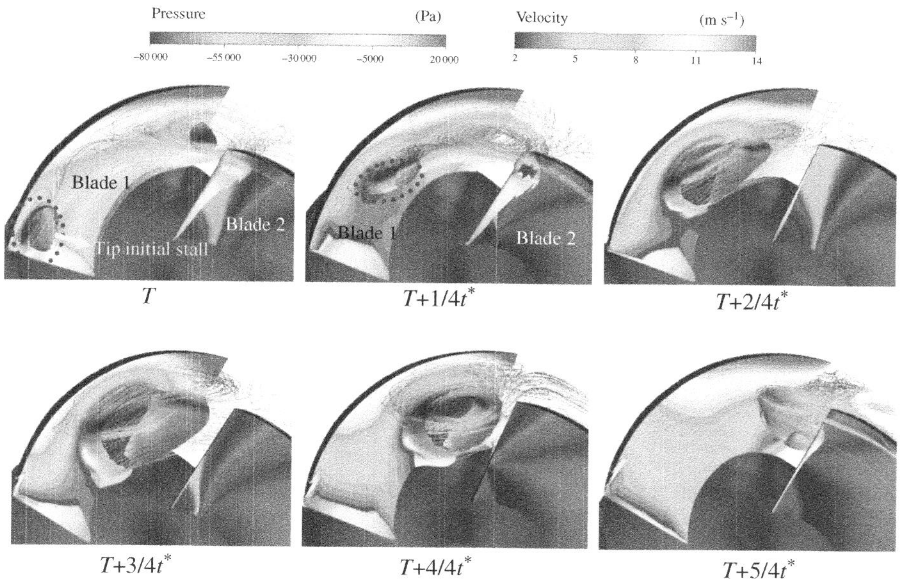


**Figure 8.32** Process of stall evolution in a single flow path (the iso-surface based on pressure where pressure = −80 kPa; transient pressure distribution contour is on SS; the color of three-dimensional streamlines disturbed by SV is determined by the velocity).

into large-scale vacuoles. At the impeller position $T + 3/4t^*$, the low-pressure region has spread to the impeller hub and occupies most of the area of the flow path. At this time, the scale of the low-pressure region has developed to the maximum. The large vortices cause the path to be completely blocked resulting in high turbulent dissipation. Then, as the low-pressure region propagates to the TE, the scale of the SV and the blockage of the flow path gradually weaken. At the impeller position $t = T + 5/4t^*$, the low-pressure region develops toward the outlet of the flow path and on the SS to form a small spherical SV core. Some of the streamlines around the SV bypass the LE of blade # 2, and the low momentum distorted fluid gradually interferes with the rim flow of the next path. This concludes the discussion of the intuitive process of generation, development, shedding, and dissipation of the SV in a single flow path of the impeller.

## 8.5 Inducements for the Circumferential Propagation of the Rotating Stall

As mentioned in Section 8.4, before a flow path of the impeller enters the stall stage, it must first transition from the stable flow state to the distorted flow state and enter the pre-stall stage. The pre-stall stage is a transition process between the steady flow and the stall flow, and the inducements for the flow to transition from the steady flow to the distorted flow and enter the pre-stall stage are discussed in this section. As shown in Fig. 8.33(a), at the impeller position $T_s + 3/8t^*$, the flow path A replaces the flow path D entering the stall stage, and slight distortions occur in the flow field at the outlet of the flow path B. With time, the degree of distortion gradually increases to form a vortex in the flow path B, which gradually enters the pre-stall stage and the inducement of the flow distortion in the formation of the vortex can be clearly seen. Figure 8.33(b) shows the backward flow in the distorted flow zone. From this figure, it can be noted that the backward flow in the

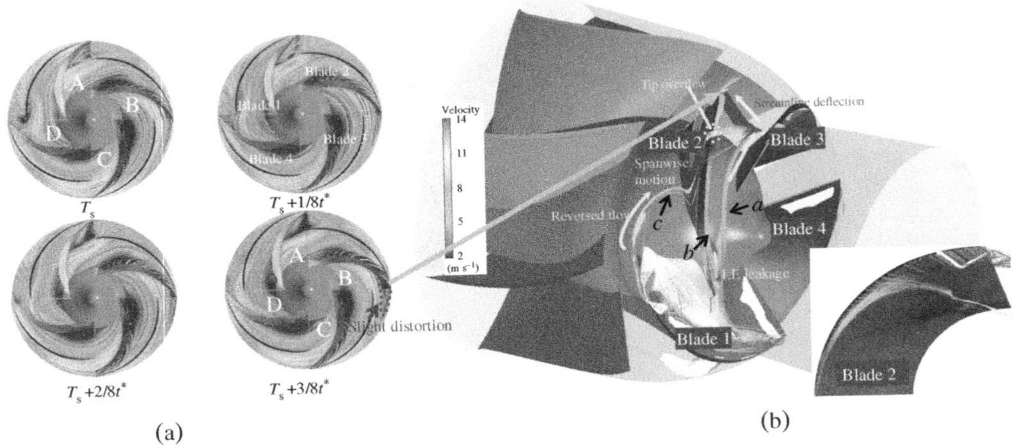


**Figure 8.33** Inducements of pre-stall stage. (a) Transient flow in a fixed section of the impeller (b) Inducements for the formation of distorted streamlines when T=$T_s$+3/8$t^*$.

distorted flow zone is composed of a LE leakage flow, the flow near the wall, and the reverse flow. First, the reverse flow passes through the low-pressure region occupied by the stall cell, deflecting the streamlines and affecting the hub area of blade #2 along the pressure gradient. Due to the transverse pressure gradient between the hub and the shroud, the fluid moves along the direction of the blade and then attaches to the tip of the blade. Then, the phenomenon of tip overflow occurs due to the pressure difference between the two sides of the blade and the streamlines deflect at 90° with high-speed flow in the flow path B. Finally, due to the entrainment effect of the leakage flow from blade #2, the streamlines deflect again at 90°, and the momentum of the fluid decreases to the lowest level, resulting in the formation of a low-speed distorted flow field. Since the flow path A is blocked by the SV, part of the fluid entrained by the SV is recirculated and flows through the LE of blade #2. Some of this fluid converges with the TLF in path B and flows downstream along the direction of the leakage flow. The other part of the fluid flows downstream and firmly adheres to the SS. The fluid attached close to the wall collides with the reverse flow at the TE, and due to the low momentum of these fluids, the two fluids converge due to the strong impact of the tip overflow. Then, the leakage flow also occurs and interferes with the former two. The deflection and convergence of the streamlines gradually form an extremely low-velocity zone at the TE, resulting in flow distortion. The three incoming flows in the distorted flow zone essentially belong to the rim flow, and the distortion in the rim flow is due to the blockage effect of the upstream SV in the flow path. The pre-stall stage caused by the distorted rim flow is a trigger sign that the flow path has entered the flow instability stage, which is a precursor for the flow path to enter the stall stage.

Similarly, the initial tip stall is the trigger sign for the flow path to enter the stall stage, and it is the transition moment for the flow to enter from the pre-stall stage to the critical stall stage. It can be seen from Figure 8.34 that when $t = T$, the low-pressure distortion zone appears at the LE, and the initial tip stall cell begins its development process in the flow path. However, the circumferential propagation process of the stall cells (the process of the SV collapse in the upstream flow path and the new stall cell initiation in the downstream path) still needs to be understood and explained next. Figure 8.34 describes the inducements for the formation of the rotor tip distortion zone when $t = T$. At this time, the stall cell of path A is located at the TE and is in the stage of imminent collapse, while the flow path B is in the flow instability stage (this is why the vortex region appears at the

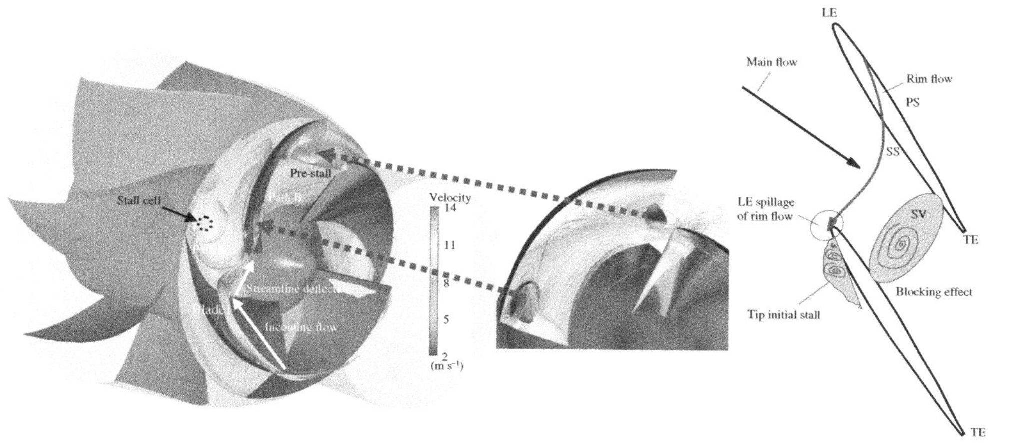


**Figure 8.34** Inducements of tip initial stall (the iso-surface is based on the turbulent kinetic energy, whose value is $5 m^2 s^{-2}$; 3D streamline is in the backward flow of leading-edge distortion zone).

TE at $t = T$), and there is a low-pressure distortion zone at the rotor tip of the flow path B. Most of the incoming flow in the LE distortion zone of path B comes from the TLF of blade #1. Due to the existence of SV, the velocity of TLF gradually decreases. There is a significant blockage effect in the path of the SV, and the flow streamlines deflect nearly 90°. The interference between the rim flow due to the deflected streamlines and the mainstream flow reduces the momentum of the rim flow. Low-energy fluid entering the downstream path tangentially leads to a sharp increase in the attack angle. Due to the severe interference between the upstream leakage flow and the mainstream flow downstream and the leakage flow, the distorted counter-current zone appears at the LE of the blade, which forms the initial tip stall cell. The interference effect between the mainstream flow and the SV gradually weakens in the upstream path due to the streamlines deflection of much of the fluid, thus the inlet attack angle decreases gradually and the stall effect decreases and disappears gradually. Like the inducements of pre-stall, flow distortion of rim flow induced by the stall vortices is the main inducement of the tip initial stall.

Pre-stall and initial tip stall induced by the distorted rim flow are important transitional stages in stall propagation; thus, the distorted rim flow contributes to the phenomenon of stall propagation. The distortion of rim flow is due to the blocking effect of SV, therefore, the fundamental reason for the circumferential propagation of rotating stall is that the existence of SV changes the trajectories of some of the fluids.

## 8.6 A Stall Prediction Model of the Mixed-flow Pump

### 8.6.1 Flow Characteristics During the Process of the Stall for Impellers with Different Blade Numbers

Figure 8.35 illustrates the external characteristic curves (head and efficiency) of the 3-blade, 4-blade, and 5-blade impellers in the mixed-flow pump. This figure shows that the head curve of the 3-blade impeller does not exhibit have a section of the positive slope. However, the head curves of the 4-blade and 5-blade impellers do show a section of positive slope, which occurs at $\Phi = 0.47$–$0.5$ and $\Phi = 0.59$–$0.61$, respectively, where $\Phi$ is the flow coefficient defined as the flow

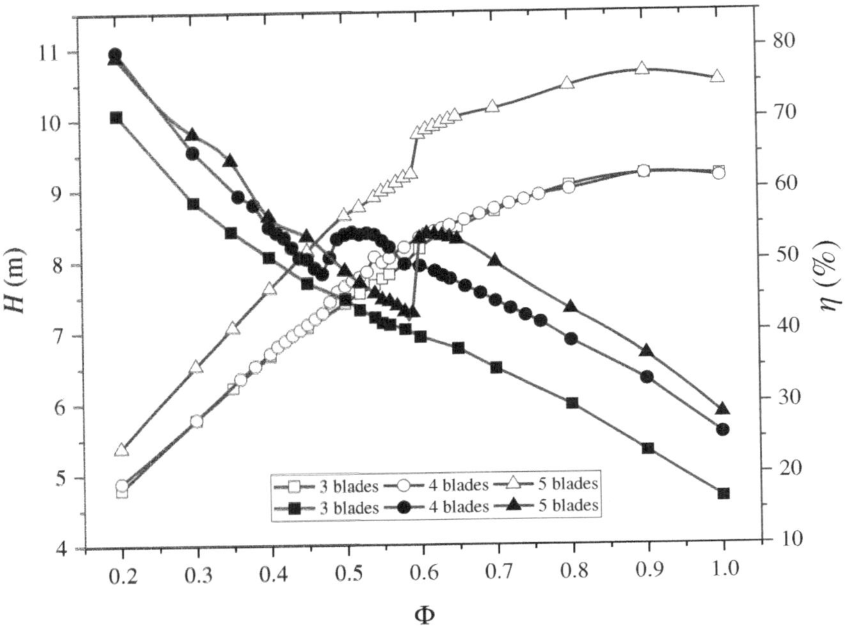


**Figure 8.35** Energy characteristic of mixed-flow pump with different number of blades.

rate normalized by the design flow rate. At $\Phi = 0.47$ and 0.59, the 4-blade and 5-blade impellers experience a deep stall, respectively. This figure indicates that the stall can occur in the 4-blade and 5-blade mixed-flow pumps, while the 3-bladed mixed-flow pump may not experience stall.

Figure 8.36 shows the velocity field in the axial cross-section of the impeller with different numbers of blades in the process of the pump stepping into stall. This figure shows the transition process of the distorted flow for the 4-blade and 5-blade impellers to form an SV in the rim region of the pump when it is stepping into the deep stall. On the other hand, the flow field of the 3-blade impeller, being closest to the pump shutdown dead center, exhibits only the flow distortion in the rim region without an SV. In the case of the 4-blade impeller, when the flow coefficient $\Phi$ is 0.49, only two flow channels out of the four exhibit the stall vortices. When $\Phi = 0.48$, the stall vortices are present in each channel. As $\Phi$ decreases to 0.47, the stall vortices in each channel become stronger. Furthermore, for the 5-blade impeller, no SV is observed when $\Phi = 0.6$ and 0.61; however, when $\Phi = 0.59$, stall vortices appear in all the five channels.

Figure 8.37 shows the velocity field on the Turbo Surface of the blade tip and on the suction surface of the blade. This figure shows that the 3-blade impeller experiences significant flow distortion in the tip region, with a small vortex structure appearing on the suction side of the inlet tip. However, there is no clear indication of an SV in the flow channel. On the other hand, the 4-blade and the 5-blade impellers exhibit an SV in the flow channel with variations in its size and shape due to the constraints of the width of the flow channel. On the suction surface of the blade, blade # 3 shows the secondary flow and the backflow from the blade root to the blade tip near the leading edge. This causes the fluid in this area to deviate from the main flow direction. The blade # 4 for the 4-blade impeller and blade # 5 for the 5-blade impeller display noticeable backflow and vortex structures on the upper right part of the blade, which corresponds to the SV in the tip area.

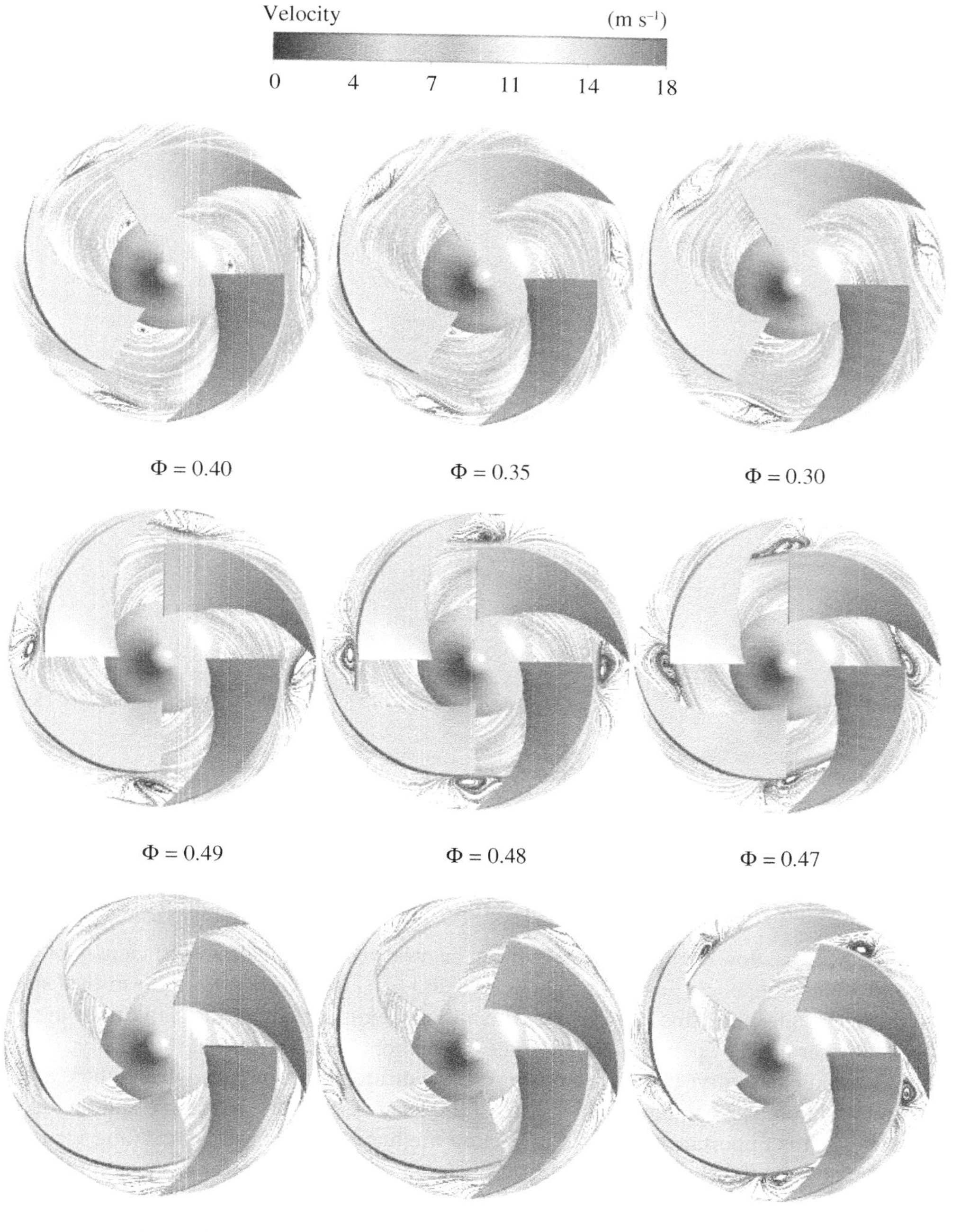


**Figure 8.36** The process of impeller entering stall with different blade numbers.

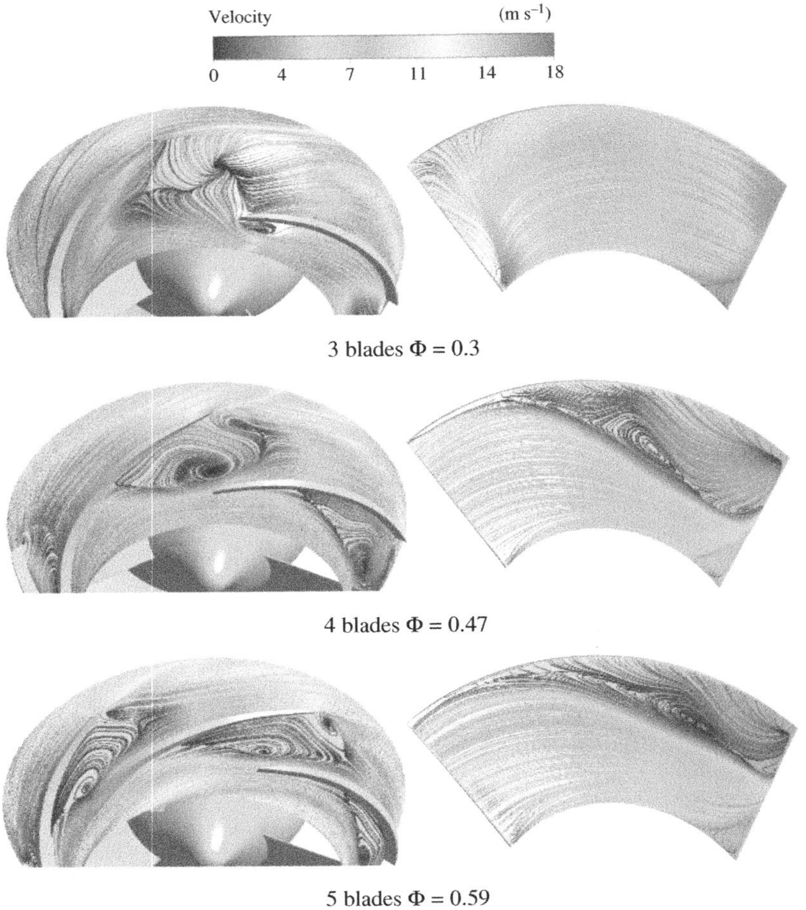


**Figure 8.37** Flow distribution of blade tip region surface and blade suction surface.

## 8.6.2 Quantitative Analysis of the Suction Surface Parameters of the Stalled Blades

Based on the head vs. flow coefficient curves, it can be observed that the flow coefficient $\Phi$ of the 4-blade and 5-blade impellers decreases only by 0.03 and 0.01, respectively, from the critical stall state to the deep stall state. This suggests that the initial operating point of the stall can be easily mistaken for other nearby operating points. Some researchers have argued that the positive slope of the head vs. flow rate curve is merely a macroscopic indicator of the pump stall, and there is no definitive correlation between the positive slope of the head vs. the flow rate curve and the deep stall. Therefore, it is important to explore prediction methods for the onset of stall and identify distinct features in the flow that represent the onset of stall. The understanding of the onset of stall phenomenon has practical significance in the development of mixed-flow pumps.

According to the discussion in the previous paragraph, the main cause for the generation of the SV can be attributed to the flow separation on the suction surface of the blade. The flow on the suction surface can be considered as the "projection" of the SV on the plane, which can be used to understand the characteristics of the SV. Therefore, the flow on the suction surface of the blade becomes an important area of investigation with the aim of finding an index that can accurately

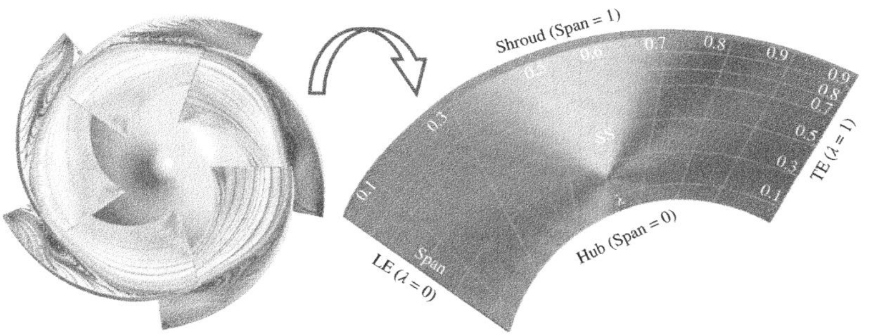


**Figure 8.38** Monitoring line for dividing different streamwise and spanwise position of suction surface.

define the onset of stall. To achieve this objective, the area on the suction surface of the blade is divided into several segments as shown in Fig. 8.38. The pressure load and the velocity distributions on these segments of the suction surface are selected for analysis. To highlight the influence of the SV, the deep stall condition point is selected. It turns out that in the deep stall condition, the pressure load and the velocity distributions on each suction surface of the impeller blades are nearly the same; thus, only a single blade suction surface for detailed analysis. In Fig. 8.38, the leading edge of the blade is defined as $\lambda = 0$, the TE as $\lambda = 1$, the blade root or hub span $= 0$, and the blade top or rim span $= 1$. It can be observed from this figure that the main body of the SV is in the flow channel near the area where the suction surface span $> 0.5$ and $\lambda > 0.5$. Therefore, the monitoring lines are densely set in this area. This phenomenon has also been noted by other researchers.

Figure 8.39 shows the pressure load distribution on the suction surface of the impeller with 4-blade and 5-blade at different positions. It is evident from this figure that in the deep stall condition, the pressure load distribution on the suction surface of both the 4-bladed and 5-bladed impellers exhibits a consistent trend in both the flow direction and the spanwise direction. Moving from the leading edge toward the TE of the blade, the pressure load on the suction side gradually increases. As the span value increases, it is expected that the pressure load should also increase gradually. However, it is observed that the pressure load at the span $= 0.5$ monitoring line on the suction surface of the 4-blade impeller is not only lower than the pressure load at span $= 0.1$ and $0.3$ after the 60% position in the flow direction but also experiences a certain decline after 85% position in the flow direction. Additionally, the pressure load at span $= 0.3$ monitoring line is also lower than the pressure load at span $= 0.1$ after the 60% position in the flow direction. The pressure load curve of the spanwise position monitoring line resembles the letter K. From the root of the blade to a blade height of approximately span $= 0.5$, the pressure load at different chord length positions on the suction surface either slightly increases or decreases or remains relatively unchanged. However, from span $= 0.5$ to the blade tip, the pressure load at different chord lengths exhibits an overall increasing trend. Moreover, the larger the value of $\lambda$, the faster the pressure rises and the higher the pressure load.

Similarly, the pressure load at the monitoring point on the suction surface of the 5-blade impeller exhibits a trend like that of the 4-blade impeller. However, the pressure load on the suction surface of the 5-blade impeller is negative. Additionally, the load curve of the 5-blade impeller changes more uniformly with variations in both the flow direction and the spanwise direction. This indicates that the 5-blade impeller imposes stronger constraints on the fluid force.

Figure 8.40 shows the velocity distributions on the monitoring lines (in the flow direction $\lambda$ for various span locations and in the spanwise direction for various values of $\lambda$) for the 4-blade and

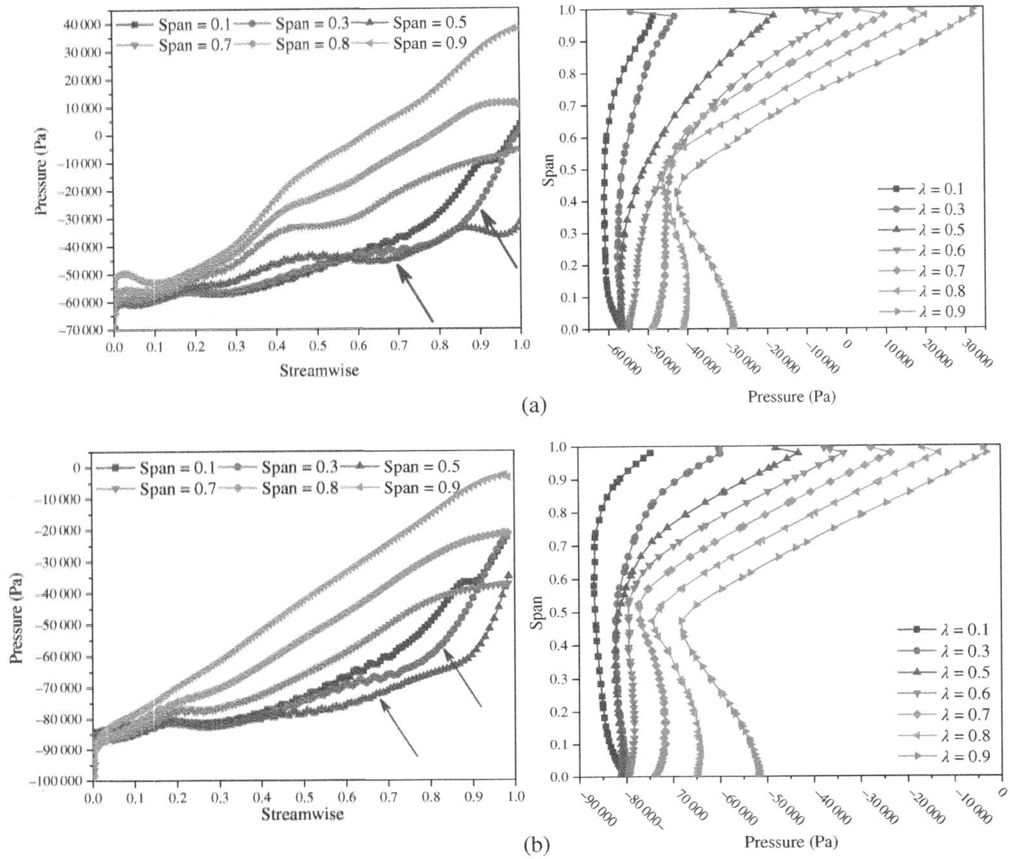


**Figure 8.39** Distribution of pressure load on blade suction surface. (a) 4-blade impeller, $\Phi = 0.47$, (b) 5-blade impeller, $\Phi = 0.59$.

5-blade impeller. This figure indicates that in the deep stall condition, the velocity distributions on the suction surface of the 4-blade and 5-blade impellers are quite similar in both the flow direction and the spanwise direction. At the leading edge of the blade, the velocity of the monitoring point along the flow direction $\lambda$ at span = 0.9 is significantly lower than the normal value, like the velocity of the monitoring point near the hub area. This is the location where the nucleus of the SV is formed. Moving from the leading edge toward the TE of the blade, the monitored velocity near the hub region initially increases, then remains relatively constant, and finally decreases. This variation in the velocity suggests that the SV has minimal influence on the flow velocity near the hub region. However, the flow velocity monitored near span = 0.5 slightly increases from the vicinity of the leading edge and then sharply decreases in the flow direction, ranging from 30% to 80% along the $\lambda$ direction. The position where the velocity drops to the lowest point along the $\lambda$ direction occurs closer to the larger values of the span. It can be noted that for the same span value, the lowest point in the velocity drop occurs closer to the leading edge for the 5-blade impeller compared to the 4-blade impeller. The lowest velocity points for both impellers are less than 1 m/s, with the minimum velocity of the 5-blade impeller approaching 0. However, the velocity on the monitoring line from LE ($\lambda = 0$) to TE ($\lambda = 1$) at span = 0.5 for a 5-blade impeller shows only a 50% drop compared to that for the 4-blade impeller. Near the blade TE, the velocity on the monitoring line for

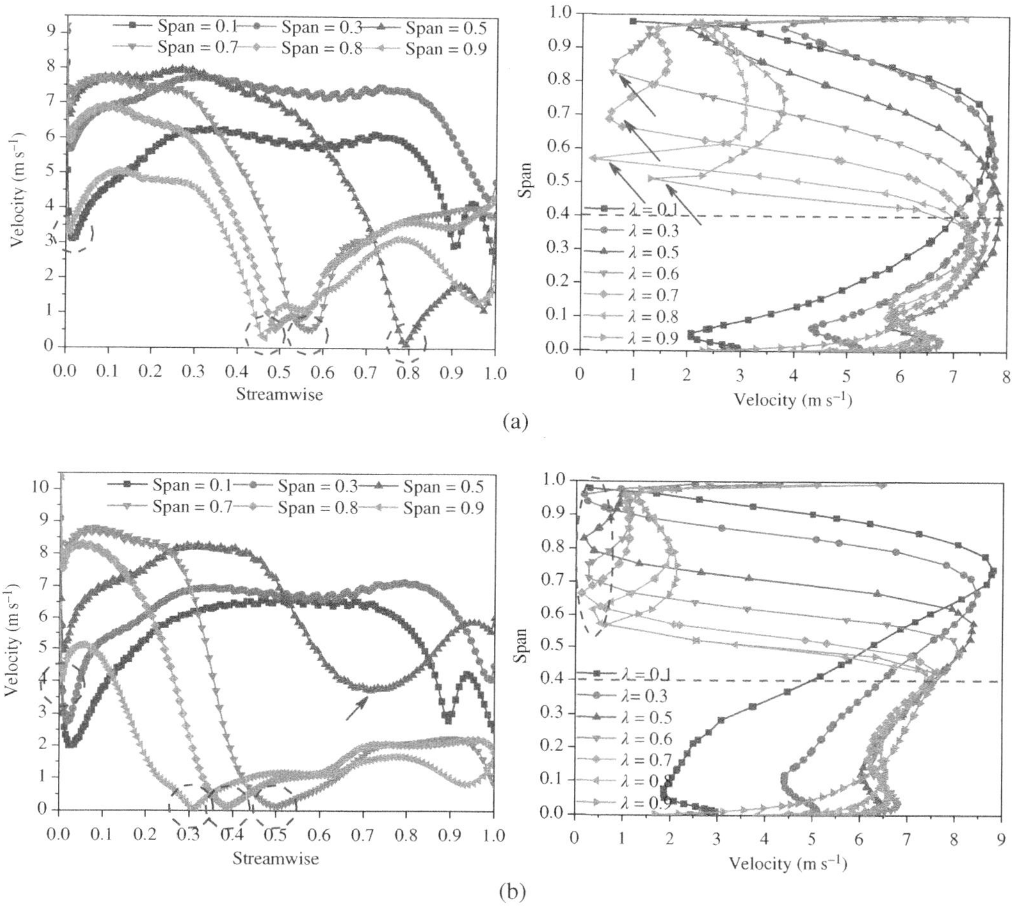


**Figure 8.40** Velocity distribution of blade suction surface. (a) 4-blade impeller, $\Phi = 0.47$, (b) 5-blade impeller, $\Phi = 0.59$.

the 5-blade impeller is generally lower than that for the 4-blade impeller. The velocity distribution curve in the spanwise direction on the monitoring line resembles the letter S. From the blade root to approximately 40% spanwise position, the velocity on the monitoring line initially decreases and then increases. From the 40% spanwise position to the blade tip, the velocity on the monitoring line shows a sharp drop. The position where the velocity drops to a minimum is farther away from the tip of the blade for larger values of $\lambda$. The minimum velocity of the 5-blade impeller is generally close to 0, which is lower than that of the 4-blade impeller, and its minimum position is closer to the tip of the blade. This observation indicates that the SV has a significant influence on the velocity with greater impact closer to the blade tip. Additionally, it suggests that the severity of the stall increases the influence of the SV on the velocity again with greater impact toward the tip of the blade.

The statement in the above paragraph not only supports the previous findings in the literature but also enhances the conclusions drawn from the data analysis presented here. It is evident that the pressure loads and the velocity distributions on the suction surface of the blade provide a more accurate and detailed representation of the SV. This information serves as a valuable reference for identifying parameters that can effectively predict the onset of stall.

### 8.6.3 The Pressure and Velocity Distributions on the Suction Side of the Blade of the Impellers with Various Blade Numbers

In Section 8.6.2 focused on the role played by the pressure and velocity distributions on the suction surface of the blade at the flow rate that results in the deep stall. However, to identify the stall inflection point, it is necessary to compare the role of these two flow variables (pressure and velocity) in other flow conditions. It can be observed from Fig. 8.40 that the velocity distribution curves along the streamwise flow direction $\lambda$ at various span locations show large variation in the stall conditions of 4-blade ($\Phi = 0.47$) and 5-blade ($\Phi = 0.59$) impeller compared to the pressure load distributions shown in Fig. 8.39 making them more of a discriminating factor for the analysis of the stall process. Additionally, the velocity difference among various streamwise positions along the $\lambda$ direction is much larger compared to that in the spanwise direction again making it a better discriminating factor for analyzing the impact of velocity distribution in the stall process. To further enhance the differences in the velocity distributions at various flow rates, they are selected on a monitoring line at span = 0.85 in the leaf top area of the blades for the 3-blade, 4-blade, and 5-blade impellers for further analysis as described below.

Figure 8.41 shows the average velocity distribution along the monitoring lines on the suction surface of the blades for impellers with 3-, 4-, and 5-blades for various flow conditions. This figure shows that the velocity distributions on the monitoring lines on the suction surface of the 3-blade impeller exhibit a consistent trend across the various flow conditions. Specifically, the velocity increases rapidly from the leading edge to the TE of the blade and then gradually decreases. Overall, as expected, the higher flow rates result in greater velocities. However, it is worth noting that the velocity value for $\Phi = 0.6$ is noticeably lower, indicating larger flow instability in the pump. In spite of this instability in the flow, the 3-blade impeller did not experience stall.

The velocity distributions on the monitoring lines on the suction sides of the blades for the 4-blade and 5-blade impellers vary significantly in various flow conditions. When operating without stall, the average velocity distributions in the streamwise flow direction for both the 4-blade and the 5-blade impeller behave like that for the 3-blade impeller, forming a "convex" curve. However, when entering a deep stall state, the average velocity distributions change to a "concave" curve. The velocity drops sharply from the TE to about 30% of its value at the TE in the flow direction and after reaching its lowest point, there is no significant increase in the velocity until the trailing edge. In the critical stall condition, the velocities near the blade TE of the 4-blade impeller at $\Phi = 0.5$ and the 5-blade impeller at $\Phi = 0.6$ are noticeably low, which distinguishes them from other operating conditions. In the deep stall condition, the velocity distributions change from "convex" to "concave" at $\Phi = 0.47$ for the 4-blade impeller and at $\Phi = 0.59$ for the 5-blade impeller. This indicates that the average velocity in the blade tip area is highly sensitive to the changes in the flow conditions, making it a valuable indicator for predicting the unstable flow in the pump leading to stall.

### 8.6.4 A Simple Stall Prediction Model

It can be seen from the above discussion that the 3-blade impeller of the mixed-flow pump will not stall, but the 4-blade and 5-blade impellers of the mixed-flow pump will stall. Now, following the ideas proposed by Aoki et al. [9], the Taylor coefficient $C_{v\_Taylor}$ can be defined as the physical quantity to measure the state of the flow in the pump and predict the stall; it is given by:

$$C_{v\_Taylor} = 6.5 - Z - \left(\frac{U_2}{U_1}\right)^2 \tag{8.14}$$

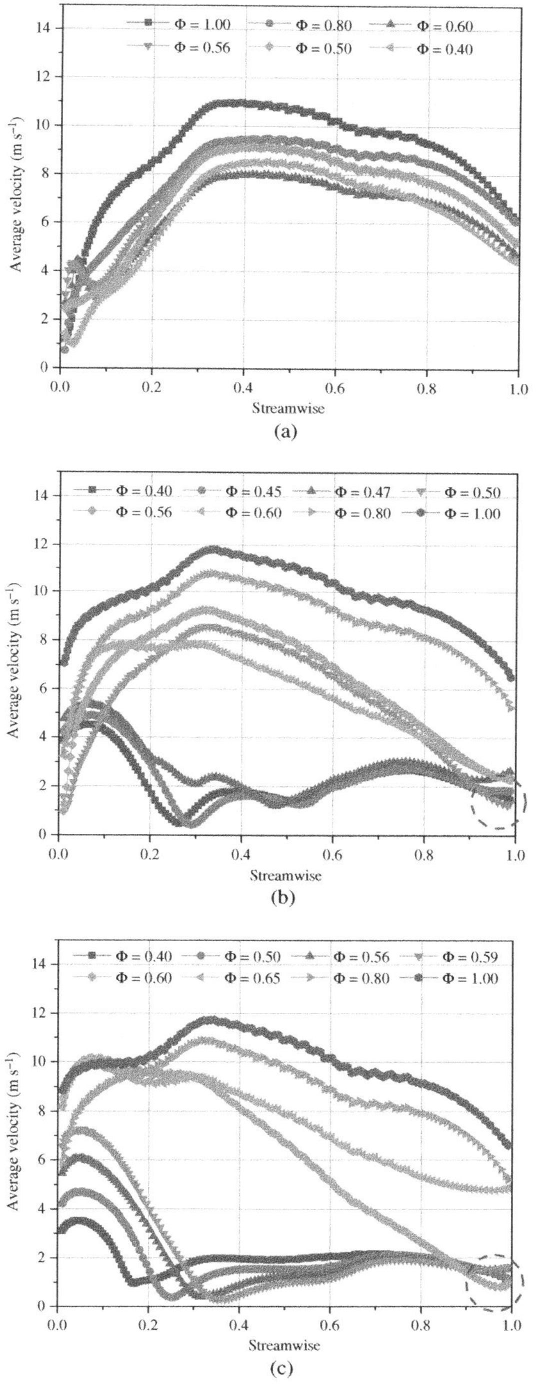


**Figure 8.41** Velocity distribution of blade suction surface monitoring line under different flowrate. (a) 3-blade impeller, (b) 4-blade impeller, (c) 5-blade impeller.

**Figure 8.42** Area of impeller inlet and outlet span≥0.85.

where $U_1$ and $U_2$ are the average flow speeds at the impeller inlet and outlet, respectively, in m/s, and $Z$ is the number of impeller blades.

The definition of $C_{v\_Taylor}$ in Eq. (8.14) includes the average velocity across the impeller inlet and outlet. However, the analysis presented in Section 8.6.3 indicates that the blade tip region has a significant impact on the critical value of stall. In the blade tip region, the relative velocity difference between the impeller inlet and the outlet is more pronounced. Therefore, both $U_1$ and $U_2$ in Eq. (8.14) are adjusted to the average velocity in the blade tip region, covering 85–100% of the blade span. Figure 8.42 shows the velocity distribution in this specific region, both at the impeller inlet and outlet.

Figure 8.43 shows the variations in $C_{v\_Taylor}$ curves with the flow coefficient in various flow conditions. It is evident from this figure that the $C_{v\_Taylor}$ values of the impellers with 3-blades, 4-blades,

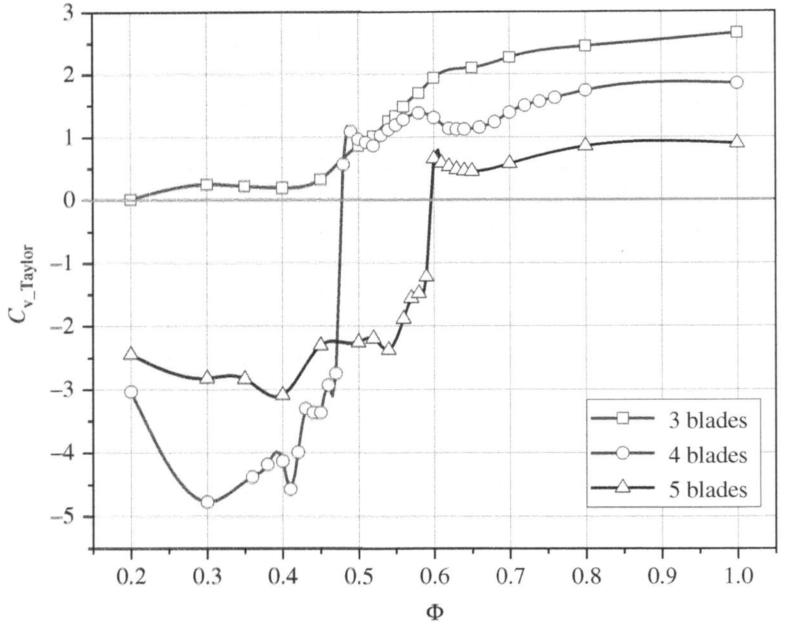


**Figure 8.43** $C_{v\_Taylor}$ curve under different flowrate.

and 5-blades all decrease as the flow rate decreases. In the operating condition where $\Phi > 0.6$, the $C_{v\_Taylor}$ value remains greater than 0 and decreases gradually. However, in the operating condition when $\Phi < 0.6$, the $C_{v\_Taylor}$ value and its downward trends for higher values of $\Phi$ differ significantly for the impellers with 3-blades , 4-blades , and 5-blades. The $C_{v\_Taylor}$ value for the 3-blade impeller consistently remains greater than 0 and exhibits a relatively moderate decline. On the other hand, the $C_{v\_Taylor}$ value of the 4-blade impeller fluctuates slightly in $\Phi = 0.49$–$0.6$ range and drops sharply in $\Phi = 0.47$–$0.49$ range At $\Phi = 0.47$, it becomes negative for the first time and fluctuates with varying amplitudes when $\Phi < 0.47$. Similarly, the $C_{v\_Taylor}$ value for the 5-blade impeller experiences a steep decline between $\Phi = 0.58$ and 0.6 and becomes negative for the first time at $\Phi = 0.59$. It shows a small fluctuation at $\Phi = 0.58$. Clearly, in the range of flow conditions where the 4-blade and 5-blade impellers transition from the critical stall to the deep stall, the $C_{v\_Taylor}$ value exhibits a cliff-like decline and becomes negative for the first time in the deep stall condition. In contrast, the $C_{v\_Taylor}$ value of the 3-blade impeller, without experiencing stall, does not demonstrate this characteristic and remains greater than 0. This indicates that the $C_{v\_Taylor}$ value can be used as reasonably accurate index to determine the occurrence of stall as well as the stall states with distinct features.

To parameterize the characteristics of this index, a physical quantity $k_T$ is defined which measures the change in the Taylor coefficient $C_{v\_Taylor}$ with respect to the flow coefficient $\Phi$. If the flow coefficient $\Phi$ is considered at many values below and above the design value, the $C_{v\_Taylor} - \Phi$ curve is a continuous smooth curve, which can be fitted into a function $C_{v\_Taylor}(\Phi)$, and its first derivative can be calculated. At the discontinuous flow point, the expression of $k_T$ can be written as:

$$k_T = \frac{dC_{v\_Taylor}}{d\Phi} \approx \frac{\Delta C_{v\_Taylor}}{\Delta \Phi} \tag{8.15}$$

Figure 8.44 shows the relationship between $k_T$ and the flow coefficient $\Phi$. It is evident that the $k_T$ curve for the 3-blade impeller, which does not experience stall, remains relatively flat. However,

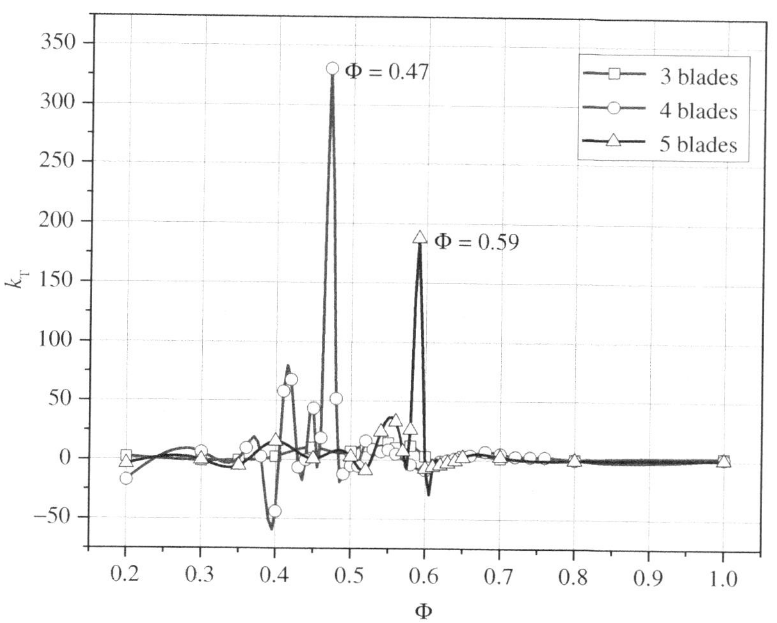


**Figure 8.44** $k_T$ curve under different flowrate.

the $k_T$ curves for the 4-blade and 5-blade impellers exhibit significant fluctuations, indicating the occurrence of stall. At $\Phi = 0.47$ (DSP for 4-blade impeller) and $\Phi = 0.59$ (DSP for 5-blade impeller), there is a noticeable "spike" in the $k_T$ curve. Additionally, this "peak" is more pronounced in the $k_T$ curve of the 4-blade impeller compared to the 5-blade impeller. This suggests that the relative velocity decline rate in the inlet and the outlet tip areas of the 4-blade impeller is greater than that for the 5-blade impeller. Consequently, the flow channel in the tip area of the 4-blade impeller is more prone to stall conditions compared to the 5-blade impeller.

In summary, the $k_T$ model, which is based on the rate of change in the relative velocity in the inlet and outlet of the impeller tip area, shows promise in predicting the occurrence of stall and its corresponding operating point in the mixed-flow pumps. The model is very simple yet quite accurate and exhibits clear characteristic curves exhibiting region of stall for pumps with different numbers of blades. However, the universality of this model cannot be confirmed at this stage due to the limited number of test cases available for verification. In this chapter, only the mixed-flow pump model considered in this book has been validated, and the number of selected operating points if the pump is also limited. The lack of sufficient number of test cases makes it difficult to determine the critical value of $k_T$.

## 8.7 Summary

The important conclusions of this chapter are summarized below:

(1) In the stall conditions, there are swirls and vortices in the inlet section of the impeller, and they are concentrated in the upper part of the blade, which not only affects the attack angle but also increases the energy loss in the inlet section and worsens the working state of the mixed-flow pump. In the deep stall condition, nearly 50% of the flow passage of the impeller is blocked by the SV, and the effective work section is further reduced. The streamlines at the LE of the SV reach the leading edge of the next blade and overflow, which has major impact on the head drop of the mixed-flow pump. When the mixed-flow pump stalls, the backflow blocks the outlet passage of the impeller resulting in the weakening of the flow capacity of the impeller passage and further increase in the energy loss. The combined effect of the TLF, SV, and backflow makes the effective flow area of the impeller outlet decrease continuously, which leads to the head drop in the saddle area of the *H*–*Q* curve, where the effect of the SV dominates, followed by that of the backflow.

(2) At the design flow rate, the TLV is stable at different tip clearances and appears in a "snail shell" shape. From the inlet to the outlet of the impeller, the core of the TLV moves away from the blade suction surface and the chamber end wall. In the rotating stall condition, the flow field in the impeller is most unstable from the middle of the blade to the outlet of the blade which affects the TLF. The TLF on the blade front is not included in the mainstream flow and forms a "flat" TLV structure. In the critical stall condition, the TLV track is approximately parallel to the circumferential direction, part of the leakage flow flows out of the leading edge of the lower blade, and the downstream leakage vortex strip crosses the rim region of the whole impeller passage, resulting in the turbulent energy dissipation in the rim region.

(3) Pre-stall is the trigger sign that indicates the flow path has entered the flow instability stage, and it is a precursor to the flow path entering the stall stage. The initial tip stall is the trigger sign that indicates that the flow path has entered the stall stage and it is in a transitional process from the instability stage to the stall stage. The circumferential propagation characteristics of the

rotating stall can be described as follows: The stable flow path is first induced by the distorted rim flow, and then it enters the pre-stall stage. Then, the initial stall cell is generated at the LE of the rotor tip and propagates along the direction of the wingspan to the TE, experiencing the stages of shedding, development, and collapse. While the stall cell collapses at the TE, another initial stall cell is generated at the LE of the downstream path. The fundamental reason for the circumferential propagation of the rotating stall is that the existence of SV changes the trajectory of some of the fluid, which is mainly reflected in the distortion of the rim flow. The distorted rim flow is one of the main inducements for pre-stall and the initial tip stall.

## References


1 Goltz, I., Kosyna, G., Stark, U. et al. (2003). Stall inception phenomena in a single-stage axial-flow pump. *Proceedings of the Institution of Mechanical Engineers, Part A: Journal of Power and Energy* 217 (4): 471–479.

2 Duc Vo, H., Tan, C.S., and Greitzer, E.M. (2008). Criteria for spike initiated rotating stall. *Journal of Turbomaehinery* 130 (1): 011023–011029.

3 Ji, L., Li, W., Shi, W. et al. (2020). Experimental study of pressure pulsation in a mixed-flow pump with different tip clearances based on wavelet analysis. *Shock and Vibration* 2020: 1–15.

4 Wei, L., Enda, L., Leilei, J. et al. (2020). Mechanism and propagation characteristics of rotating stall in a mixed-flow pump. *Renewable Energy* 153: 74–92.

5 Emmons, W., Pearson, C.E., and Grant, H.P. (1955). Compressor surge and stall propagation. *Transactions of the ASME* 127 (2): 455–469.

6 Ren, X., Fan, H., Xie, Z. et al. (2019). Stationary stall phenomenon and pressure fluctuation in a centrifugal pump at partial load condition. *Heat and Mass Transfer* 55.

7 Yin, H. and Jin, Q.Y. (2018). Exploration of three-dimensional unsteady model for compressor post-stall under inlet distortion. *Journal of Aerospace Power* 33 (6): 1393–1402.

8 Tan, L., Xie, Z., Liu, Y. et al. (2018). Influence of T-shape tip clearance on performance of a mixed-flow pump. *Proceedings of the Institution of Mechanical Engineers, Part A: Journal of Power and Energy* 32: 386–396.

9 Aoki, M. and Goto, A. (1996). positively sloped head-flow characteristics and inlet recirculation of a mixed-flow pump. *Transactions of the JSME* 62 (596): 162–167.

# 9

# Passive Suppression of Rotating Stall in Mixed-flow Pump

## 9.1 Introduction

As an archetypal vane-type rotary apparatus, the mixed-flow pump showcases a plethora of merits including its capacity for achieving high flow rates and commendable efficiency, rendering it ubiquitously applicable across a multifarious spectrum of engineering domains [1, 2]. In scenarios characterized by low-flow rates, the mixed-flow pump is markedly susceptible to experiencing rotational stall, which not only culminates in a discernible attenuation of performance but also harbors the latent potential to incite fluid-induced vibrations, thereby proffering a constellation of safety hazards to the associated equipment. Rotational stall, in this context, pertains to the phenomenon where, as flow rates decrease, the angle of attack at the leading edge of the impeller blades gradually increases. Consequently, it results in the separation of water flow within the inter-blade flow passages. Over the course of this evolution, the region of flow separation metamorphoses into a convoluted assemblage known as a stall cluster, thereby imperiously occluding the flow passages and propagating at a specific velocity between various flow passages within the pump, ultimately disrupting the operational stability of the pump [3]. This phenomenon routinely afflicts the impeller-type rotating machines that serve as conduits for both gaseous and liquid fluid mediums. In epochs antecedent to the contemporary era, investigations into the intricacies of internal rotational stall in rotating machinery predominantly hinged upon experimental methodologies. However, in tandem with the relentless march of progress in the realm of computational fluid dynamics, numerical simulations have progressively ascended to pre-eminence as the prevailing modality for delving into the abstruse mechanisms that govern the genesis of internal stall due to its low cost and high efficiency [4, 5].

Some scholars have studied a variety of suppression methods in order to reduce the effects of stall. These stall control methods can be divided into two categories, the active suppression of stall and the passive suppression of stall. Active suppression of stall uses external energy input to control the flow by adding an appropriate disturbance to the flow field and coupling it with the internal flow regimes. Passive suppression does not need external energy input and only depends on fixed devices to change the energy distribution of the fluid to control the flow. In terms of active suppression, Tong [6] proposed a micro-jet to control the flow field of the unsteady tip leakage vortex (TLV) so as to suppress the rotating stall and improve the stability of the pump. Goto [7] installed a jet device at the inlet of the impeller of the mixed-flow pump to change the secondary flow inside the pump. The instability and surge of the mixed-flow pump were effectively suppressed. Shi and Li [8] used the method of reverse "forced pre-rotation" to improve the reflux at the inlet section of the centrifugal pump. Zhang et al. [9] weakened the reflux intensity at the inlet of the centrifugal

*Mixed-flow Pumps: Modeling, Simulation, and Measurements*, First Edition.
Wei Li, Leilei Ji, Ramesh Agarwal, Weidong Shi, and Ling Zhou.

pump in small flow rate conditions and improved the unsteady flow in the pump by filling the impeller inlet with high-pressure water.

However, since the cost of active suppression is much higher than that of passive suppression and the improvement in the efficiency of the pump is not obvious, currently passive suppression is widely used in the rotating machinery. Krishnababu et al. [10] compared the tip leakage flow (TLF) and heat transfer characteristics of the turbomachinery using the flat tip structure and the two-side squealer and one-side squealer tip structure and found that the cavity structure of the tip can weaken the TLF and the average heat transfer to the rim area. Ameri and Bunker et al. [11] confirmed that the tip cavity structure can not only reduce the TLF by 14% but also improve the corner separation of the rim through the fillet structure by numerical calculation. Tallman and Lakshminarayana [12] used the structure of the blade LE chamfer, TE chamfer, and full rim chamfer to study the flow field in the rim area of the turbomachinery and found that the TE chamfer is better than the LE chamfer in weakening the strength of the TLV. Saha et al. [13] studied the flow field in the rim clearance with winglets at the tip. By adding winglets under the flat tip structure and squealer tip structure, it was found that the heat transfer coefficient under the flat tip structure decreased by 30%, and the effect of winglets under the squealer structure was insufficient. Harvey and Ramsden [14] added a large-size winglet structure on both sides of the blade tip and found that the efficiency of the 2% blade height in the turbine rim area increased by 1.2–1.8%. Dey and Camci [15] also studied the sensitivity of tip winglet width and cross-section shape in the rim flow field of turbomachinery and found that the winglet on the suction side has little effect on the tip leakage, while the winglet on the pressure side has a greater effect on TLF.

In addition, there have been suggestions to incorporate spokes parallel to the pressure gradient, referred to as "J-Groove" in fluid machinery to suppress various abnormal flow phenomena. Saha et al. [16, 17] initially installed the "J-Groove" upstream of the impeller leading edge on the casing wall of a mixed-flow pump. They found that this approach completely mitigated the instability in the performance curve, allowing the pump to operate stably across the entire flow range. Subsequently, they also employed the "J-Groove" on either the casing wall or the diffuser wall and demonstrated through theoretical analysis and experimentation that this method effectively suppressed the rotational stall within the impeller diffusers. Shabbir and Adamczyk [18] investigated the flow mechanisms of circumferential casing grooves in an axial compressor to enhance the stall margin. Choi et al. [19], by installing the "J-Groove" on the casing wall near the blade tip, observed that it almost completely suppressed the occurrence of rotating backflow cavitation at low-flow rates and cavitation surge near the best efficiency point. Shimiya and Choi et al. [20], by introducing the "J-Groove" on the pipe wall, controlled the leakage flow through the blade tip clearance, thus ameliorating the flow instability. Kurokawa et al. [21] investigated the suppressive effect of the "J-Groove" on swirling flow, and the results indicated that the installation of the "J-Groove" significantly reduced the intensity of swirl within the diffuser. Moreover, this configuration also reduced the pressure pulsation amplitude caused by the rotation of the vortex core near the inlet of the diffuser. Furthermore, Kurokawa [22] suggested that the placement of the "J-Groove" near the impeller inlet effectively suppressed the surge without compromising the maximum efficiency of mixed-flow and axial pumps. Lee et al. [23] conducted a study within the stator blades of a power generation turbine to investigate the influence of the depth of the blade tip recesses on the flow structure near the blade tips. They found that regardless of the depth of the recesses, the leakage flow region was narrower compared to a flat blade tip. Du et al. [24] investigated the influence of a single-groove structure on the stability of the pump and the flow in the blade tip clearance. They found that the modified single-groove structure improved the long-scale disturbances in the stall flow field near the suction surface. Lu et al. [25], using a modified axial transonic compressor rotor as their subject of study, confirmed that introducing grooves on the casing wall could weaken the intensity of blade TLV under near-stall conditions, consequently enhancing the overall performance of the compressor. Yang et al. [26] introduced axial grooves on the wall surface in the inlet section of an axial-flow pump to enhance its head-flow ($H$–$Q$) curve. The results indicated that at low-flow operating conditions, these axial grooves effectively reduced the inlet swirl and the blade leading-edge angle, thereby effectively eliminating the hump phenomenon in the axial-flow pump's head-flow curve. Chen et al. [27] investigated the influence of U-shaped and S-shaped grooves on stall margin and unit efficiency. They found that U-shaped grooves improved the maximum stall margin by 60% but incurred a 2.0% loss at the best efficiency point. S-shaped grooves enhanced the stall margin by 36% without affecting the best efficiency point. Wu et al. [28] investigated the influence of multi-location circumferential single-groove casings on blade tip flow stability and their control mechanisms. The research revealed varying degrees of reduction in blade TLF, flow losses, and blade tip flow blockage, resulting in enhanced flow stability. Luo et al. [29] were the first to apply the J-groove structure to tailwater tube optimization. They employed entropy theory to determine the specific locations of energy losses in the impeller and diffuser sections, concluding that "J-Groove" plays a beneficial role in suppressing the stall vortices and reducing the energy dissipation.

This chapter explores the more economical passive suppression methods to improve the performance of a mixed-flow pump operating in the stall range conditions, and to delay or suppress the occurrence of stall. By constructing different rim geometric structures and a pioneering circumferential spoke structure, termed "O-Spoke," with the help of computational fluid dynamics (CFD), their performance ability in improving the energy characteristics of the mixed-flow pump near-stall condition is comprehensively compared.

## 9.2 Impeller Blade Rim Structures for Stall Suppression

### 9.2.1 Geometries of Various Blade Rim Structures

In this section, the rim structure of the original impeller with a tip size of 0.5 mm is modified, and the geometries of four different modified rim structures are presented as shown in Fig. 9.1. For the convenience of description, four new rim structures are named as Groove_Tip, Bulge_Tip, Double_Rib_Tip, and Single_Rib_Tip. Among them, the Groove_Tip is equipped with a 2 mm

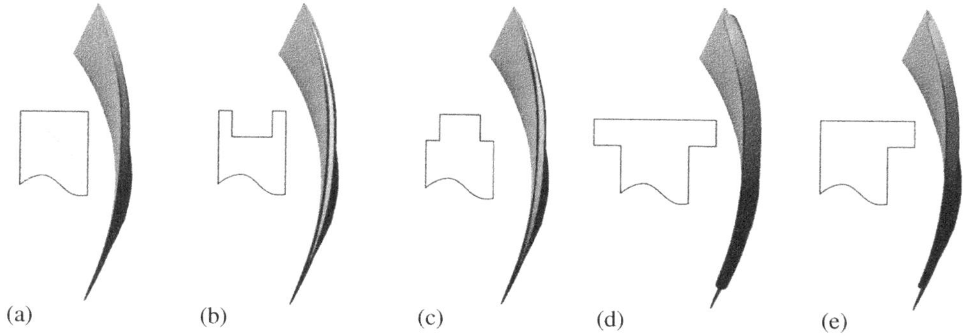


**Figure 9.1** Various types of blade rim structures of the impeller blade. (a) Original_Tip, (b) Groove_Tip, (c) Bulge_Tip, (d) Double_Rib_Tip, and (e) Single_Rib_Tip.

depth groove located in the middle of the blade rim area, and the distances between the two sides of the groove away from the pressure surface and the suction surface of the blade are 1 mm. The two sides of the groove near the blade inlet and blade outlet are connected by the round corners. In case of Bulge_Tip, 2 mm deep and 1 mm wide areas are cut-off on both sides of the pressure surface and suction surface of the blade rim, so that a 2 mm high and 3 mm thick bulge is formed in the middle of the blade rim. In case of Double_Rib_Tip, a winglet structure of 2 mm width and 2 mm thickness is added on both sides of the pressure surface and suction surface of the blade rim. The Single_Rib_Tip represents the addition of 2 mm winglets in width and thickness on the suction side of the blade rim.

## 9.2.2 Results of Numerical Calculations for Various Blade Rim Structures

### 9.2.2.1 Comparison of the External Characteristics of the Pump for Various Blade Rim Structures

The head and efficiency of the mixed-flow pump with four different rim clearance structures described in Section 9.2.1 are compared with the original experimental data with original blade rim as shown in Figs. 9.2 and 9.3. In these figures, the operating point corresponding to the first decline in the head curve within the stall interval is called the "critical stall operating point," while the operating point corresponding to the final recovery of the head curve is called the "deep stall operating point." Table 9.1 shows these two operating points of the "saddle area" for different blade rim structures.

It can be seen from these figures that for the design flow rate, the head of the mixed-flow pump with Groove_Tip structure is about 1.8% higher than that of the original impeller, and the head of the mixed-flow pump with Single_Rib_Tip structure remains basically the same as that of the original impeller, while the head of the pump with Bulge_Tip and Double_Rib_

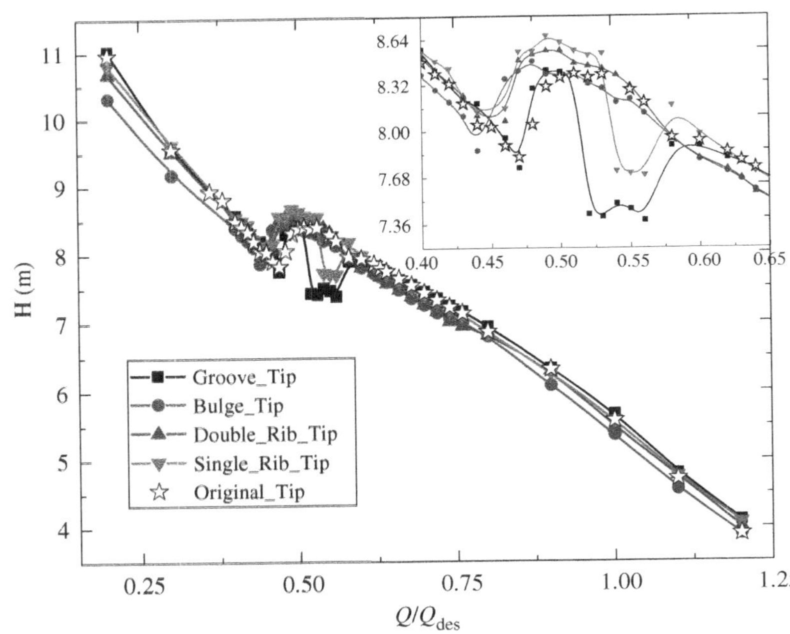


**Figure 9.2** Comparison of the head curves for various blade rim structures.

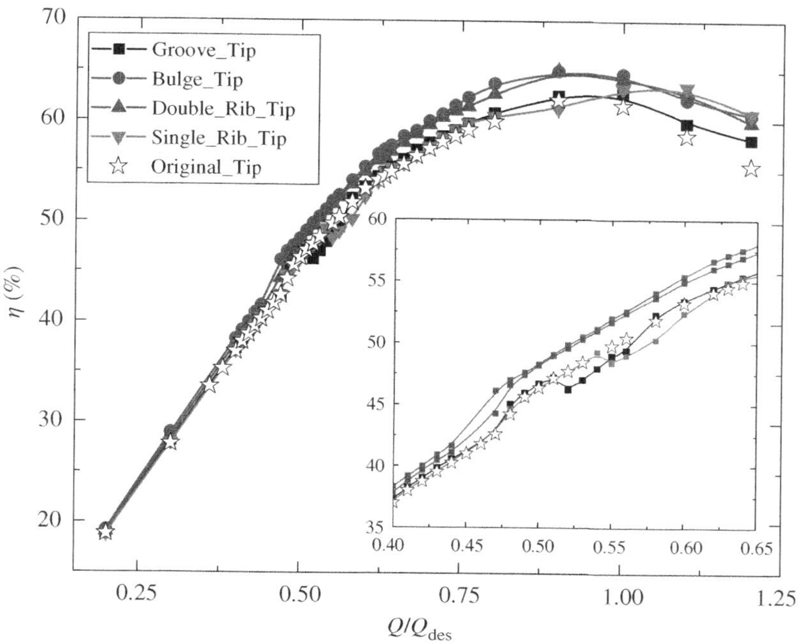


**Figure 9.3** Comparison of the efficiency curves for various blade rim structures.

**Table 9.1** The flow rates under critical and deep stall for different blade tip geometries.

| Blade rim structure | Critical stall condition | Deep stall condition |
|---|---|---|
| Original_Tip | $0.51Q_{des}$ | $0.47Q_{des}$ |
| Groove_Tip | $0.58Q_{des}$ | $0.47Q_{des}$ |
| Bulge_Tip | $0.48Q_{des}$ | $0.44Q_{des}$ |
| Double_Rib_Tip | $0.49Q_{des}$ | $0.45Q_{des}$ |
| Single_Rib_Tip | $0.58Q_{des}$ | $0.44Q_{des}$ |

Tip structure is 2.0% and 3.8% lower than that of the original impeller, respectively. Compared to the original impeller, the efficiency of the mixed-flow pump with all four different blade rim structures has increased, and the order from largest to smallest increase is: Bulge_Tip > Double_Rib_Tip > Single_Rib_Tip > Groove_Tip increasing by 5.2%, 4.3%, 2.8%, and 2.1%, respectively. Under the stall conditions, the head and the efficiency curves of the four different blade rim geometries are quite different from the original curves. There is an obvious phenomenon of secondary stall in both the Groove_Tip structure and the Single_Rib_Tip structure, that is, in the stall interval, there are two sharp reduction segments in the head curve, and the downward trend of the efficiency curve in the corresponding flow segment is also significantly accelerated. Under the other two blade rim structures, the positive slope characteristic of the head curve only appears once in the stall region, while the pump efficiency under the same flow rate is slightly higher than that in the original blade rim structure. Therefore, if only the performance within the stall interval of the four different blade rim structures is considered, the Groove_Tip structure and the Single_Rib_Tip structure show even a negative impact on the stall region. Compared to the original impeller, the critical stall and deep stall operating points of the Bulge_Tip and the Double_Rib_Tip structure are slightly reduced, and the stall range is narrowed, which can improve the "saddle zone" performance of the mixed-flow pump to some extent.

#### 9.2.2.2 Influence of Blade Rim Geometry on the Flow Field in Rim Clearance Region

In order to compare the effects of different blade rim geometries on the flow characteristics of the TLF, some flow parameters are compared in this section. The position parameters of the blade rim along the chord length and the axial direction are nondimensionalized by the blade rim chord length $C$, which are denoted as $\lambda = S/C$ and $\gamma = (Z - Z_{min})/(Z_{max} - Z_{min})$, where $S$ is the coordinate of the local position which is related to the chord length of the blade rim, and $Z$ is the axial coordinate of the local position; $Z_{max}$ and $Z_{min}$ are the axial coordinates of the corresponding positions of the inlet edge and outlet edge of the blade rim respectively. Figure 9.4 describes the position parameters at the blade rim. $L$ denotes the corresponding thickness at the blade rim as and $\zeta$ indicates the position parameter from the pressure surface to the suction surface of the blade rim. Thus, the position parameters in the rim gap can also be nondimensionalized using the blade rim thickness $L$ as $\zeta/L$. Figure 9.5 shows the definitions of the position parameters in the rim clearance region for different blade rim structures. Using the above defined position parameters, the velocity components $U_{str}$ and $U_{cir}$ along the streamwise location and circumferential direction for different span parameters in the rim clearance region are synthesized to obtain the size distribution of velocity $U_{span}$ in the span section, where $U_{span} = (U_{str}{}^2 + U_{cir}{}^2)^{1/2}$. At the same time, the distribution of turbulent kinetic energy $k$ in the rim gap for different span parameters is obtained. The velocity $U_{span}$

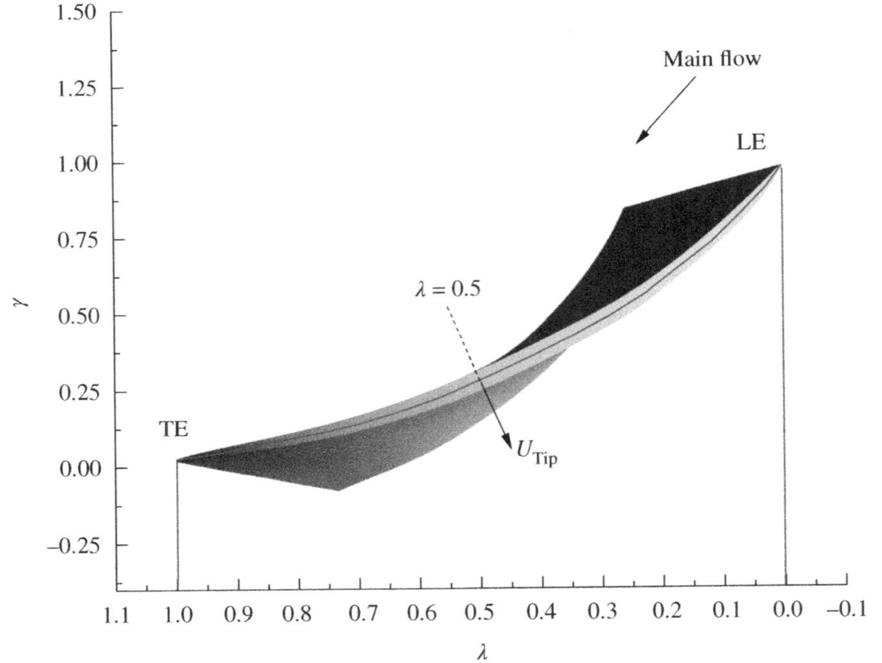


**Figure 9.4** Definition of the position parameters at the blade rim.

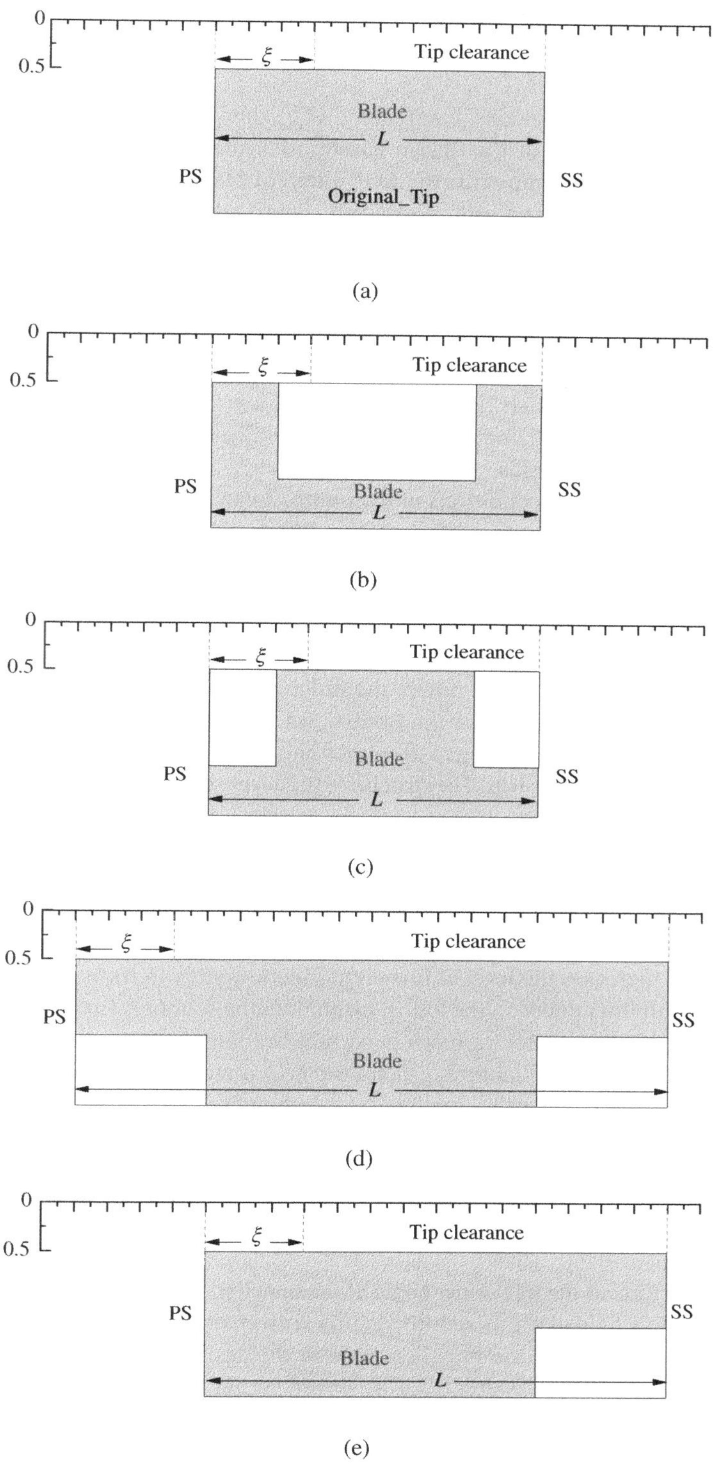


**Figure 9.5** Definitions of the dimensionless position parameters in the rim clearance for various blade rim structures. (a) Original_Tip, (b) Groove_Tip, (c) Bulge_Tip, (d) Double_Rib_Tip, and (e) Single_Rib_Tip.

is normalized using the circumferential velocity $U_{\mathrm{Tip}}$ at the blade rim, while the turbulent kinetic energy is normalized using the square of $U_{\mathrm{Tip}}$; $U_{\mathrm{Tip}} = \pi n D_{\mathrm{Tip}}/60$ where $D_{\mathrm{Tip}}$ is the diameter of the rim area.

Figures 9.6–9.10 show the distribution of velocity component $U_{\mathrm{span}}$ and turbulent kinetic energy $k$ in each span section of the rim gap in the design condition, critical stall condition, and deep stall condition for the mixed-flow pump with the four different blade rim structures. It can be seen from these figures that the distribution of velocity field and turbulent kinetic energy in the rim gap area varies significantly with different blade rim structures. No matter what the working condition of the pump is, the difference in the velocity distribution between the Groove_Tip structure and the Bulge_Tip structure along the radial direction in the gap is obvious, and the turbulent kinetic energy distribution also shows a large difference. However, the Double_Rib_Tip structure and Single_Rib_Tip structure have a rectification effect due to increase in the time required for the fluid to pass through the tip clearance region, and the difference in fluid velocity in the direction of the inner diameter of the gap gradually decreases, and the difference in turbulent kinetic energy also decreases.

In addition, comparing the different working conditions of the pump, it can be noticed that for the two stall conditions, with a decrease in the flow rate, the velocity difference in the radial direction of the original and four different rim clearances decreases, but the turbulent kinetic energy level in the clearance changes significantly for different blade rim structures especially in the deep stall condition and the big change in turbulent kinetic energy is obvious. For the Groove_Tip structure, the turbulent kinetic energy level in the middle of the gap rises sharply, which is caused by the unstable flow in the groove in the middle of the gap, while in the Bulge_Tip structure, the turbulent kinetic energy level in the gap decreases rapidly. For the Double_Rib_Tip and Single_Rib_Tip structures, the level of turbulent kinetic energy in the gap decreases along the flow direction, but overall, due to the rectification effect of Double_Rib_Tip structure, the level of turbulent kinetic energy in the rim clearance region decreases faster.

It can be noted that different rim structures have an impact on the performance of mixed-flow pump under different working conditions of the pump. By comparing with the velocity field and turbulent kinetic energy distribution in the rim gap of the original impeller, it is found that the Groove_Tip structure not only increases the unstable flow in the gap under the design condition and the stall condition but also increases the level of turbulent kinetic energy in the gap; the Bulge_Tip structure has a certain inhibitory effect on the loss of turbulent kinetic energy in the gap under stall conditions; the Double_Rib_Tip structure has a good rectification effect on the flow field in the gap with the velocity field gradually stabilizing along the flow direction and reducing the turbulent kinetic energy loss with the decrease in the flow rate. Although the Single_Rib_Tip structure also has a certain rectification effect, the effect is weak and also has certain limitations. These results show that by selecting the appropriate blade rim structure, the velocity field and turbulent kinetic energy loss in the gap can be optimized for the purpose of favorably optimizing the stall flow conditions and associated flow fields in the mixed-flow pump.

#### 9.2.2.3 Influence of Blade Rim Structure on the Inlet Flow Angle of the Impeller

In this section, the influence of the flow field of the mixed-flow pump with change in the blade rim structure on the inlet flow angle distribution of the impeller blade in the design flow condition, critical stall condition, and deep stall condition is described as shown in Fig. 9.11. The circumferential average value of the angle at the blade inlet from hub to shroud (normalized streamwise parameter = 0.15) for the four-blade rim structures is selected, and the velocity triangle at the blade inlet rim is combined to explain the flow field, as shown in Fig. 9.12. The velocity vectors at the blade inlet under deep stall conditions are shown in Fig. 9.13.

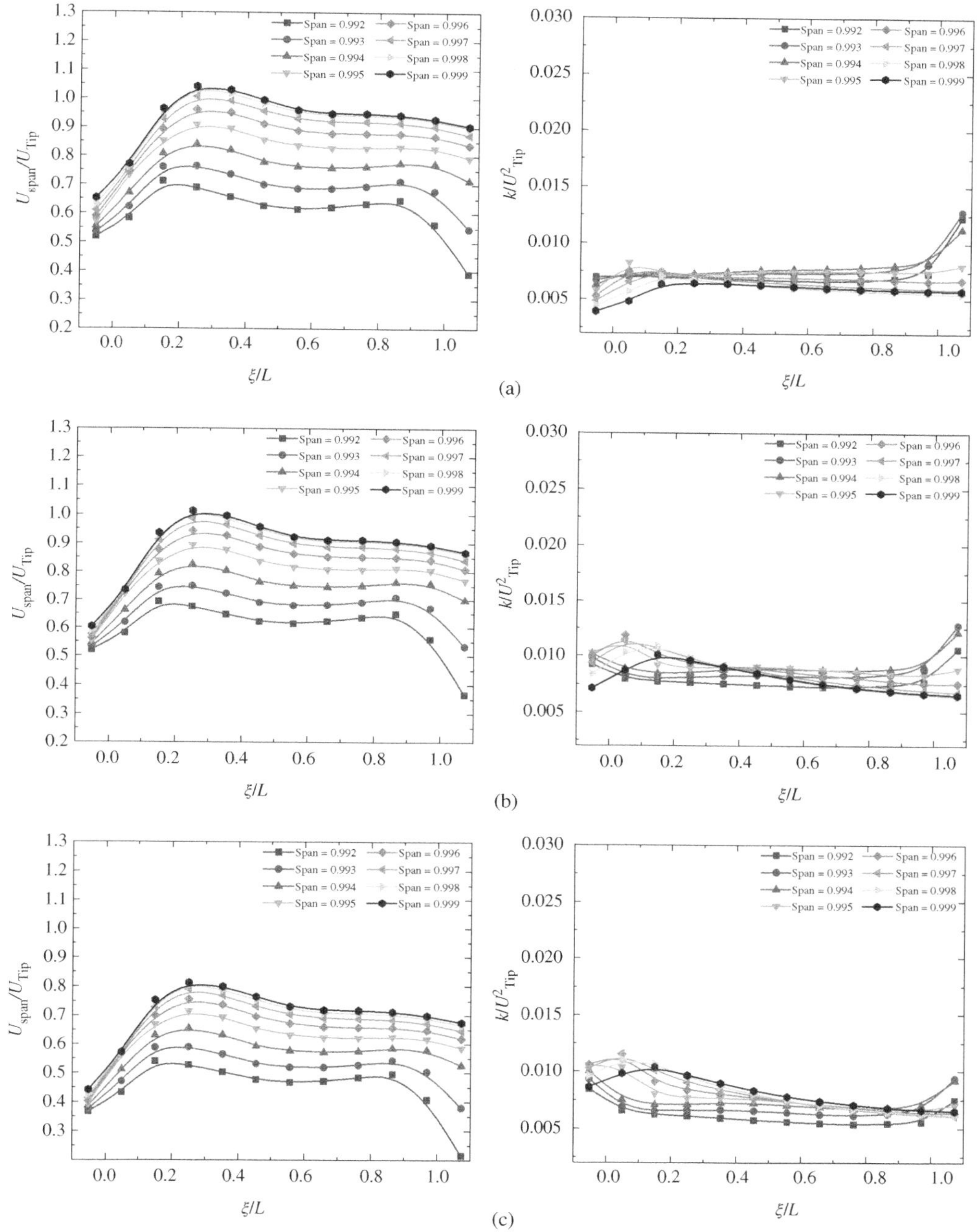


**Figure 9.6** Distributions of the velocity and the turbulent kinetic energy in the rim gap of the Original_Tip impeller blade. (a) Design condition, (b) Critical stall condition, and (c) Deep stall condition.

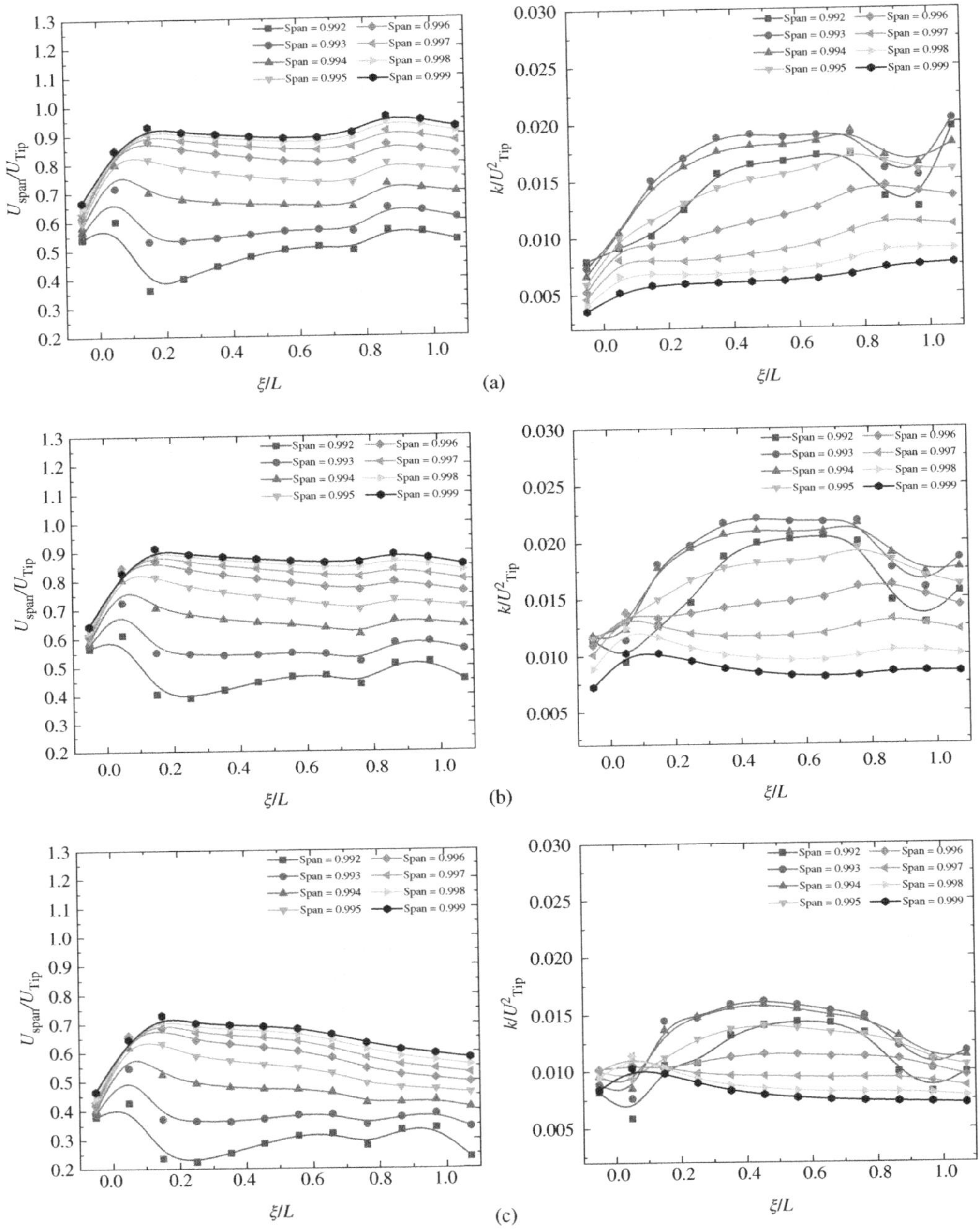


**Figure 9.7** Distributions of the velocity and the turbulent kinetic energy in the rim gap of the Groove_Tip impeller blade. (a) Design condition, (b) Critical stall condition, and (c) Deep stall condition.

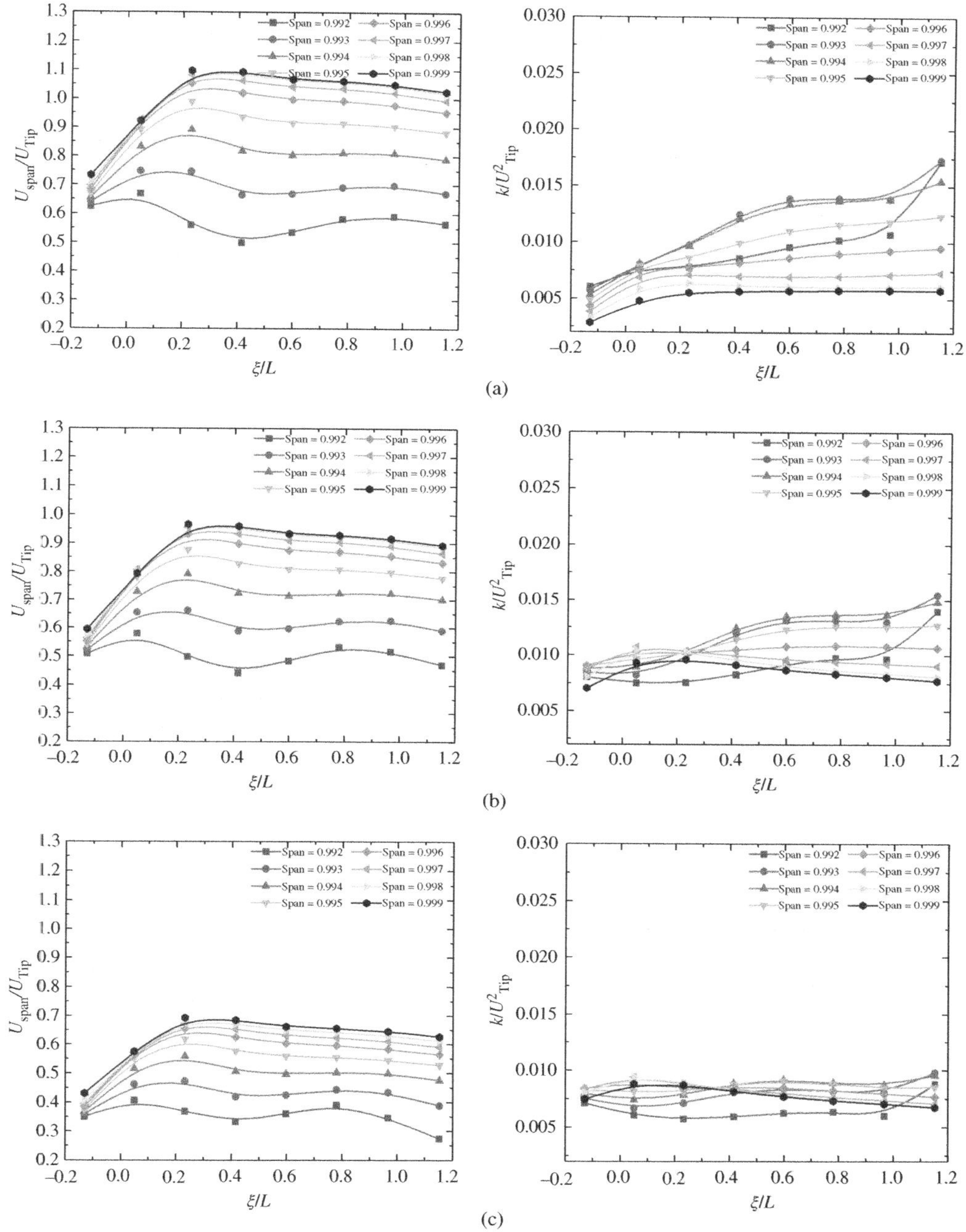


**Figure 9.8** Distributions of the velocity and the turbulent kinetic energy in the rim gap of the Bulge_Tip impeller blade. (a) Design condition, (b) Critical stall condition, and (c) Deep stall condition.

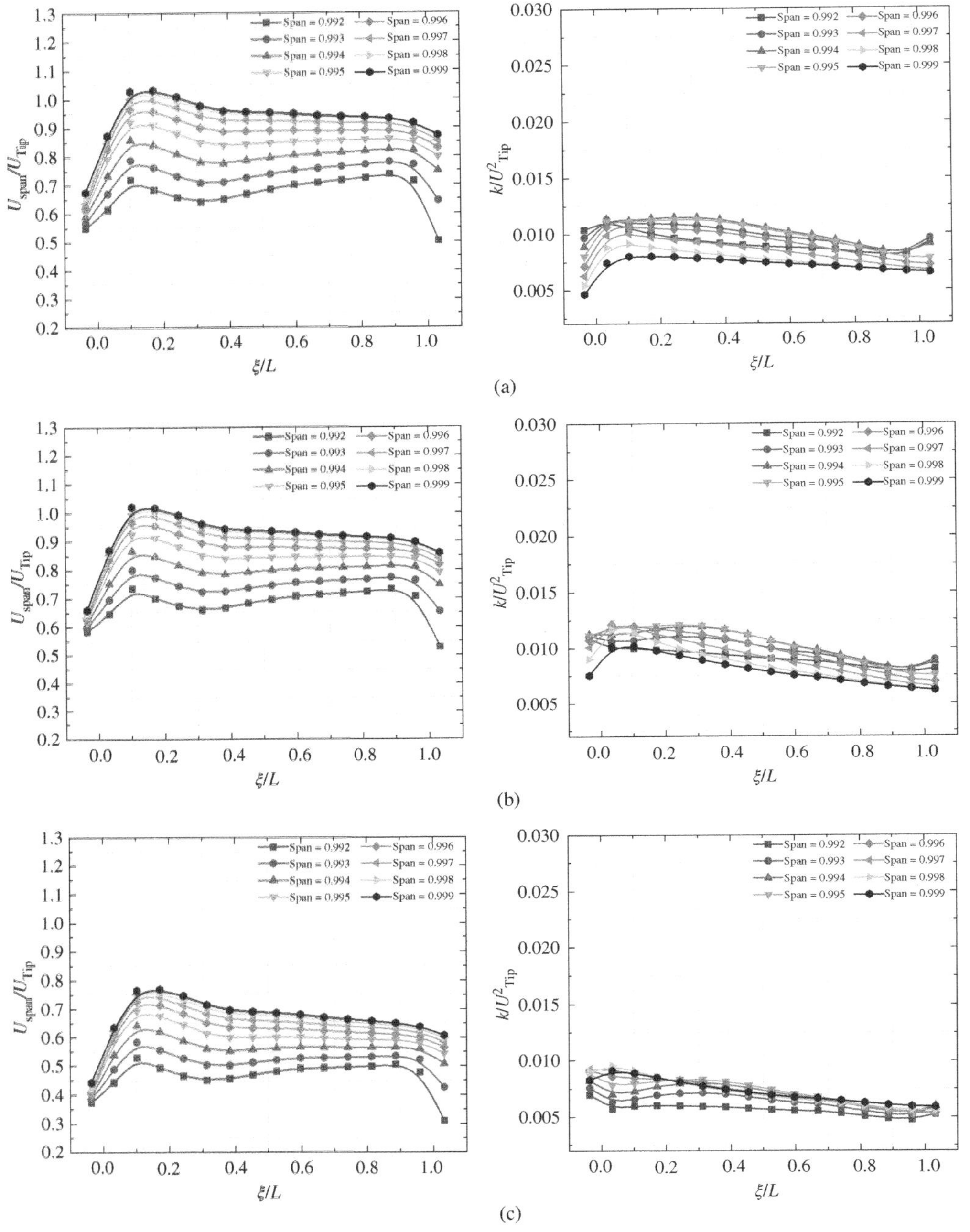


**Figure 9.9** Distributions of the velocity and the turbulent kinetic energy in the rim gap of the Double_Rib_Tip impeller blade. (a) Design condition, (b) Critical stall condition, and (c) Deep stall condition.

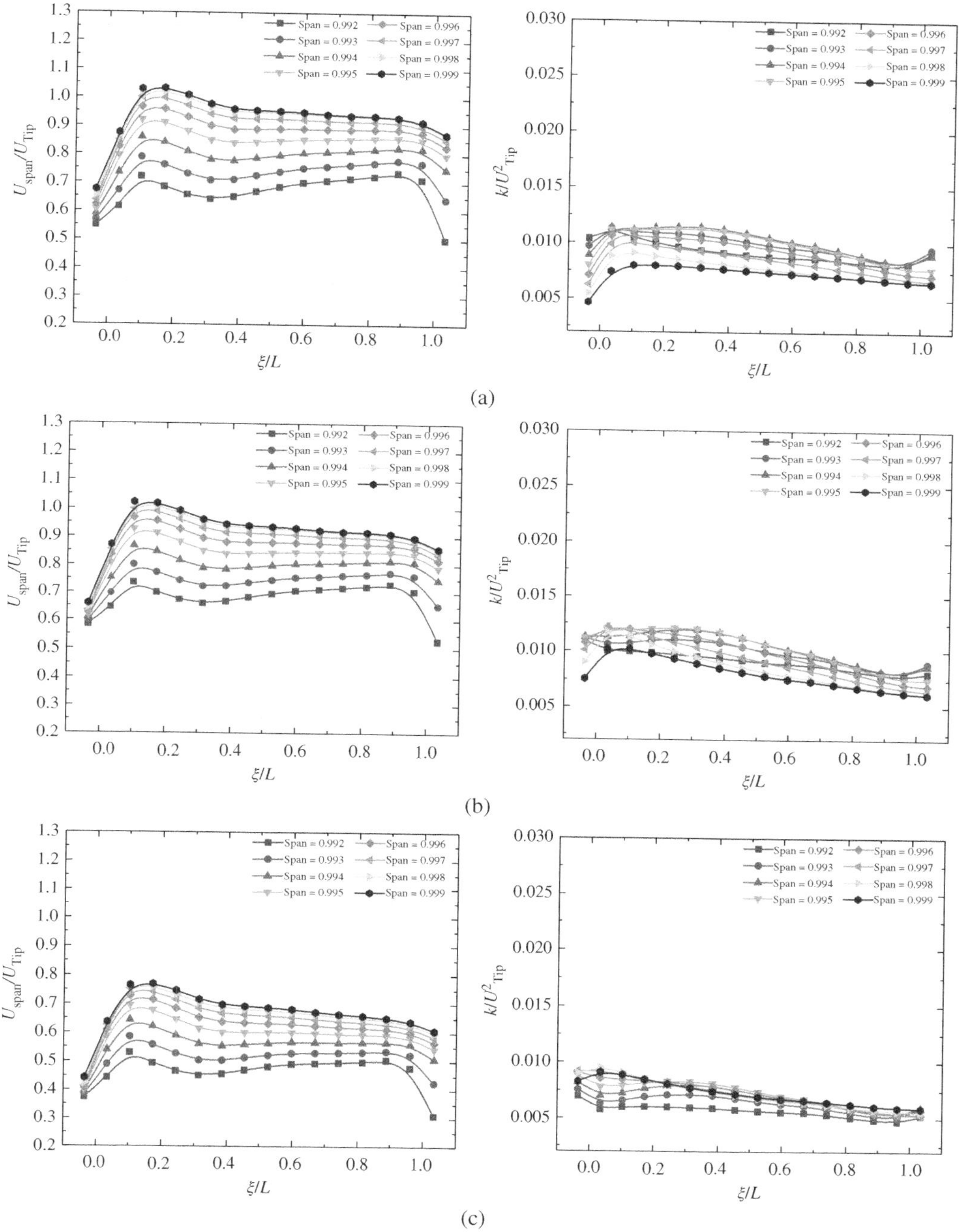


**Figure 9.10** Distributions of the velocity and the turbulent kinetic energy in the rim gap of the Single_Rib_Tip impeller blade. (a) Design condition, (b) Critical stall condition, and (c) Deep stall condition.

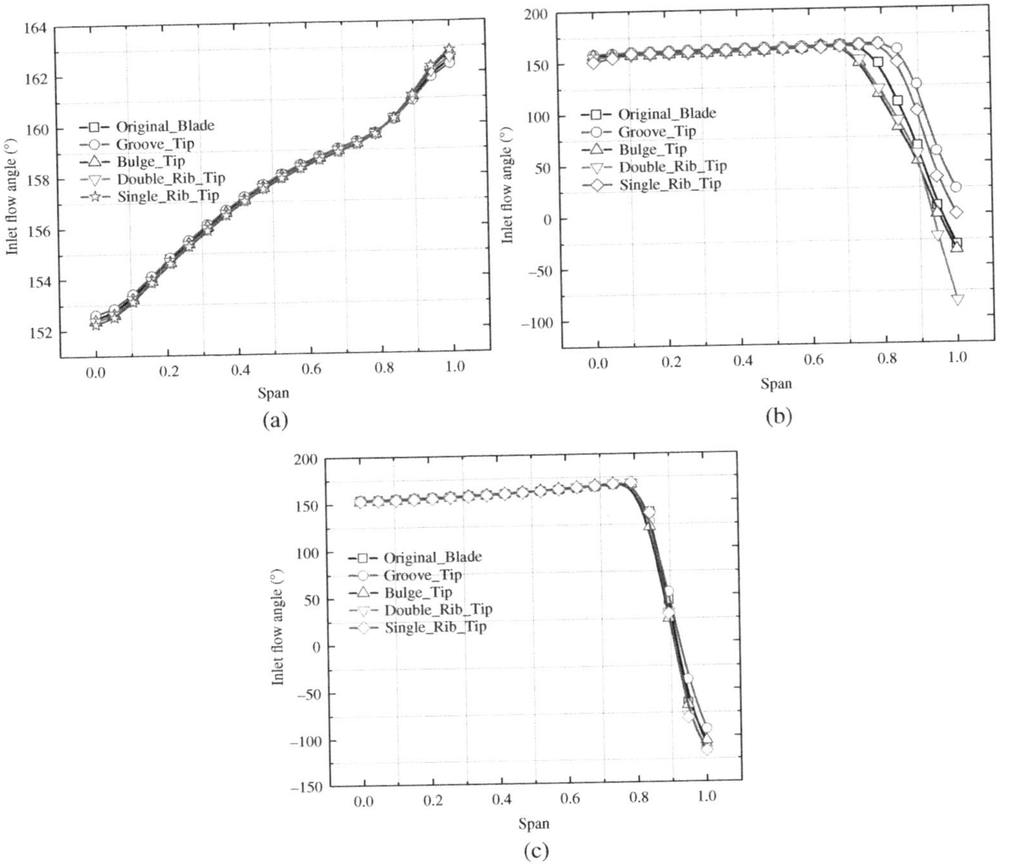


**Figure 9.11** Inlet angle distributions of the blades for various blade rim structures. (a) Design condition, (b) Critical stall condition, and (c) Deep stall condition.

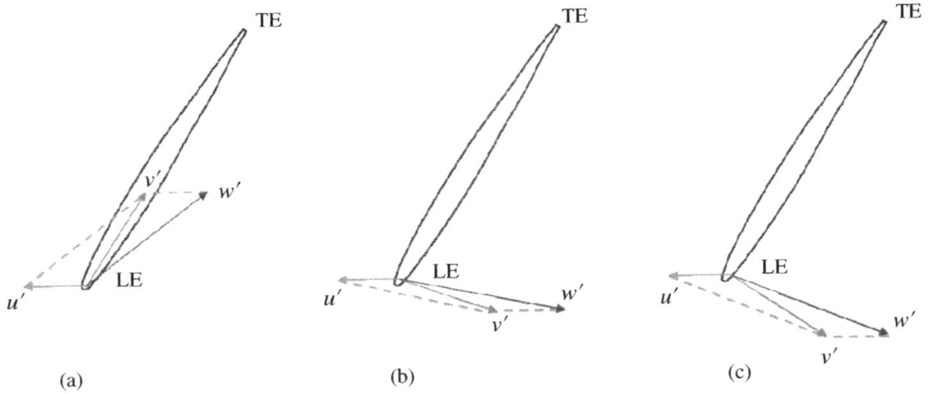


**Figure 9.12** The velocity triangles of the fluid at the blade inlet rim at various pump working conditions. (a) Design condition, (b) Critical stall condition, and (c) Deep stall condition.

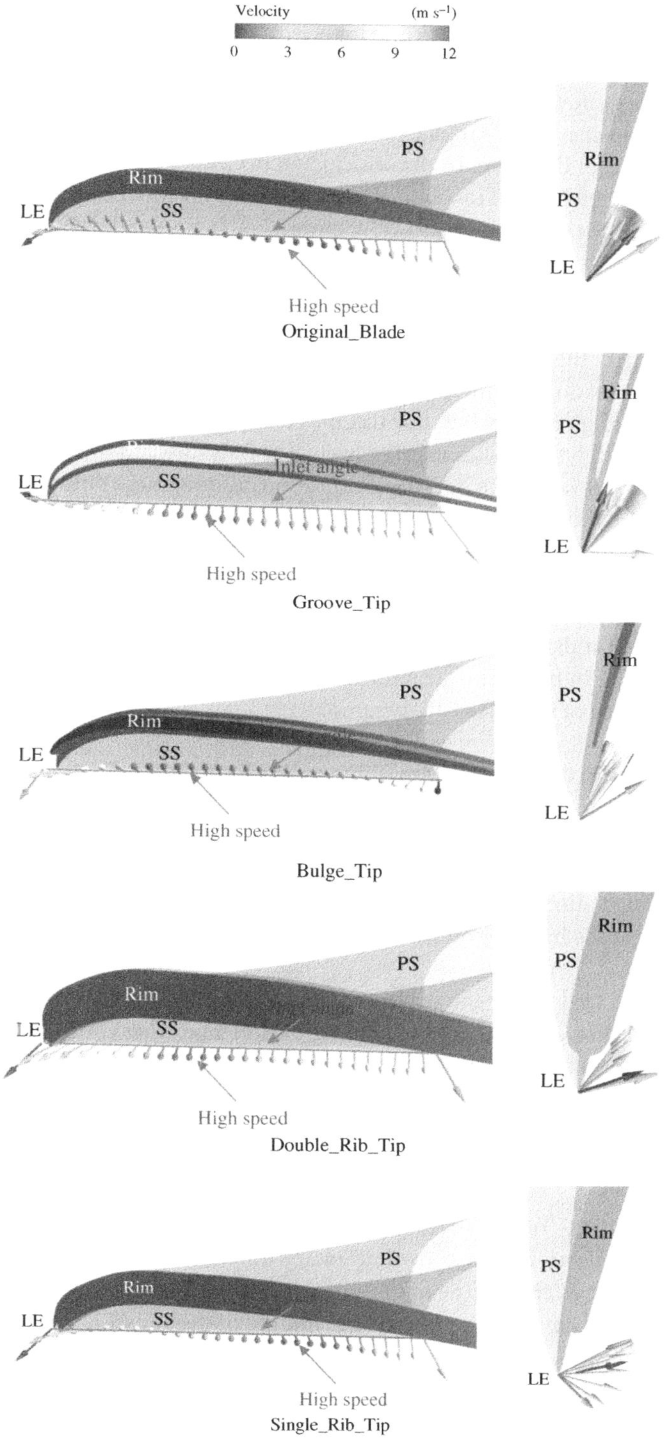


**Figure 9.13** Inlet angle distributions of the blades for various rim structures in the deep stall condition.

As shown in Fig. 9.11, in the design flow rate condition, the inflow angles at the blade inlet for the five-blade rim structures show an increasing trend from the hub to the rim, reaching the maximum value near the rim. Furthermore, the fluid velocity at the blade inlet gradually decreases from the rim to the hub in all five cases; it has the same characteristics as in the original impeller in the design condition, indicating that the change in the rim clearance structure has little effect on the inlet angle of the mixed-flow pump at the design flow rate. In the critical and deep stall conditions, the flow field in the impeller rim area is affected by the TLF of the upper blade and the swirl flow at the pump inlet, thus affecting the flow field of the whole mixed-flow pump. For example, the velocity triangle at the blade inlet rim is impacted by the TLF from the upper blade, and the absolute velocity may even show a negative angle, as shown in Fig. 9.12; these results are consistent with the previous research reported in the literature [30–34]. Thus, the change in the blade rim structure can lead to a change in the TLF, which can then affect the fluid inlet angle at the blade inlet. In the critical stall condition, near the rim area, the fluid inlet angle of the Groove_Tip and Single_Rib_Tip structures is significantly larger than that of the original impeller, but the Bulge_Tip and Double_Rib_Tip structures show an opposite trend. Although there is some influence of the flow rate, but the impact is small since the operating point is very close. In the deep stall condition, although the difference in the flow conditions of the original and four-blade rim structures increases, their influence range and the reduction in the rim inlet angle remain essentially the same. Four new rim structures not only change the inlet angle of the rim area but also move the high-speed region to the middle of the blade inlet. It can be concluded that the change in the blade rim structure can affect the flow field structure of the pump inlet under stall condition, affecting the overall performance of the mixed-flow pump. Therefore, the change in blade rim structure can be considered as a key factor to improve the "saddle area" performance of the mixed-flow pump.

#### 9.2.2.4 Distribution of Vortex Structures in the Impeller with Different Rim Structures

In order to further analyze the influence of different rim structures on the flow field of the mixed-flow pump and the mechanism for suppressing the stall, the internal vortex structure distribution of the four rim structures is obtained, as shown in Figs. 9.14–9.16. The $Q$ criterion method is adopted for the vortex visualization, and the threshold value is set at $90\,000\,s^{-1}$. At the same time, four sections S1–S4 are set at the impeller outlet to observe the changes in the turbulent kinetic energy.

It can be seen from Figs. 9.14 to 9.16 that under the design condition, the four new blade rim structures have formed a stable TLV structure in the impeller rim area and a passage vortex (PV) structure in the impeller hub area. In the four sections S1–S4, only section S3 is affected by the TLV, and there is high turbulent kinetic energy (TKE) near the end wall region of the impeller; however, the performance of the internal flow field due to the four new rim structures is not different from that of the original impeller, which is also the reason why the external characteristics are similar under the design condition.

In the critical stall condition, the TLF of the five rim structures is affected to varying degrees, depicting an unstable state. Due to decrease in flow in the impeller passage, a large curvature-bounded vortex (BV) structure is formed near the pressure surface of the blade and close to the outlet. The vortex begins to form near the hub in the middle of the pressure surface of the blade and then extends to the outlet of the impeller to converge and coalesce with the PV structure. Then, it forms an unsteady secondary vortex (SV) structure in the stator-rotor interaction flow field, which further affects the flow field at the inlet of the guide vane. At the same time, a secondary tip leakage vortex structure appears at the tail of the blade rim, which intensifies the energy loss in the end wall region of the impeller outlet. In addition, the stall vortex

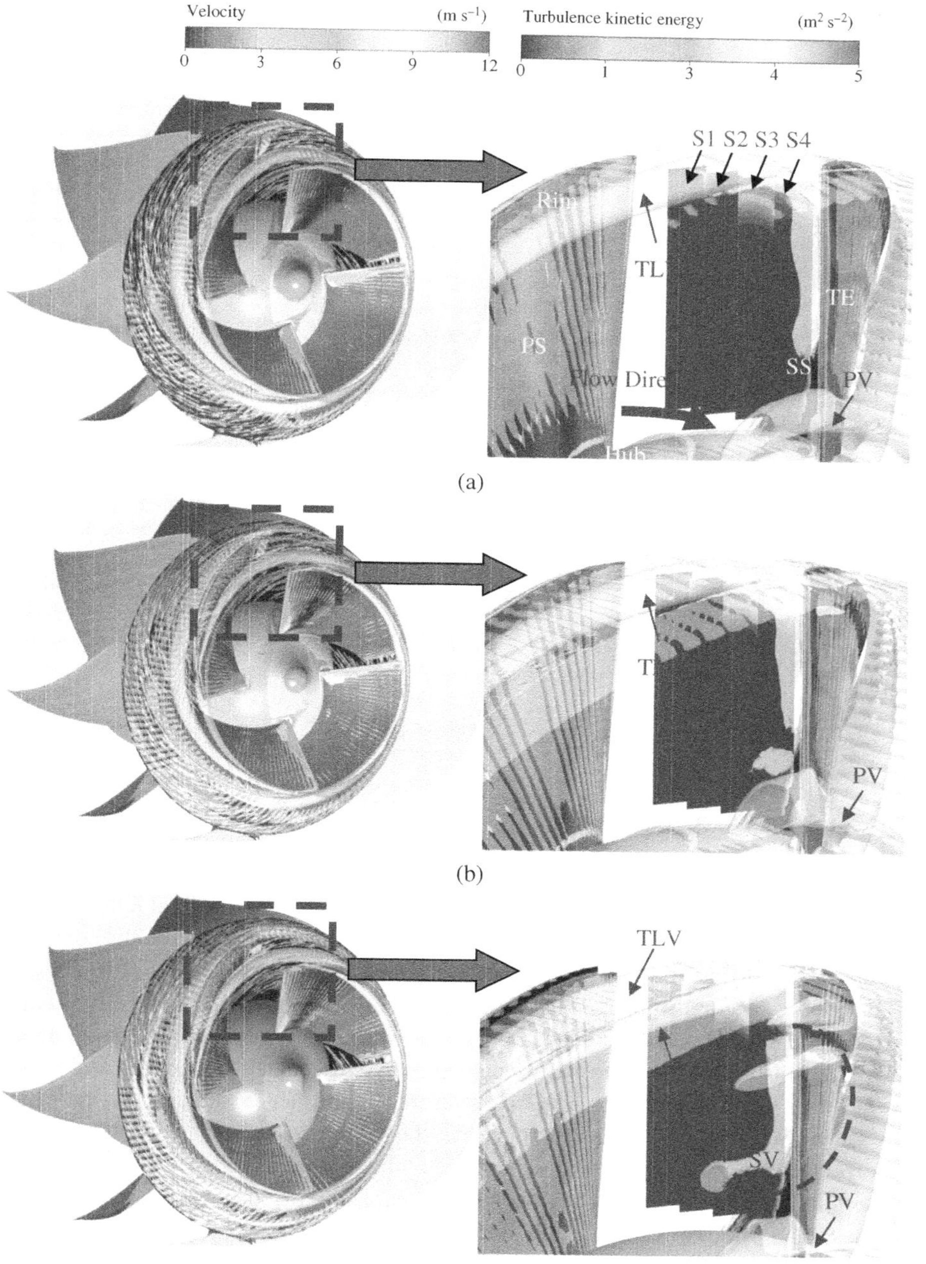


**Figure 9.14** Vortex structures in each impeller and the turbulent kinetic energy distributions at the impeller outlet in the design condition. (a) Original_Tip, (b) Groove_Tip, (c) Bulge_Tip, (d) Double_Rib_Tip, and (e) Single_Rib_Tip.

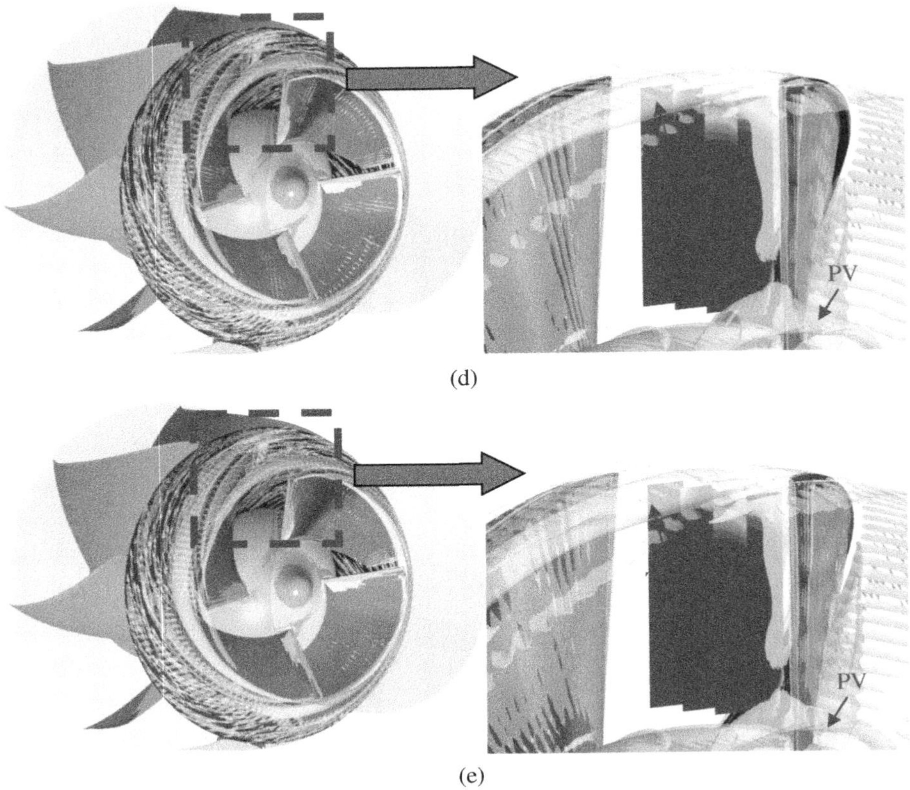


**Figure 9.14** *(Continued)*

in the impeller passage begins to form (marked by the dotted black line and capital letter A), and the TKE in the four sections S1–S4 increases significantly. However, comparing the different rim structures, it can be noted that the flow field of the four new rim structures has also changed compared to that of the original impeller. At the impeller outlet of the Groove_Tip structure, a slender SV structure is formed and extruded, shifting toward the center of the impeller channel; this is also one of the reasons for the performance degradation of the mixed-flow pump under stall condition. In the Bulge_Tip and Double_Rib_Tip structure, BV and PV coalesce early, and their starting points are in the impeller passage. At this time, the SV structure is not only closer to the end wall region but also the strength of the SV tail is notably reduced, which is also the factor that these two rim structures can improve the performance of the mixed-flow pump under stall conditions. In the Single_Rib_Tip structure, BV and PV in the impeller have delayed coalescence, which intensifies the turbulence at the impeller outlet and causes more energy losses.

In flow condition of the deep stall, the region of higher TKE (black dotted box) appears in the four sections S1–S4 at the outlet of the impeller channel for the original and the four-blade rim structures, indicating that the stall vortex leads to high energy dissipation. At the same time, compared to the critical stall condition, the strength of the PV is slightly reduced, which hardly affects the formation of the SV, while the influence range of the SV is significantly larger than that in the critical stall condition. Compared to the original impeller, the vortex in the Groove_Tip and Single_Rib_Tip structures changes little; The BV in the Bulge_Tip structure decreases, and the starting point has

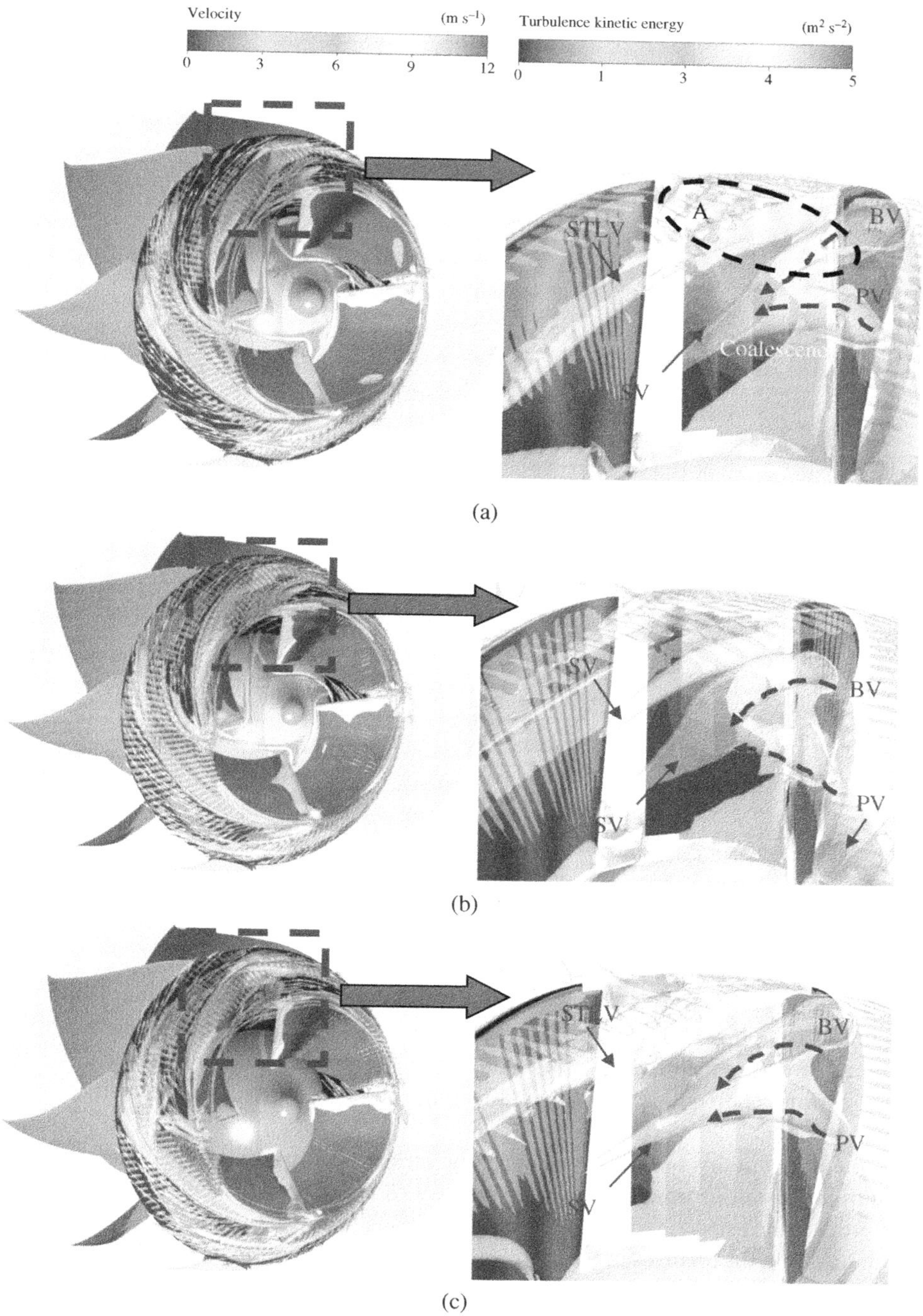


**Figure 9.15** Vortex structures in each impeller and the turbulent kinetic energy distributions at the impeller outlet in the critical stall condition. (a) Original_Tip, (b) Groove_Tip, (c) Bulge_Tip, (d) Double_Rib_Tip, and (e) Single_Rib_Tip.

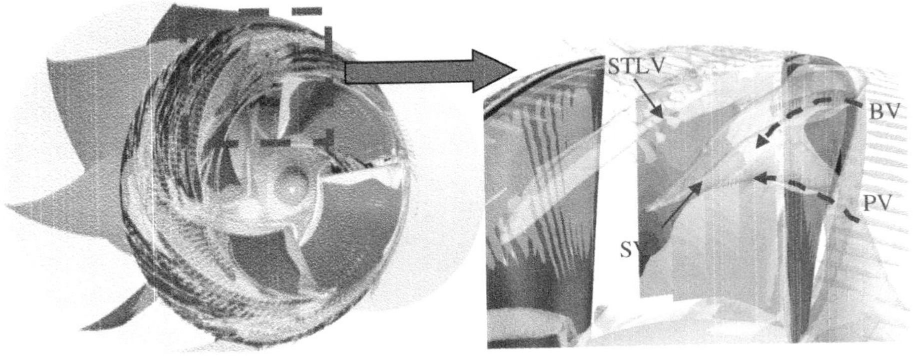


(d)

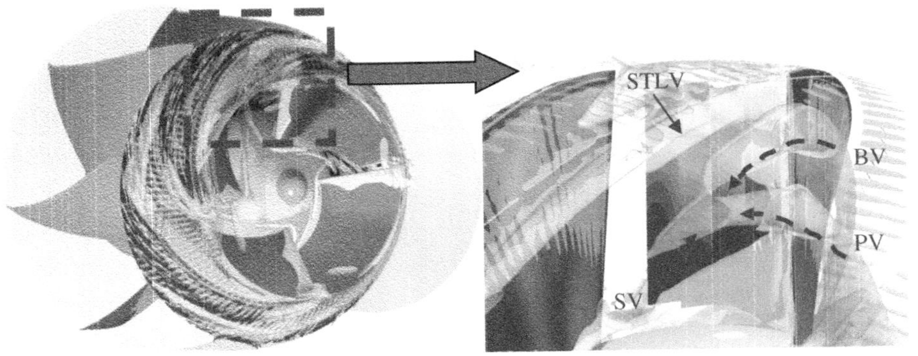


(e)

**Figure 9.15** (*Continued*)

the trend of moving toward the impeller outlet. The SV strength in the Double_Rib_Tip structure is weakened, which further shows that the Bulge_Tip and Double_Rib_Tip structures play a positive role in the flow field in the deep stall condition.

In summary, when the structure of the rim clearance is changed, the Bulge_Tip and Double_Rib_Tip structures of the blade can influence the convergence of the BV structure on the blade surface and the PV structure near the hub, thus limiting the location and strength of the unsteady SV structure at the impeller outlet, ultimately affecting the energy performance of the whole mixed-flow pump.

In Fig. 9.17, the distributions of stall vortex cores for various rim structures are compared, and the line distributions of stall vortex cores are given in Fig. 9.18. It can be seen from these two figures that the stall vortex in the impeller with the five different blade rim structures is still located at the impeller outlet; however, the changes in the rim clearance affect the distributions of the stall vortex core. The influence range of the stall vortex in the Groove_Tip structure is significantly increased, and it can be seen from the vortex core line that the vortex core is not only in the $r$ direction but is also in the $\theta$ direction. Both directions are wide, which further proves that the Groove_Tip blade rim structure has a negative effect on the performance of the mixed-flow pump under stall condition. The stall vortex core in the Single_Rib_Tip structure also exhibits a similar phenomenon. However, in the Bulge_Tip structure and the Double_Rib_Tip structure, the stall vortex hardly changes compared to in the original impeller, and the connection line of the stall vortex core is similar to that in

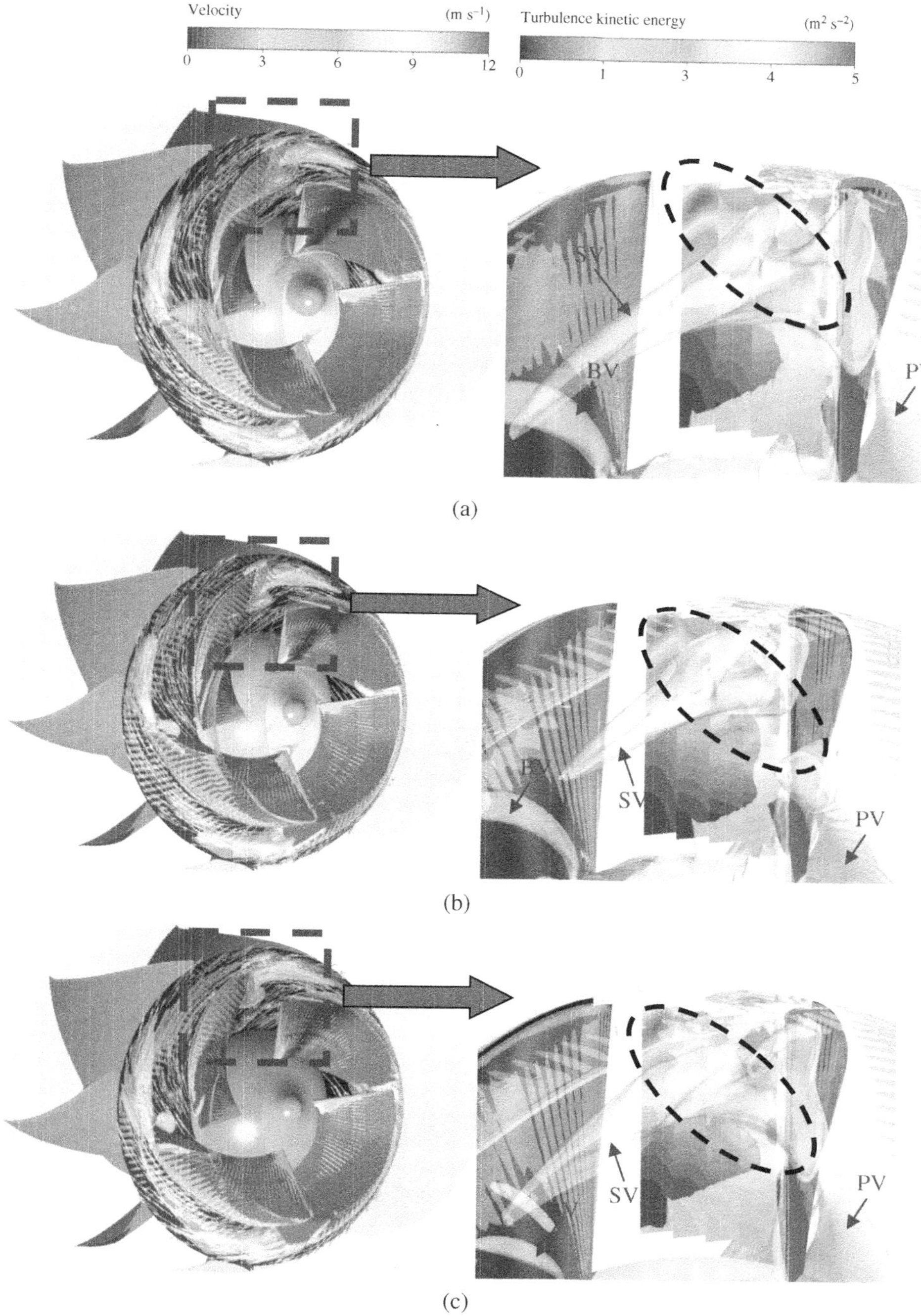


**Figure 9.16** Vortex structures in each impeller and the turbulent kinetic energy distributions at the impeller outlet in deep stall condition. (a) Original_Tip, (b) Groove_Tip, (c) Bulge_Tip, (d) Double_Rib_Tip, and (e) Single_Rib_Tip.

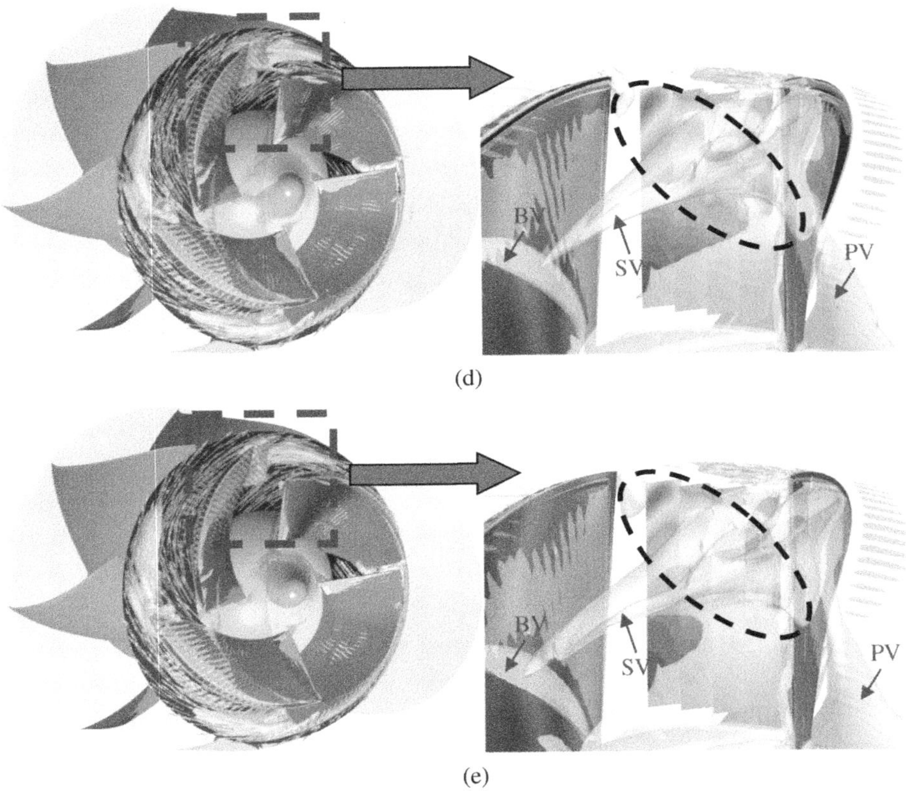


Figure 9.16 (*Continued*)

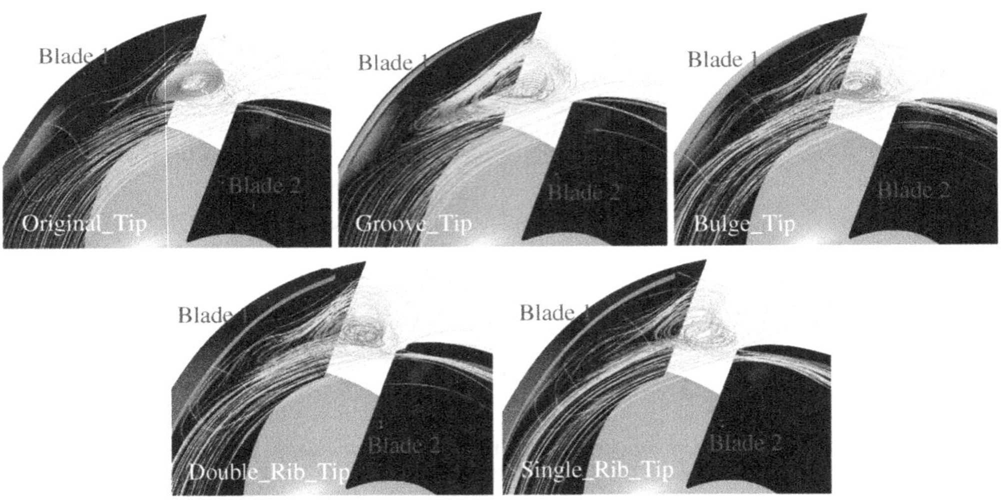


**Figure 9.17** Distributions of stall vortex structures in the impeller in deep stall condition for various rim structures.

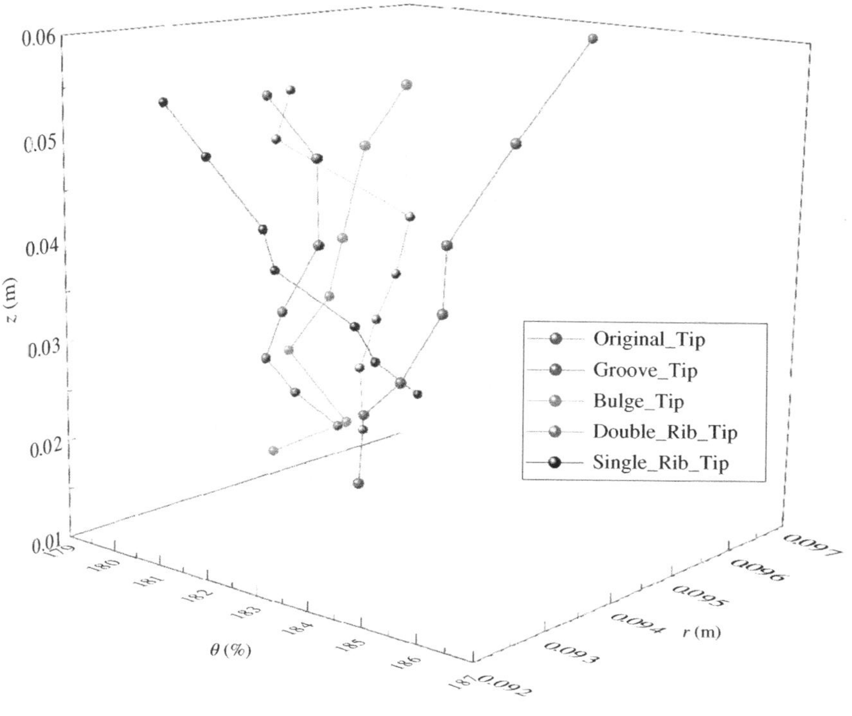


**Figure 9.18** Distributions of stall vortex core for various blade rim structures.

the original impeller indicating that these two rim structures will not promote the flow instability in the impeller channel.

In conclusion, the blade rim structures have a great influence on the stall flow field, especially on the strength and distribution of the stall vortex. In terms of overall performance, both the Bulge_Tip structure and Double_Rib_Tip structure can shorten the range of stall conditions without deteriorating the stall flow field in the impeller and optimize the flow field in the rim gap and the impeller inlet and outlet. Thus, these two-blade rim structures have the potential for suppression of stall and are candidates for improved design of mixed-flow pump.

## 9.3 Influence of the Circumferential Spokes

### 9.3.1 The Structure of "O-Spoke"

Prior studies have demonstrated that the utilization of grooving techniques substantially improves the flow conditions at the impeller inlet, thereby aiding in the attenuation of adverse effects in the hump region of the head vs. flow rate curve of the mixed-flow pump. In this section, an "O-Spoke" configuration is introduced as illustrated in Fig. 9.19, which includes the circumferential "O-Spoke" implemented according to three distinct schemes labeled as A, B, and C. These spokes are positioned 2 mm upstream of the impeller inlet leading edge. The essential design parameters are provided in Table 9.2.

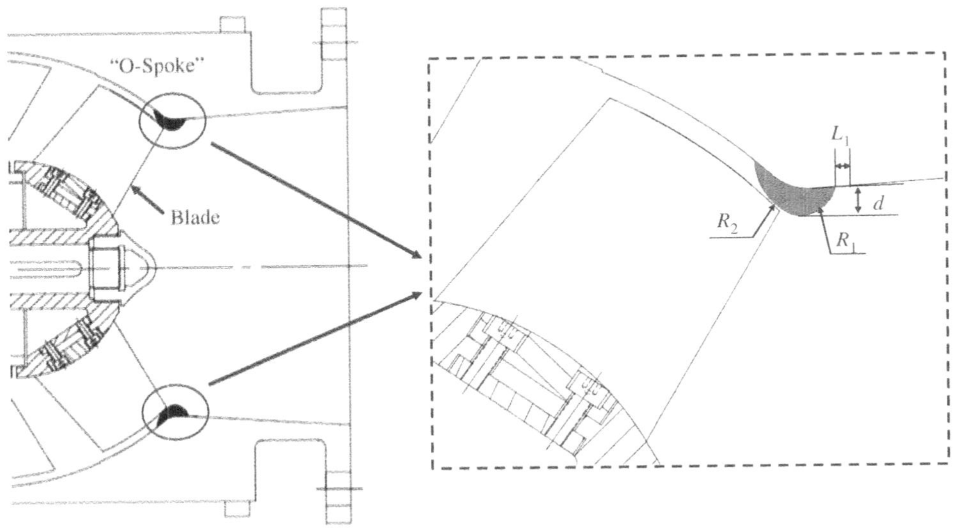


**Figure 9.19** Schematic diagram of the "O-Spoke".

**Table 9.2** Key dimensions of the "O-Spoke".

| "O-Spoke" | Depth $d$ (mm) | $L_1$ (mm) | $R_1$ (mm) | $R_2$ (mm) |
|---|---|---|---|---|
| A | 4 | 2 | 6 | 13 |
| B | 5 | 2 | 9 | 16 |
| C | 6 | 2 | 10 | 11 |

## 9.3.2 Analysis of Results

### 9.3.2.1 External Characteristics Analysis

Figure 9.20 depicts the head and efficiency curves of the mixed-flow pump as a function of the flow rate. At lower flow rates, the head curve exhibits a "saddle region," signifying the onset of rotational stall. To facilitate the analysis, multiple sampling points are introduced in the "saddle region," and the data is curve-fitted. When compared to the numerical simulation results, it becomes evident that the stall region for the mixed-flow pump falls within the range of $0.4Q_{\text{des}}$ to $0.6Q_{\text{des}}$ flow rates. In proximity to the stall conditions, the inclusion of the "O-Spoke" leads to substantial deviations in the head and efficiency curves compared to the original configuration without "O-Spoke." The flow rate condition associated with the maximum head just prior to a sharp decrease is termed the "critical stall condition," while the flow rate condition linked to the minimum head immediately before the rise is termed as the "deep stall condition.".

Figure 9.20 shows that, as the flow rate decreases, the mixed-flow pump with original blade rim structure experiences a sharp drop in head at a specific flow rate. On further reduction in the flow rate, the head curve gradually begins to rise again. On the other hand, the mixed-flow pump equipped with "O-Spoke" maintains a relatively smooth head curve within the stall region. At the deep stall condition ($0.54Q_{\text{des}}$), the efficiency of the mixed-flow pump with "O-Spoke" at depths of 4, 5, and 6 mm increases by 2.45%, 3.96%, and 2.34%, respectively, compared to the

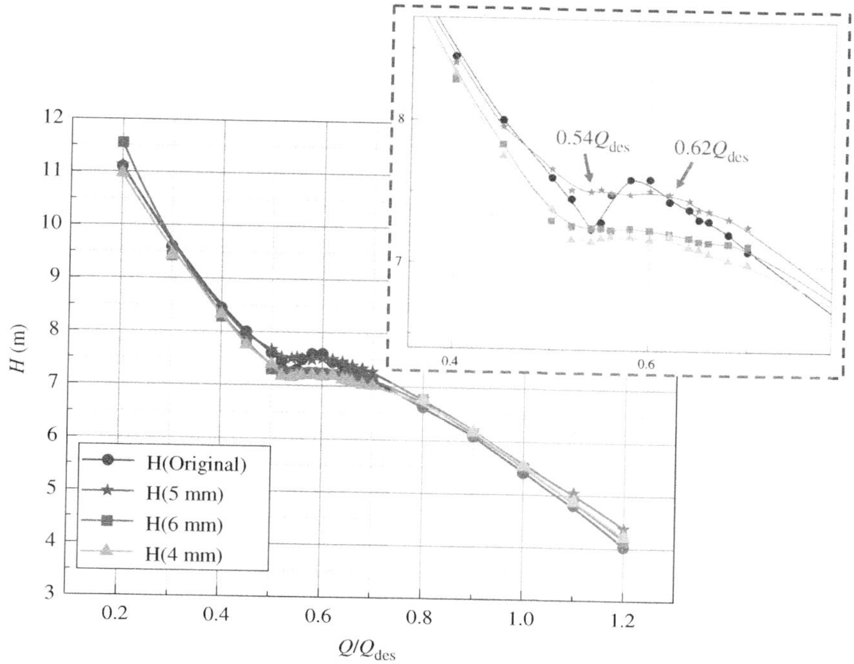


(a)

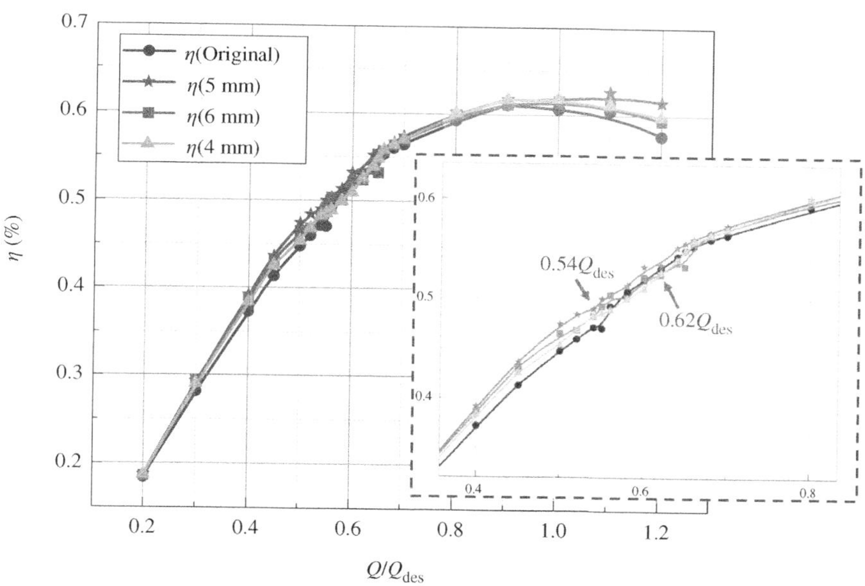


(b)

**Figure 9.20** External characteristic curves. (a) Head curve and (b) Efficiency curve.

original mixed-flow pump. The head of the mixed-flow pump increases by 3.62% and 0.10% with the addition of 5 and 6 mm "O-Spoke," respectively, but decreases by 1.25% with 4 mm "O-Spoke." At the critical stall condition ($0.62Q_{des}$), the efficiency of the mixed-flow pump increases by 0.59% with 5 mm "O-Spoke" but decreases by 0.85% and 1.28% with 4 and 6 mm "O-Spoke," respectively. The head of the mixed-flow pump increases by 0.67% with 5 mm "O-Spoke" but decreases by 3.26% and 3.03% with 4 and 6 mm "O-Spoke," respectively. At the design condition ($1.0Q_{des}$), the efficiency of the mixed-flow pump increases by 1.98%, 1.27%, and 1.61% with 4, 5, and 6 mm "O-Spoke," respectively compared to the efficiency of the original mixed-flow pump. At the design condition ($1.0Q_{des}$), the head increases by 2.31%, 2.18%, and 2.10% with 4, 5, and 6 mm "O-Spoke," respectively compared to the head of the original mixed-flow pump. It is evident that at the design condition, the addition of "O-Spoke" of different depths results in varying degrees of improvement in both the head and efficiency of the mixed-flow pump. Furthermore, the mixed-flow pump with the 5 mm "O-Spoke" configuration (B scheme) also exhibits improved head and efficiency under stall conditions, reducing the hydraulic losses in the stall region. This result highlights the effective stall mitigation capability of the "O-Spoke" structure.

#### 9.3.2.2 Analysis of the Flow Field in Various Cross-Sectional Areas of the Pump

To compare the flow fields inside the impeller at various operating conditions, five Aligned Turbo Surface flow cross-sections are employed along the flow path from the inlet to the outlet of the impeller, as shown in Fig. 9.21. Figure 9.22(a)–(d) depicts the velocity distribution profiles in the impeller for both the original mixed-flow pump and the mixed-flow pump featuring "O-Spoke" with varying depths. These velocity distributions are presented at the design flow condition of $1.0Q_{des}$, the critical stall condition of $0.62Q_{des}$, and the deep stall condition of $0.54Q_{des}$. Figure 9.23(a)–(d) shows the pressure distribution maps in the impeller for the same pump configurations with "O-Spoke" and operating conditions used for velocity distributions in Fig. 9.22(a)–(d).

The graphical representations illustrate that, at the design flow condition ($1.0Q_{des}$), the velocity and pressure distribution profiles at the impeller flow cross-sections remain consistent after the incorporation of different depths of "O-Spoke" when compared to the original mixed-flow pump model without "O-Spoke." At the critical stall condition ($0.62Q_{des}$), the low-velocity region in the impeller flow cross-section of the mixed-flow pump equipped with 4 mm "O-Spoke"

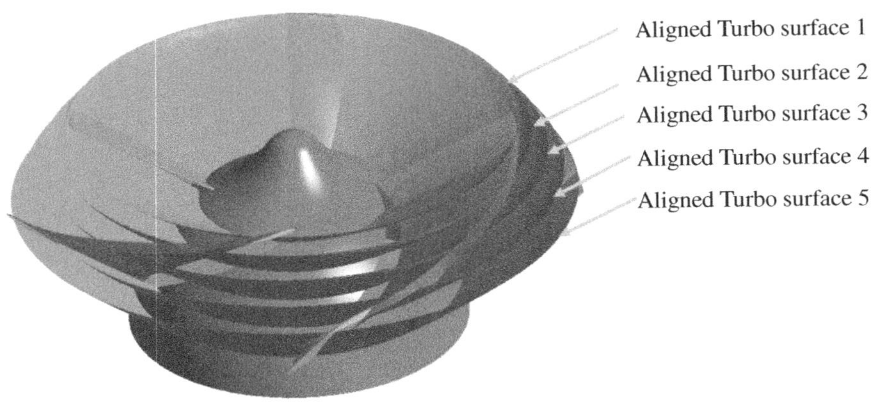


**Figure 9.21** Aligned Turbo Surface flow cross-section.

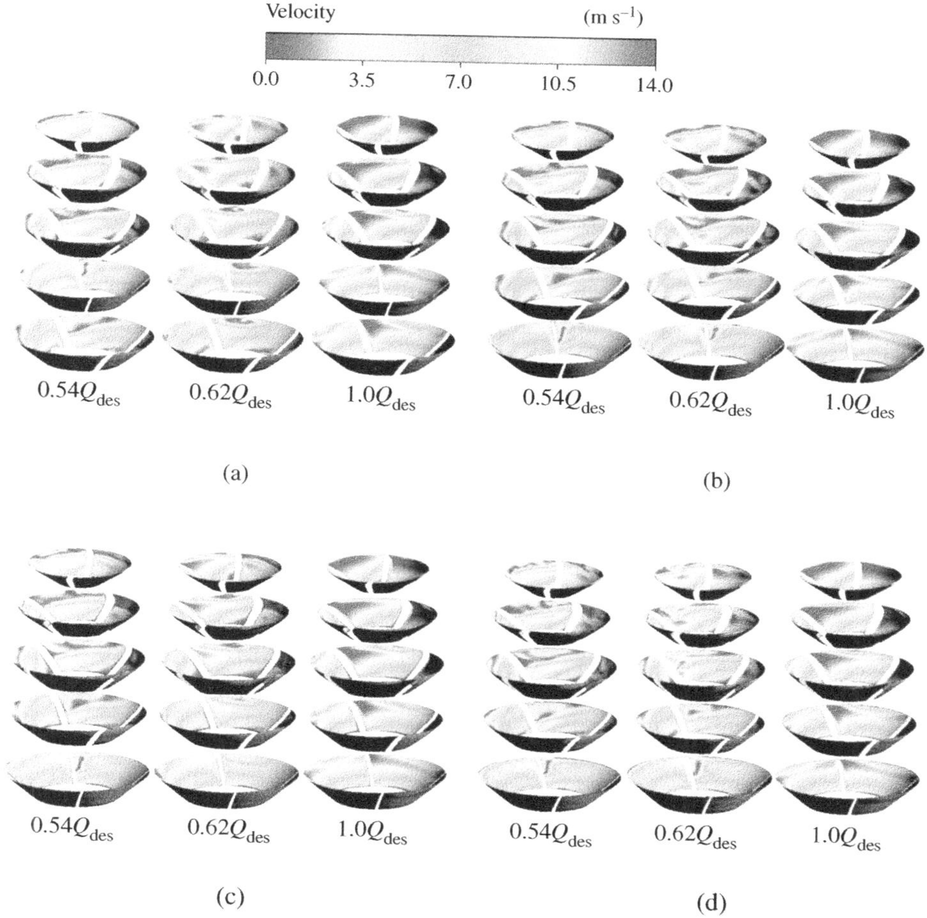


**Figure 9.22** Velocity distribution maps in the impeller cross-section. (a) The original pump, 4 mm "O-Spoke," (c) 5 mm "O-Spoke," and (d) 6 mm "O-Spoke."

expands significantly compared to the original mixed-flow pump. Similarly, the inclusion of 6 mm "O-Spoke" results in a marginal increase in the low-velocity region. In contrast, the presence of 5 mm "O-Spoke" results in a noticeable improvement in the low-velocity region when compared to the original mixed-flow pump. In the deep stall condition (0.54$Q_{des}$), the variations in the velocity distributions among different "O-Spoke" schemes are not particularly significant.

Scrutinizing the pressure variations across different operating conditions, it becomes apparent that the mixed-flow pump fitted with varying depths of "O-Spoke" exhibits a more gradual transition in pressure fluctuations in comparison to the original impeller without "O-Spoke." In essence, the magnitude of the pressure oscillations in the impeller gets diminished. At the deep stall condition (0.54$Q_{des}$), the incorporation of 4 and 6 mm "O-Spoke" in the impeller yields akin variations in the low-pressure zone. On the other hand, the impeller featuring 5 mm "O-Spoke" demonstrates a more uniform distribution of pressure areas. This signifies that the "O-Spoke" design described here has a discernible mitigating effect on pressure fluctuations that occur during the rotational stall of the mixed-flow pump.

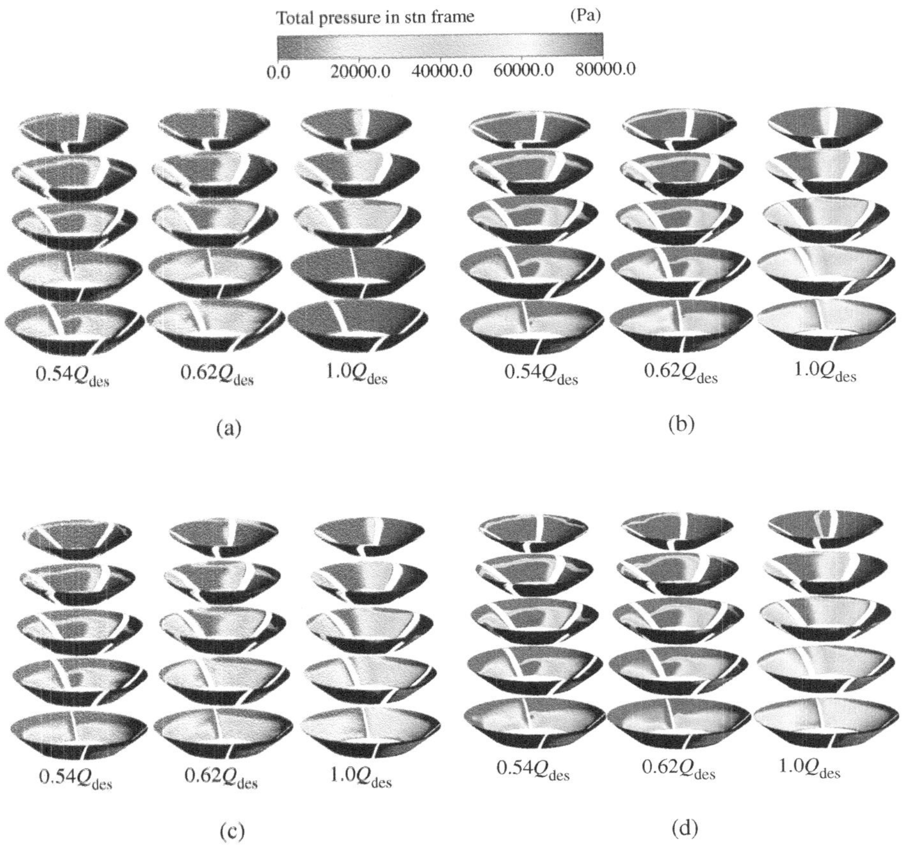


**Figure 9.23** Pressure distribution maps in the impeller cross-section. (a) The original pump, 4 mm "O-Spoke," (c) 5 mm "O-Spoke," and (d) 6 mm "O-Spoke."

#### 9.3.2.3 Flow Field Analysis in the Impeller Shroud Region

Figures 9.24 and 9.25 depict the flow fields in a single passage and the associated vortex structures based on the $Q$ criterion ($Q = 80\,000\,\text{s}^{-2}$) in the impeller shroud region of both the original mixed-flow pump and the modified mixed-flow pump impellers with varying depths of "O-Spoke" under stall conditions. In the mixed-flow pumps, the TLF usually originates near the leading edge and primarily resides in the primary flow passage of the pump, and the stall vortex extends vertically across the rear section of the flow passage.

Analysis of Fig. 9.24(a) and 9.25(a) reveals that, during the deep stall f condition (0.54$Q_{des}$), the mixed-flow pump impeller equipped with the 4 mm "O-Spoke" exhibits a noticeable improvement in the TLF and a slight reduction in the stall vortex compared to the original mixed-flow pump. The 5 mm "O-Spoke" configuration results in a modest enhancement of the TLF and a significant reduction in the stall vortex compared to the original mixed-flow pump. In contrast, the 6 mm "O-Spoke" configuration generates more TLF and exacerbates the stall vortex structure compared to the original mixed-flow pump.

Figures 9.24(b) and 9.25(b) provide insight into the flow field in the critical stall condition (0.62$Q_{des}$). In this scenario, the mixed-flow pump impeller with 4 mm "O-Spoke" exhibits reduced TLF and a minor improvement in the stall vortex compared to the original mixed-flow pump. The

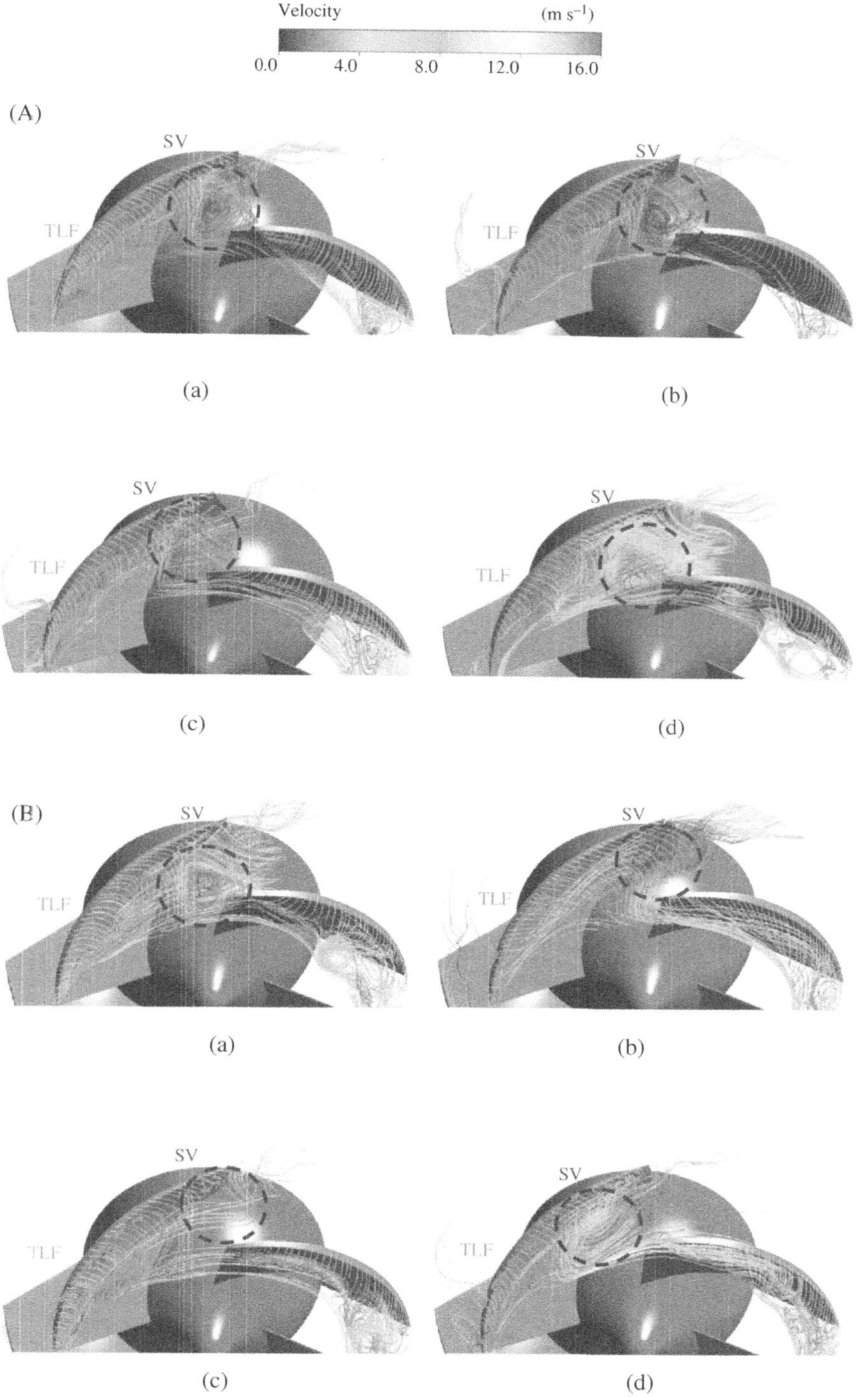


**Figure 9.24** Impeller shroud region flow field plots. (A) $0.54Q_{des}$ (a) The original pump, 4 mm "O-Spoke," (c) 5 mm "O-Spoke," and (d) 6 mm "O-Spoke." (B) $0.62Q_{des}$ (a) The original pump, 4 mm "O-Spoke," (c) 5 mm "O-Spoke," and (d) 6 mm "O-Spoke."

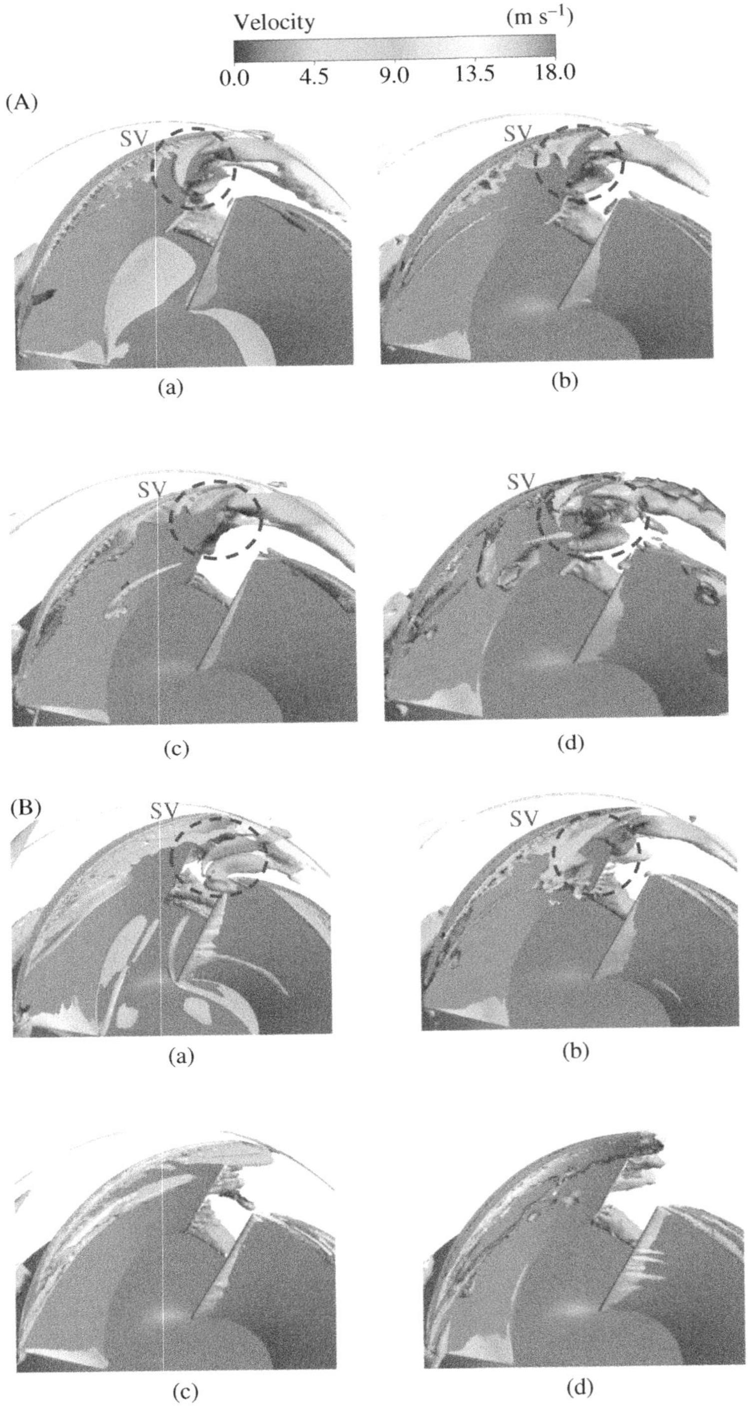


**Figure 9.25** Impeller single passage vortex plots. (A) $0.54Q_{des}$ (a) The original pump, 4 mm "O-Spoke," (c) 5 mm "O-Spoke," and (d) 6 mm "O-Spoke." (B) $0.62Q_{des}$ (a) The original pump, 4 mm "O-Spoke," (c) 5 mm "O-Spoke," and (d) 6 mm "O-Spoke."

configuration with the 5 mm "O-Spoke" not only demonstrates a significant reduction in the TLF compared to the original mixed-flow pump but also effectively mitigates the generation of the stall vortex. The mixed-flow pump impeller with 6 mm "O-Spoke" also shows a slight reduction in TLF compared to the original mixed-flow pump, leading to an improved stall vortex structure.

#### 9.3.2.4 Energy Performance Analysis Based on the Entropy Generation Theory

Illustrated in Figure 9.26 are the local entropy production distribution maps for both the original mixed-flow pump and mixed-flow pumps equipped with various depths of "O-Spoke" across different flow conditions. These visual representations reveal that with the decreasing flow rates, entropy production in the mixed-flow pump impeller increases, signifying the increased turbulence effects and the elevated hydraulic losses in the impeller. Notably, a high-entropy production region emerges along the impeller shroud pinpointing the TLF as the predominant contributor to the energy dissipation within the impeller.

Under the critical and deep stall conditions, the stall vortex generated near the impeller's outlet is the principal source of the increased entropy production. Furthermore, at the critical stall condition ($0.62Q_{des}$), energy losses within the shroud region surpass those observed at the deep stall flow condition ($0.54Q_{des}$). Significantly, at the critical stall condition, the entropy production in the shroud region of the mixed-flow pump equipped with "O-Spoke" substantially diminishes in comparison to that of the original mixed-flow pump. This provides additional evidence that the "O-Spoke" design described in this section possesses the capability to partially mitigate the stall vortex formation, thereby ameliorating the rotational stall phenomenon within the mixed-flow pump impeller.

#### 9.3.2.5 Analysis of the Blade Height Section Flow Field

This analysis underscores the effectiveness of introducing "O-Spoke" structure at the inlet of the mixed-flow pump at low-flow rate conditions in mitigating the formation of the stall zone. When operating at low-flow rate conditions, the circumferential velocity of the mixed-flow pump increases resulting in an augmented impeller inlet impingement angle. Consequently, the flow separation on the suction side of the blade becomes pronounced, causing a reduction in incoming flow. Furthermore, this recirculation extends toward the blade inlet giving rise to vortices in conjunction with the incoming flow shock waves. These vortices in the impeller blade passage cannot promptly dissipate or expel from the impeller passage. This phenomenon leads to their gradual accumulation and development eventually obstructing the passage.

Figure 9.27 depicts the streamlines at various span sections of both the original mixed-flow pump and the mixed-flow pump equipped with the "O-Spoke" of 5 mm depth operating at a flow rate of $0.62Q_{des}$ (critical stall condition). This figure reveals the generation of a stall vortex near the impeller outlet, specifically adjacent to the suction side of the blade. This phenomenon primarily accounts for the energy losses in the impeller passage. At Span = 0.8, two vortices manifest along the impeller path of the original mixed-flow pump, with the vortex near the impeller inlet rotating clockwise and the vortex near the impeller outlet rotating counterclockwise. In contrast, the stall vortex produced near the outlet of the mixed-flow pump impeller equipped with 5 mm depth "O-Spoke" is considerably smaller. At Span = 0.85, the two vortices in the original mixed-flow pump's impeller begin to separate and weaken. Concurrently, the stall vortex within the impeller passage of the mixed-flow pump equipped with the 5 mm depth "O-Spoke" gradually diminishes. Upon reaching Span = 0.9, the stall vortex near the impeller outlet of the original mixed-flow pump further diminishes, while the vortex near the impeller inlet nearly dissipates. In sharp contrast, the stall vortex within the impeller passage of the mixed-flow pump with 5 mm depth "O-Spoke" is

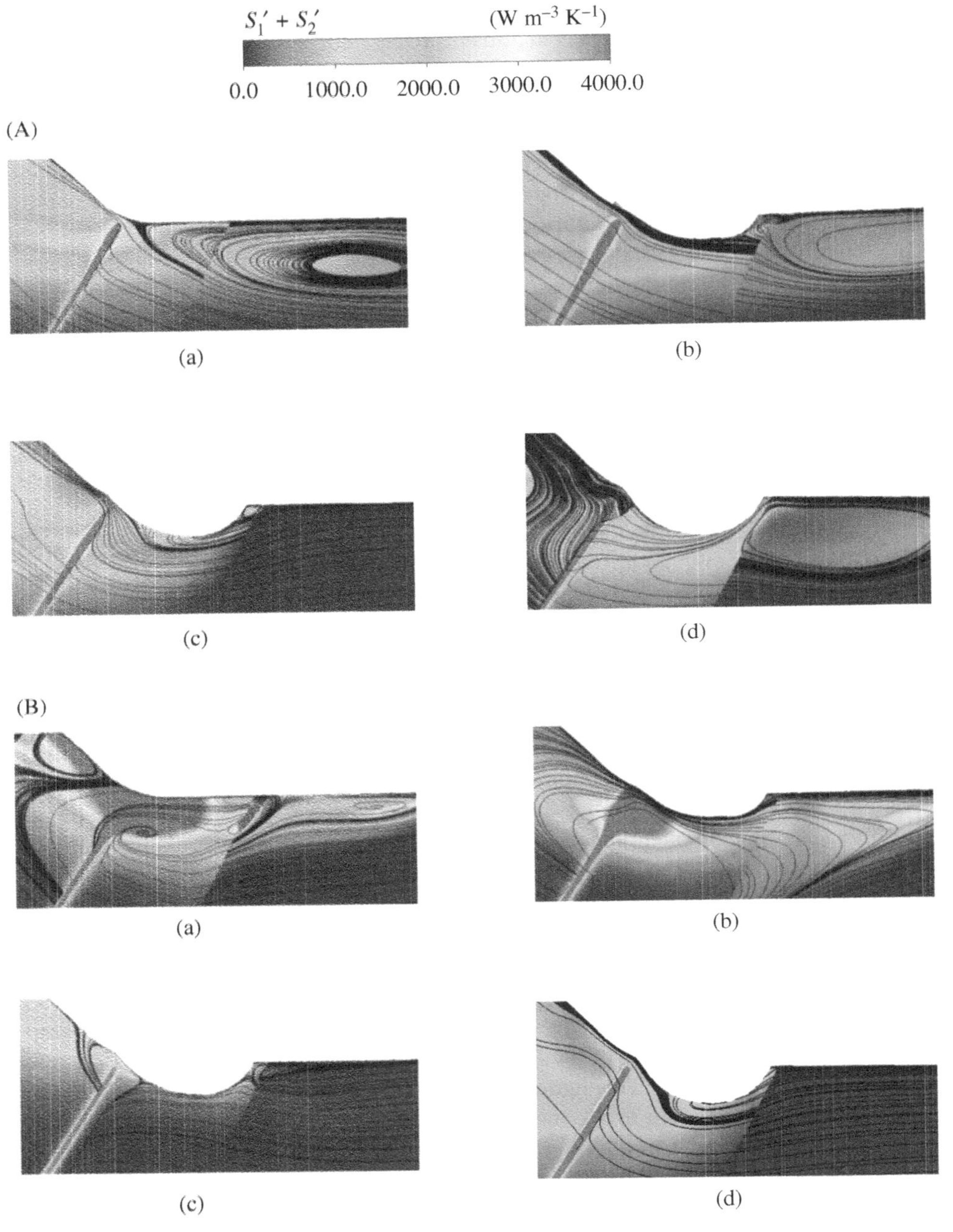


**Figure 9.26** Local entropy generation distributions at various operating conditions. (A) $0.54Q_{des}$ (a) The original pump, 4 mm "O-Spoke," (c) 5 mm "O-Spoke," and (d) 6 mm "O-Spoke." (B) $0.62Q_{des}$ (a) The original pump, 4 mm "O-Spoke," (c) 5 mm "O-Spoke," and (d) 6 mm "O-Spoke."

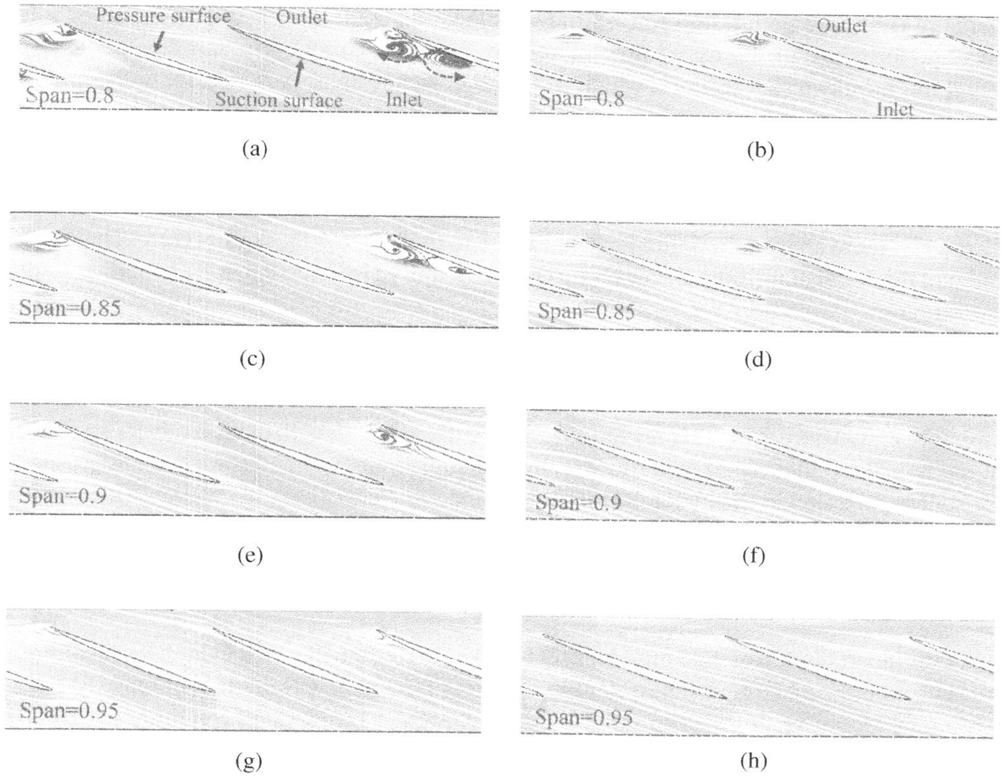


**Figure 9.27** The flow streamlines in various sections at $0.62Q_{des}$. (a) Span = 0.8, original pump, (b) Span = 0.8, 5 mm "O-Spoke," (c) Span = 0.85, original pump, (d) Span = 0.85, 5 mm "O-Spoke," (e) Span = 0.9, original pump, (f) Span = 0.9, 5 mm "O-Spoke" (g) Span = 0.95, original pump, and (h) Span = 0.95, 5 mm "O-Spoke."

almost entirely dissipated. By Span = 0.95, the stall vortex near the impeller outlet of the original mixed-flow pump gradually dissipates, with no vortex structure observed in the impeller passage after installing the 5 mm depth "O-Spoke." A more intuitive representation can be gleaned by examining Fig. 9.27(b), (d), (f), and (h), which highlight the effectiveness of the 5 mm depth "O-Spoke" in eliminating the stall vortex generated in the impeller passage when installed on the impeller inlet wall.

## 9.4 Summary

This chapter described the influence of the bade rim structure and the circumferential spoke structure on suppression of stall in the mixed-flow pump. The main conclusions are as follows:

(1) The blade rim structure has an influence on the energy performance curve of the mixed-flow pump. In the design flow rate condition, the Groove_Tip structure increases the head by about 1.8%, while the Bulge_Tip and Double_Rib_Tip structures reduce the head by 2.0% and 3.8%, respectively, and the head of the Single_Rib_Tip structure remains almost unchanged. In the stall condition, Bulge_Tip structure and Double_Rib_Tip structure make the critical

stall and deep stall points shift toward the low-flow rate point, which improves the "saddle area" performance of the mixed-flow pump; however, the Groove_Tip and Single_Rib_Tip structures have two stall sections, which worsen the "saddle area" performance of the mixed-flow pump.

(2) The structure of the rim gap affects the distribution of the fluid velocity field and the loss of TKE in the gap. The Bulge_Tip structure and the Double_Rib_Tip structure can restrain the TKE loss in the gap under stall condition to a certain extent, but Groove_Tip structure increases the loss, while there are certain limitations that relate to the Single_Rib_Tip structure. Therefore, the stall flow field of the mixed-flow pump can be optimized by selecting the appropriate blade rim structure.

(3) The rim clearance structure can not only affect the inlet angle distribution of the blade but also further affect the vortex structure inside the impeller by influencing the flow field in the rim area. Among them, Bulge_Tip and Double_Rib_Tip structures can affect the coalescence of BV structure and PV structure, thereby limiting the location and strength of unsteady SV structure at the impeller outlet. These two rim structures demonstrate a potential application prospect for stall suppression in the hydraulic machinery.

(4) Among the three proposed configurations of "O-Spoke" in this chapter, the 5 mm-depth "O-Spoke" (B scheme) exhibits superior performance, both in enhancing the external characteristics of the pump and in suppressing the leakage flow and stall vortices in the impeller passage.

(5) After the incorporation of the 5 mm-depth "O-Spoke" in the impeller (B scheme), the head-flow rate curve of the mixed-flow pump exhibits a smooth transition within the stall range ($0.4Q_{des}$ ~$0.6Q_{des}$). Moreover, at the deep stall condition ($0.54Q_{des}$), critical stall condition ($0.62Q_{des}$), and design condition ($1.0Q_{des}$), the pump efficiency is improved by 3.96%, 0.59%, and 1.27%, respectively, when compared to the original mixed-flow pump. The head also showed enhancements of 3.62%, 0.67%, and 2.18%, respectively. These results indicate that the inclusion of a 5 mm-depth "O-Spoke" (B scheme) effectively mitigates the stall in the mixed-flow pump impeller.

(6) The 5 mm-depth "O-Spoke" structure (B scheme) effectively reduces the leakage flow within the mixed-flow pump impeller and eliminates the formation of the stall vortices in the impeller passage. Consequently, this results in a reduced entropy generation distribution in the mixed-flow pump impeller compared to the original design. Furthermore, this structure also enhances the low-speed region in the cross-sectional area of the mixed-flow pump impeller in the stall range compared to the original impeller. Additionally, it leads to a smoother transition of pressure changes across different cross-sections of the Aligned Turbo Surface areas under the same flow conditions compared to the original mixed-flow pump impeller.

## References


**1** Li, W., Huang, Y., Ji, L. et al. (2023). Prediction model for energy conversion characteristics during transient processes in a mixed-flow pump. *Energy* 271: 127082.

**2** Li, W., Liu, M., Ji, L. et al. (2023). Study on the trajectory of tip leakage vortex and energy characteristics of mixed-flow pump under cavitation conditions. *Ocean Engineering* 267: 113225.

**3** Suh, J.W., Yang, H.M., Kim, Y.I. et al. (2019). Multi-objective optimization of a high efficiency and suction performance for mixed-flow pump impeller. *Engineering Applications of Computational Fluid Mechanics* 13 (1): 744–762.

4 Wang, M., Li, Y., Yuan, J. et al. (2022). Effects of different vortex designs on optimization results of mixed-flow pump. *Engineering Applications of Computational Fluid Mechanics* 16 (1): 36–57.

5 Sinha, M., Pinarbasi, A., and Katz, J. (2001). The flow structure during onset and developed states of rotating stall within a vaned diffuser of a centrifugal pump. *Journal of Fluids Engineering* 123 (3): 490–499.

6 Zhiting, T. (2006). Experimental study on unsteady correlation of tip leakage vortex, stall precursor and tip micro jet in axial compressor. Graduate School of Chinese Academy of Sciences (Institute of Engineering Thermophysics).

7 Goto, A. (1993). Suppression of mixed-flow pump instability and surge by the active alteration of impeller secondary flows. In: *ASME 1993 International Gas Turbine and Aeroengine Congress and Exposition*, May 24–27 1993, 93-GT-298. Cincinnati, OH.

8 Hongchao, S. and Yimin, L. (2000). Study on inlet reflux and its control of centrifugal pump. *Water Pump Technology* 03: 27–29. 37.

9 Zhangjinfeng, Yun, L., Yuanjianping et al. (2012). Numerical simulation of centrifugal pump inlet reflux flow field and its control method. *Journal of Jiangsu University (Natural Science Edition)* 33 (4): 402–407.

10 Krishnababu, S.K., Newton, P.J., Dawes, W.N. et al. (2009). Aerothermal investigations of tip leakage flow in axial flow turbines—Part I: effect of tip geometry and tip clearance gap. *Journal of Turbomachinery* 131 (1).

11 Ameri, A.A., Steinthorsson, E., and Rigby, D.L. (1998). Effect of squealer tip on rotor heat transfer and efficiency. 120: 753–759.

12 Tallman, J. and Lakshminarayana, B. (2001). Methods for desensitizing tip clearance effects in turbines. In: *ASME Turbo Expo 2001: Power for Land, Sea, and Air. American Society of Mechanical Engineers Digital Collection.*

13 Saha, A.K., Acharya, S., Bunker, R. et al. (2006). Blade tip leakage flow and heat transfer with pressure-side winglet. *International Journal of Rotating Machinery* 2006: 017079. (1–15).

14 Harvey, N.W. and Ramsden, K. (2001). A computational study of a novel turbine rotor partial shroud. *Journal of Turbomachinery* 123 (3): 534–543.

15 Dey, D. and Camci, C. (2001). Aerodynamic tip desensitization of an axial turbine rotor using tip platform extensions. In: *ASME Turbo Expo 2001: power for land, sea, and air. American Society of Mechanical Engineers Digital Collection.*

16 Saha, S.L., Kurokawa, J., Matsui, J. et al. (2000). Suppression of performance curve instability of a mixed flow pump by use of J-groove. *Journal of Fluids Engineering* 122 (3): 592–597.

17 Saha, S.L., Kurokawa, J., Matsui, J. et al. (2001). Passive control of rotating stall in a parallel-wall vaned diffuser by J-grooves. *Journal of Fluids Engineering* 123 (3): 507–515.

18 Shabbir, A. and Adamczyk, J.J. (2005). Flow mechanism for stall margin improvement due to circumferential casing grooves on axial compressors. *Journal of Turbomachinery* 127 (4): 708–717.

19 Choi, Y.D., Kurokawa, J., and Imamura, H. (2007). Suppression of cavitation in inducers by J-Grooves. *Journal of Fluids Engineering* 129 (1): 15–22.

20 Shimiya, N., Fujii, A., Horiguchi, H. et al. (2008). Suppression of cavitation instabilities in an inducer by j groove. *Journal of Fluids Engineering* 130 (2): 021302. (1–7).

21 Kurokawa, J., Imamura, H., and Choi, Y.D. (2010). Effect of J-groove on the suppression of swirl flow in a conical diffuser. *Journal of Fluids Engineering* 132 (7): 071101.

22 Kurokawa, J. (2011). J-groove technique for suppressing various anomalous flow phenomena in turbomachines. *International Journal of Fluid Machinery and Systems* 4 (1): 1–13.

**23** Lee, S.W. and Kim, S.U. (2010). Tip gap height effects on the aerodynamic performance of a cavity squealer tip in a turbine cascade in comparison with plane tip results: part 1—tip gap flow structure. *Experiments in Fluids* 49: 1039–1051.

**24** Juan, D., Jichao, L., Lipeng, G. et al. (2016). The impact of casing groove location on stall margin and tip clearance flow in a low-speed axial compressor. *Journal of Turbomachinery* 138 (12): 121007.

**25** Lu, H., Han, X., Zhang, Y. et al. (2017). Numerical simulation of transonic compressor rotor performance affected by the casing circumferential groove suction. *Journal of Engineering Thermophysics* 38 (6): 1191–1195.

**26** Yang, K.F., Feng, J.J., Zhu, G.J. et al. (2018). Study on improvement of hump characteristic of an axial flow axial pump by grooving inlet wall. In: *IOP Conference Series: Earth and Environmental Science*, vol. 163, 012090. IOP Publishing.

**27** Chen, H., Koley, S.S., Li, Y. et al. (2019). Systematic experimental evaluations aimed at optimizing the geometry of axial casing groove in a compressor. In: *Turbo Expo: Power for Land, Sea, and Air. American Society of Mechanical Engineers*, vol. 58554, V02AT39A021.

**28** Wu, X., Liu, B., Zhang, B. et al. (2021). Effect of circumferential single casing groove location on the flow stability under tip-clearance effect in a transonic axial flow compressor rotor. *Energies* 14 (19): 6143.

**29** Luo, Z., Nie, C., Lv, S. et al. (2022). The effect of J-groove on vortex suppression and energy dissipation in a draft tube of Francis turbine. *Energies* 15 (5): 1707.

**30** Li, W., Li, E., Ji, L. et al. (2020). Mechanism and propagation characteristics of rotating stall in a mixed-flow pump. *Renewable Energy* 153: 74–92.

**31** Ji, L., Li, W., Shi, W. et al. (2021). Effect of blade thickness on rotating stall of mixed-flow pump using entropy generation analysis. *Energy* 236: 121381.

**32** Ji, L., Li, W., Shi, W. et al. (2020). Diagnosis of internal energy characteristics of mixed-flow pump within stall region based on entropy production analysis model. *International Communications in Heat and Mass Transfer* 117: 104784.

**33** Li, W., Ji, L., Li, E. et al. (2021). Effect of tip clearance on rotating stall in a mixed-flow pump. *Journal of Turbomachinery* 143 (9): 1–39.

**34** Ji, L., Li, W., and Shi, W. (2020). Influence of tip leakage flow and inlet distortion flow on a mixed-flow pump with different tip clearances within the stall condition. *Proceedings of the Institution of Mechanical Engineers, Part A: Journal of Power and Energy* 234 (4): 433–453.

# 10

# Cavitation in a Mixed-flow Pump

The numerical results used to describe the special flow features/characteristics of the mixed-flow pump in this chapter were obtained by employing the computational methodology described in Chapter 3 which includes the pump model, mesh generation, turbulence models and their near wall treatment, and the boundary conditions. Additional details specific to this chapter are provided inside the text of the chapter.

## 10.1 Numerical Model

### 10.1.1 Cavitation Model

The homogenous mixture model is generally used to numerically simulate the two-phase cavitation flow. The cavitation process is controlled by the mass transfer equation with the conservation of mass fraction of liquid phase and gas phase, which can be defined as:

$$\frac{\partial(\rho_v\alpha_v)}{\partial t}+\frac{\partial(\rho_v\alpha_v u_j)}{\partial x_j}=\dot{m}^+-\dot{m}^- \tag{10.1}$$

where $\alpha_v$ is the volume fraction of the gas phase, and $\dot{m}^+$ and $\dot{m}^-$ represent the mass transfer during evaporation and condensation respectively.

Assuming that the gas phase region is composed of many tiny bubbles, the vapor volume fraction can be expressed as:

$$\alpha_v=\frac{4}{3}n_b\pi r_b^3 \tag{10.2}$$

where $n_b$ is the number of frontal vacuoles per unit volume and $r_b$ is the cavitation radius with value of $1\times10^{-6}$ m.

Here, the Zwart Gerber Belamari model is selected as the mass transfer model. The model is simplified by the Rayleigh-Plesset equation. Because of its high accuracy, this model is widely used in the cavitation simulation [1]. Rayleigh-Plesset equation is given as follows:

$$r_b\frac{d^2r_b}{dt^2}+\frac{3}{2}\left(\frac{dr_b}{dt}\right)^2+\frac{2\sigma}{r_b}=\frac{p_v-p}{\rho_l} \tag{10.3}$$

Ignoring the influence of the second-order term and surface tension term, the equation can be simplified to:

$$\frac{dr_b}{dt}=\sqrt{\frac{2}{3}\frac{|p_v-p|}{\rho_l}} \tag{10.4}$$

*Mixed-flow Pumps: Modeling, Simulation, and Measurements*, First Edition.
Wei Li, Leilei Ji, Ramesh Agarwal, Weidong Shi, and Ling Zhou.

The evaporation term and condensation term in the cavitation model can be expressed as:

$$\dot{m}^{+} = C_{\text{vap}} \frac{3\alpha_{\text{nuc}}(1-\alpha_{\text{v}})\rho_{\text{v}}}{R_{\text{b}}} \sqrt{\frac{2}{3} \frac{\max(p_{\text{v}} - p, 0)}{\rho_l}} \tag{10.5}$$

$$\dot{m}^{-} = C_{\text{cond}} \frac{3\alpha_{\text{v}}\rho_{\text{v}}}{R_{\text{b}}} \sqrt{\frac{2}{3} \frac{\max(p_{\text{v}} - p, 0)}{\rho_l}} \tag{10.6}$$

where $\alpha_{\text{nuc}}$ is the nucleation volume fraction, and $C_{\text{vap}} = 50$ and $C_{\text{cond}} = 0.01$ represent the evaporation coefficient and condensation coefficient respectively.

### 10.1.2 Numerical Simulation Setup and Boundary Conditions

The computations reported here were conducted using the commercial CFD software ANSYS-CFX. The details of mesh generation and the solution independence of mesh employed are not presented in detail here, the reader should refer to the previous literature for details [2]. The pressure-velocity direct coupling method is used to solve the governing fluid dynamics equations. For numerical calculation, the total pressure inlet is set as the inlet boundary condition and the turbulence intensity is set at 5%. The volume fraction of liquid and vapor are set 1 and 0 respectively. The mass flow outlet is adopted as the outlet boundary condition in the flow solver. The steady cavitating flow simulation starts from a steady non-cavitating flow field, and the transient cavitating flow simulation starts from the computed steady cavitating flow.

The physical parameters used in the simulation at assumed room temperature of $T = 298$ K are water density $\rho_{\text{l}} = 997\ \text{kg/m}^3$, water vapor density $\rho_{\text{v}} = 0.02308\ \text{kg/m}^3$, and saturated vapor pressure $\rho_{\text{v}} = 3169$ Pa. The impeller domain is in the rotating coordinate system with rotational speed $n = 1450$ r/min. The intersection of each hydraulic component is the Grid Generation Interface (GGI); the dynamic-static interface is set to Frozen Rotor type in steady simulation and the transient simulation is set to Transient Rotor Stator type. The rest of the geometric wall surfaces employ the no-slip boundary condition. The total time of transient simulation is set to be 0.62 seconds, and the time step is set at $1.149 \times 10^{-4}$ seconds. The time step is set in the transient simulation keeping in mind both the accuracy and computational efficiency. The convergence criterion is set as $1 \times 10^{-5}$ for all flow quantities and governing equations.

### 10.1.3 Cavitation Test Procedure

The experiment reported in this section has been conducted in the closed-circuit mixed-flow pump test rig of the Research Center of Fluid Machinery Engineering and Technology at Jiangsu University, China. The MPM-type pressure sensors are used to measure the inlet pressure and outlet pressure of the mixed-flow pump. They have 0.5% full-scale test accuracy. The accuracy of the LWGY-type tubing flowmeter is 0.5%, and that of the ZJ-type torque tachometer is 0.2%. The HSJ-2010 tester is linked to a computer and is used to capture and process all physical data collected by the sensors. High-speed photography is performed with a Photon high-speed camera with high photographic performance. The high-speed camera uses a high-sensitivity CMOS sensor with 100 million pixels at 20 000 fps, and its resolution size is $1024 \times 1024$. The light source is a halogen lamp with a power of 750 W. The experiment sets the sampling frequency to 20 000 Hz and the exposure time to 107 μs. At the start of the test, the high-speed camera is adjusted to a horizontal position using a level ruler, and the location of the light source is adjusted so that the camera can clearly record the flow condition within the impeller as shown in Fig. 10.1.

**Figure 10.1** High-speed photography experiment.

The experiment is carried out at an ambient temperature of 25 °C. The water level in the storage tank is adjusted to the suitable level at the start of the experiment; the inlet and outlet valves are opened, the motor is turned on and the speed is set to 1450 r/min, and the outlet valve is adjusted until the flowmeter display stabilizes at 380 $m^3/h$. The experiment employs the active cavitation approach, which involves continually lowering the input pressure in order to force the impeller runners to cavitate. When the pump pressure meter indicates a steady input pressure, the high-speed camera begins recording the cavitation pattern and collecting the external characteristics data.

### 10.1.4 Comparison of Numerical Simulations and Cavitation Test Results

When the total pump inlet pressure is reduced to a certain level, cavitation occurs in the flow path. The net positive suction head ($NPSH_a$) is defined as

$$NPSH_a = \frac{P_{\text{inlet}} - P_{\text{v}}}{\rho g} \tag{10.7}$$

The cavitation number is defined as

$$\sigma = \frac{P_{\text{inlet}} - P_{\text{v}}}{\frac{1}{2}\rho_l u^2} \tag{10.8}$$

where $P_{\text{inlet}}$ is the inlet pressure of the pump and $u$ is the circumferential velocity of the impeller.

For the design condition, the results of computations are compared in this section with the experimental data of the model mixed-flow pump described in Chapter 5 for different inlet pressure conditions. The head attenuation characteristic curves of the pump for various inlet pressures are obtained as shown in Fig. 10.2. In the cavitation test, the flow rate at the outlet is kept constant, and the head of the pump is obtained for different $NPSH_a$ values by adjusting the inlet pressure. Both the experimental and numerical results show that the pump head first increases slightly with decrease in $NPSH_a$, and then suddenly decreases after reaching the critical cavitation value. The reason for the small increase is that at the beginning of cavitation, the cavitation bubbles attach to the blade surface, leading to an increase in the smoothness of the blade, reducing the friction loss. As cavitation increases, the attached cavities break off and interact with the vortex, causing blockage of

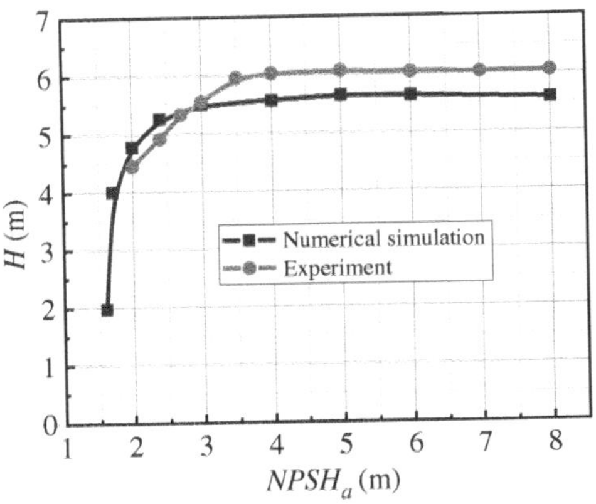


**Figure 10.2** Numerical and experimental comparison of cavitation performance curves.

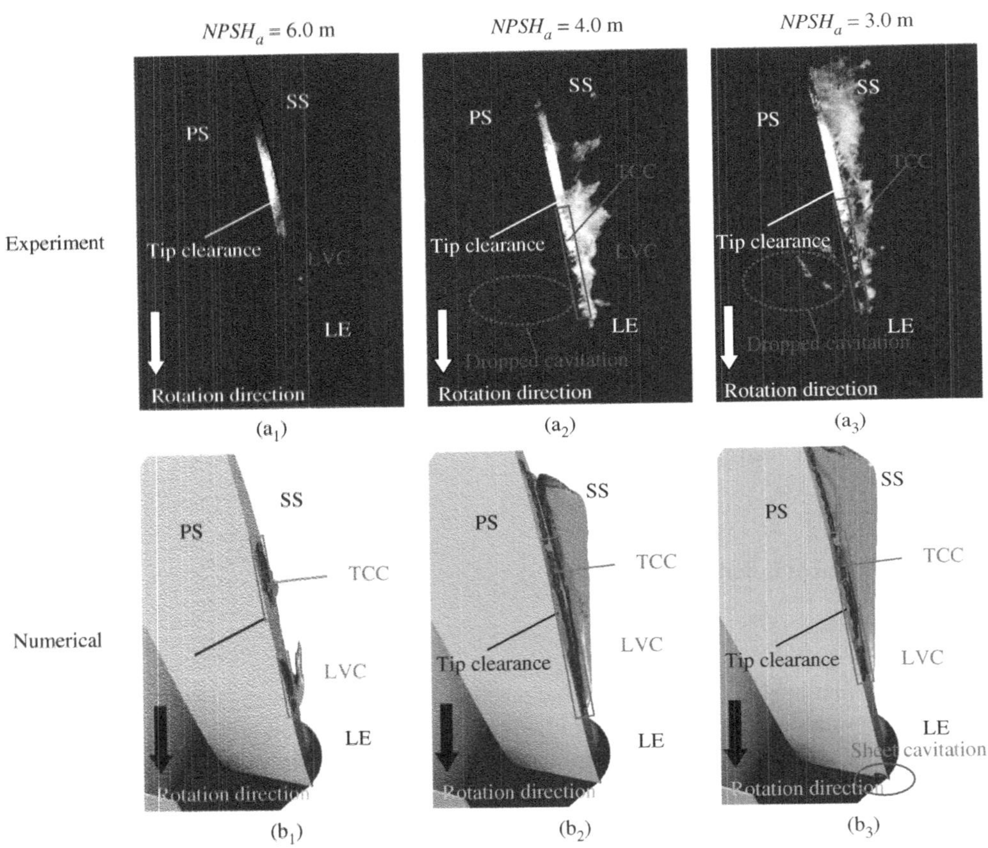


**Figure 10.3** Cavitation structure in impeller passage for various cavitation conditions.

the flow channel resulting in a sudden drop in pump performance. The comparison of computed and experimental results shows that the numerical simulation model and the method described in Chapter 6 have high accuracy in predicting both the external characteristics and cavitation characteristics of the pump.

To further demonstrate that the numerical simulation model and method can accurately predict the cavitation structure, Fig. 10.3 depicts the numerical simulation and experimental comparison of the cavitation flow structure in the flow passages under different cavitation conditions. Under slight cavitation condition, small local vortical cavitation (LVC) occurs near the leading edge (LE) of the blade on the suction side (SS). The numerical results accurately predict the small structure of LVC; however, the numerical results also show the tip clearance cavitation (TCC), which was not observed in the experiment. As the cavitation increases, the LVC develops toward the trailing edge (TE), and the TCC region expands. In addition, the pressure drops in the passage due to cavitation, which is distinctly visible, as shown in Figs. 10.3($a_2$) and ($a_3$). When $NPSH_a = 3.0$ m, the sheet cavitation attached to the pressure side (PS) appears near the blade LE in Fig. 10.3($b_3$), which is difficult to capture in the experiment. Overall, it can be concluded that the numerical simulations can predict the cavitation structure reasonably well in the tip region of the pump, which provide certain degree of confidence in the simulations.

## 10.2 Flow Characteristics of the Mixed-flow Pump Under Cavitation

### 10.2.1 Blade Surface Load Distribution

To better understand the pressure load distribution on SS under cavitation, the pressure coefficients at normalized blade span = 0.2, 0.5, and 0.8 are presented in Fig. 10.4(a)–(c), respectively. In these figures, the $X$-axis represents the dimensionless length of a blade from the LE to TE, the $Y$-axis represents the cavitation stage, and the $Z$-axis represents the pressure coefficient ($C_\mathrm{p}$) on SS. The enclosed area represents the distribution of load on the suction surface SS. The pressure coefficient of the suction surface can be expressed as

$$C_\mathrm{p} = \frac{P_\mathrm{static}}{\frac{1}{2}\rho_l u^2} \tag{10.9}$$

where $P_\mathrm{static}$ denotes the static pressure on the blade surface.

As can be noted from Fig. 10.4, when $NPSH_a = 6.0$ m, the pressure coefficient on the SS does not vary with span. On the other hand, when the $NPSH_a$ value decreases to 3, the pressure load in the span region from 0.3 to 0.6 decreases streamwise with an increase in the span and has a similar trend under other cavitation conditions. However, in the same span region, the load distribution on SS is inconsistent under different cavitation conditions. The numerical results show that the stronger the cavitation, the lower the pressure load on the SS. In addition, the region of low-pressure load expands with the intensity of cavitation. It can be noted that at a low $NPSH_a$ value, there is a large range of low-pressure regions on the SS, and the low-pressure region expands toward the LE, the TE, and the hub with decrease in $NPSH_a$. According to the study of Yun Long, these results are consistent with the trend in cavitation distribution caused by the low-pressure regions [3]. Comparing the three-span areas, the load distribution of SS in the 0.8 span region is significantly lower than that in the other two regions, and the region near the tip clearance has more

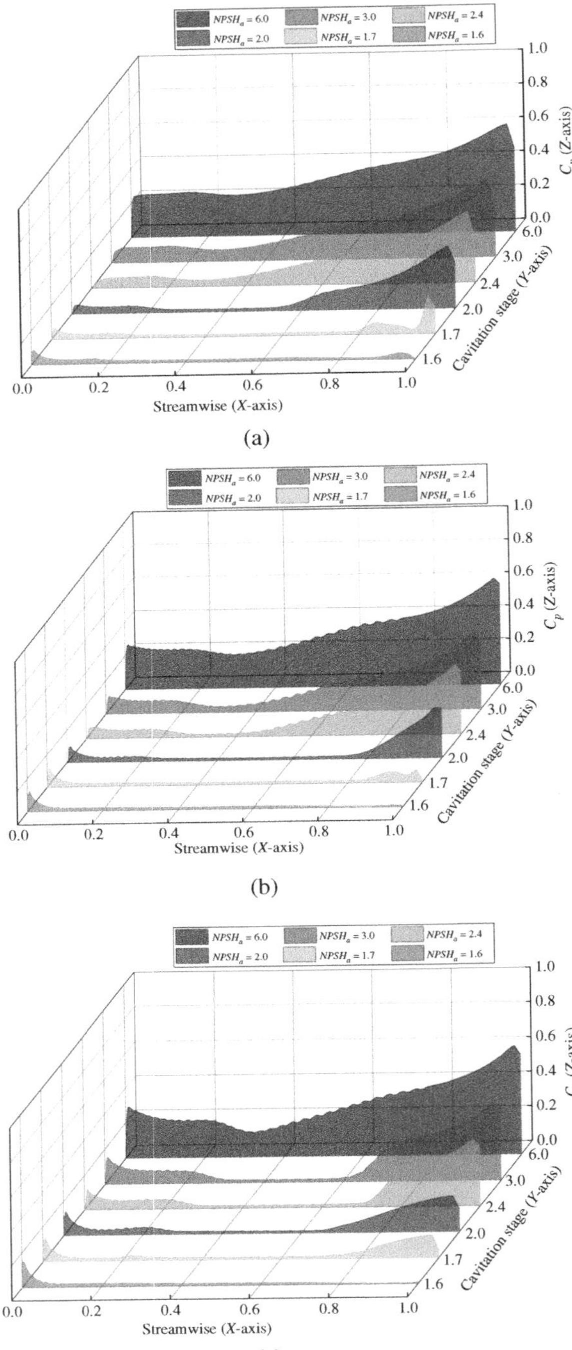


**Figure 10.4** Pressure load distribution on the suction side (SS) of the blade for various $NPSH_a$ values. (a) 0.2 span, (b) 0.5 span, and (c) 0.8 span.

serious cavitation. In conclusion, the intensity of cavitation greatly affects the load distribution on the SS, and the variation in the distribution of low-pressure region on the SS is related to the location of cavitation.

### 10.2.2 Trajectory of TLV for Different Values of $NPSH_a$

Due to the existence of tip clearance, the occurrence of tip leakage flow and tip leakage vortex (TLV) caused by the pressure gradient between PS and SS is inevitable. They are not only affected by the clearance size and the flow rate, but also are related to the state of cavitation. Figure 10.5 shows the streamlines at the impeller outlet for different $NPSH_a$ values and the vortex structures based on the $Q$ criterion, where $Q = 300\,000\,\text{s}^{-2}$.

The geometry of the TLV at the tip clearance of the impeller is altered dramatically for different values of $NPSH_a$ as shown in Fig. 10.5. As the $NPSH_a$ value decreases, the starting location of the TLV shifts from the TE to the LE, and then back to the TE. When there is slight cavitation, the TLV is mostly localized around the PS of the lower blade, and the TLV trajectory is almost complete. However, when the $NPSH_a$ value decreases, the TLV location shifts away from the PS of the lower blade toward the LE of the current stage of the impeller. A considerable number of vortex filaments develop at the tip clearance. As cavitation increases, a separation vortex develops toward the TE, and a high number of trailing edge cavitation (TEC) structures congregate near the outflow of the impeller passage, causing substantial obstruction. As a result, the development of cavitation not only affects the form and structure of the TLV at the tip clearance but also enhances the vortex structure in the impeller. The presence of cavitation obstructs the impeller passage and lowers the efficiency.

Figure 10.6 shows the flow field structure in the impeller of the pump for various values of $NPSH_a$. The velocity distribution and the three-dimensional streamlines emanating from the tip clearance region are shown in this figure. Seven sections are divided and marked as Sections 1–7. Because of large flow at the impeller, velocity is large, and the size of the low-speed zone is tiny; thus, the forward cloud scale is not suitable for the study and therefore the reverse cloud scale is employed. As can be seen from this figure, the TLF is entrained into the TLV structures at different positions in the impeller passage under various cavitation conditions. The TLV develops at the LE and continues to develop toward the TE. With increase in cavitation, the TLV entrainment is enhanced and the TLV area expands. When $NPSH_a = 6.0$ m, due to slight cavitation, the influence range of the TLV structure is small, and the degree of entrainment is small. In addition, because the TLV with strong mainstream flow is extruded farther, the flow field on the PS of the lower blade is affected. With the decrease in $NPSH_a$, the influence of the TLV on the PS becomes more prominent and the influence of TLV on PS and SS increases, and the vortex core on the suction surface gradually approaches the tip clearance. In addition, part of the upper blade TLF includes the secondary TLV (STLV) which is sucked into the tip clearance of this stage forming a new TLF. In summary, with the increase in cavitation, the flow field structure in the impeller passage becomes more complex, and the influence range of TLV increases. In addition, the secondary TLV (STLV) appears as the cavitation becomes stronger.

### 10.2.3 Evolution of the Transient Flow Trajectory of the TLV and Tip Cavitation Flow

To understand the relationship between TLV and its induced cavitation morphology, the transient evolution process of TLV and tip cavitation is captured by both the numerical simulation and the high-speed photography as shown in Fig. 10.7. The number of blades considered is 4, the rotation

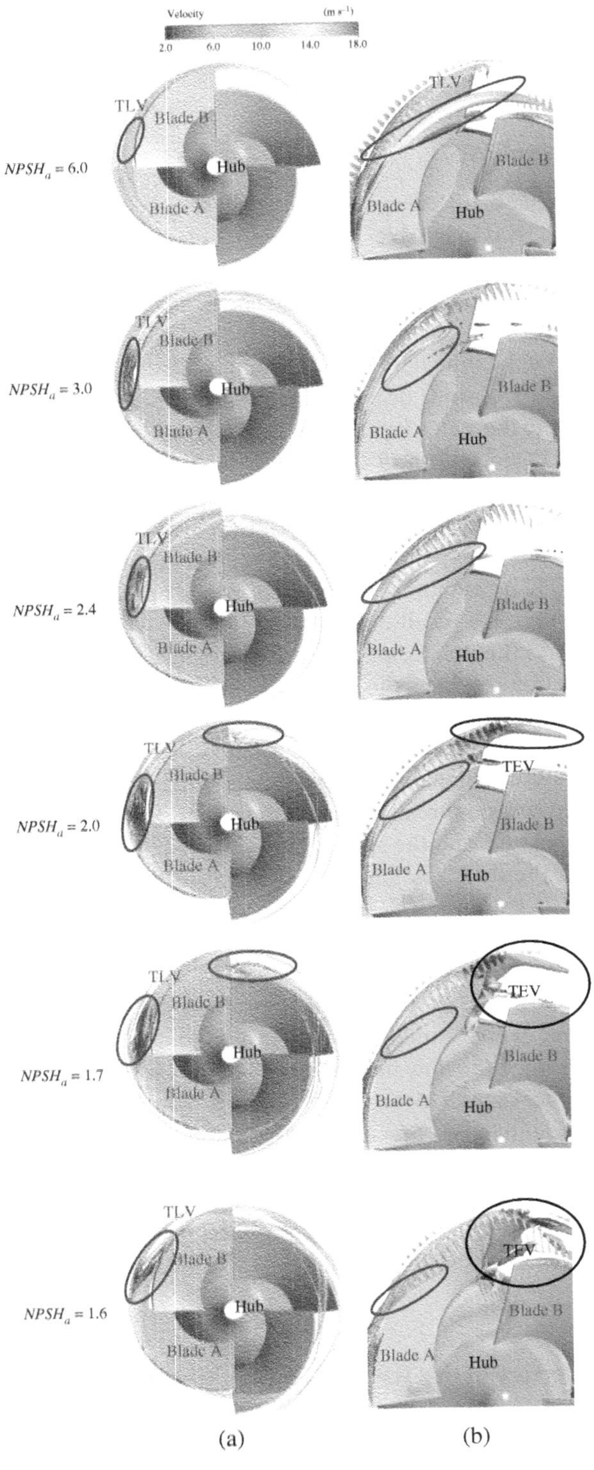


**Figure 10.5** Impeller outlet streamlines diagram and vortex core distribution for various $NPSH_a$ values. (a) Streamline diagram and (b) Vortex structure diagram.

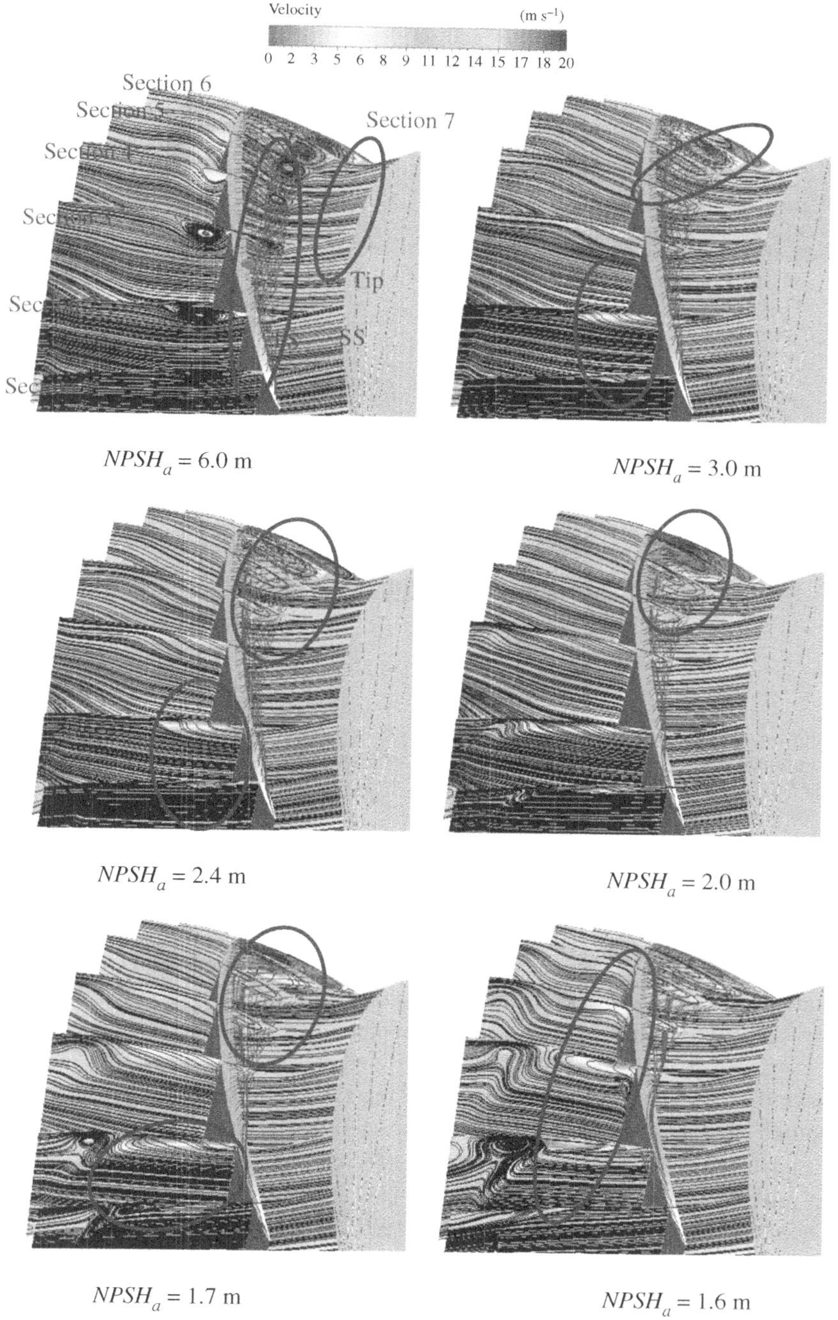


**Figure 10.6** Tip clearance vortex core distribution for various $NPSH_a$ values.

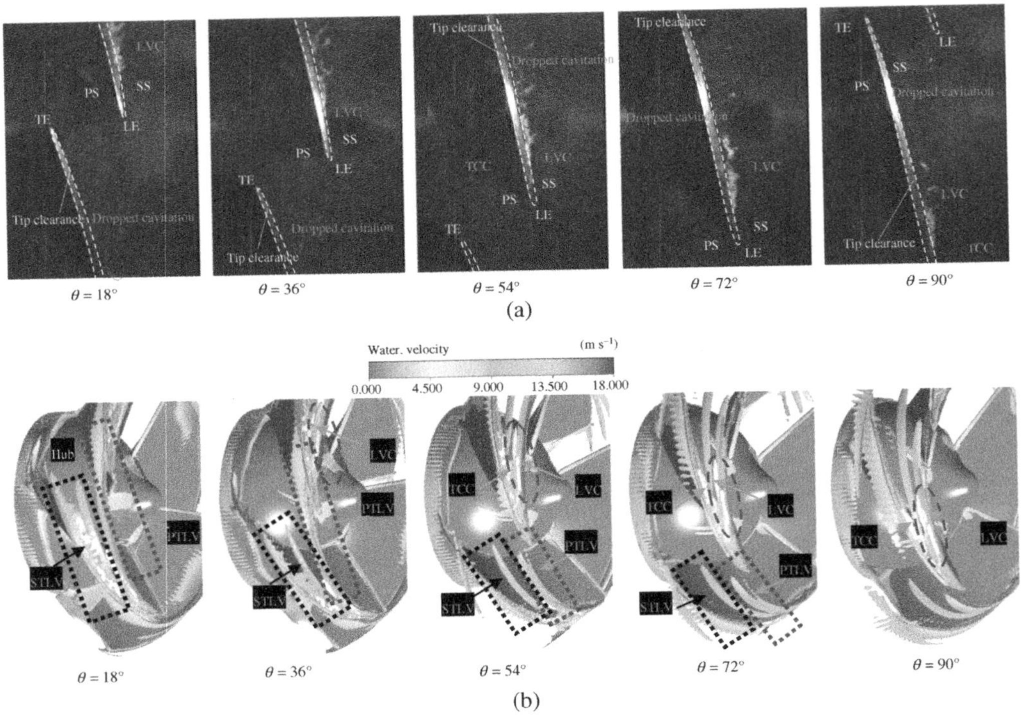


**Figure 10.7** Time evolution of leakage vortices for $NPSH_a = 5$ m. (a) Transient process of tip cavitation flow captured by high-speed photography. (b) Transient process of tip cavitation flow captured by numerical simulation ($\alpha v = 0.1$, $Q = 320\,000\ s^{-2}$).

angle of a single blade is 90° when the impeller rotates one circle, and five images are selected at equal angle intervals to show the structure of cavitation flow, that is one image is taken every 18°. Figure 10.7(a) shows the cavitation morphology of the tip area observed at different instances of time by high-speed photography. The evolution of cavitation over time can be visualized in this figure. The cavitation area starts close to the blade LE. Over time, the cavitation morphology and the location of its occurrence hardly changes and is accompanied by the dropped cavitation structures.

Figure 10.7(b) shows the transient process of capturing the tip cavitation flow by numerical simulation; the computed cavitation shape and area are consistent with the experimental results. The $Q$ criterion is used to describe the computed TLV trajectory. There are two typical TLV structures in the tip region, which include the primary TLV (PTLV) and the secondary TLV (STLV) structures. The red dotted box in the figure represents the PTLV, and the black dotted box represents the STLV. At a rotation angle $\theta = 18°$, the PTLV propagates toward the pressure surface (PS) of the lower stage blades, and a strong STLV is formed on the SS of the blades at this time. When $\theta = 54°$, there is significant vortex diffusion in the PTLV and STLV structures compared to that at $\theta = 18°$. At this time, the PTLV propagation distance is significantly shortened, and STLV is slightly diffused but retains its structure. During the forward advance of the impeller, the main STLV continuously advances toward the LE, and its starting point is close to the starting position of the PTLV. The interaction between the PTLV and STLV disturbs the mainstream flow causing the pump head to drop, which is consistent with the trend of the curve shown in Fig. 10.2.

## 10.3 Analysis of Cavitation Energy Characteristics

### 10.3.1 Analysis of Pump Energy Characteristics

Figure 10.8 shows the variation in the head coefficient Ψ along the *Z*-axis for various values of $NPSH_a$ at the design flow condition. The head coefficient Ψ is defined as

$$\Psi = \frac{P_{\mathrm{ave}} - P_{\mathrm{inlet}}}{\rho_1 g} \tag{10.10}$$

where $P_{\mathrm{ave}}$ is the average pressure of any section of the mixed-flow pump passage in Pa.

As can be seen from Fig. 10.8, the variation in the head coefficient is basically the same from the pump inlet to the guide vane outlet at different $NPSH_a$ values, but the head coefficient shows a decreasing trend as the $NPSH_a$ value decreases. In the pump zone A, the head coefficient Ψ is the same not only in value but also in the trend since cavitation hardly occurs in the inlet pipe. However, in the pump zone B, as the value of $NPSH_a$ decreases, the head coefficient Ψ shows a decreasing trend which can be noted especially when $NPSH_a = 1.6$; it is due to the significant cavitation on the impeller resulting in the blockage of the flow passage thereby reducing the work capacity of the impeller. In the C region of the guide vane, the head coefficient Ψ again shows a decreasing trend like in the case of the impeller with the decreasing value of $NPSH_a$ indicating that the over-flow capacity inside the guide vane is seriously affected due to the bubbles generated by the cavitation. In summary, with the increase in cavitation, not only the work capacity of the impeller is weakened, but also the flow in guide vane is affected by the cavitation resulting in a sharp reduction in the energy performance of the pump, which is the main reason for the head drop and operating instability of the pump.

### 10.3.2 Analysis of Energy Loss in the End Wall Region of the Pump Caused by the TLV-Induced Cavitation

The results in Section 10.3.1 show that the head drop of the model pump is mainly due to the energy loss caused by the cavitation; however, the fluid mechanical sources of energy loss were

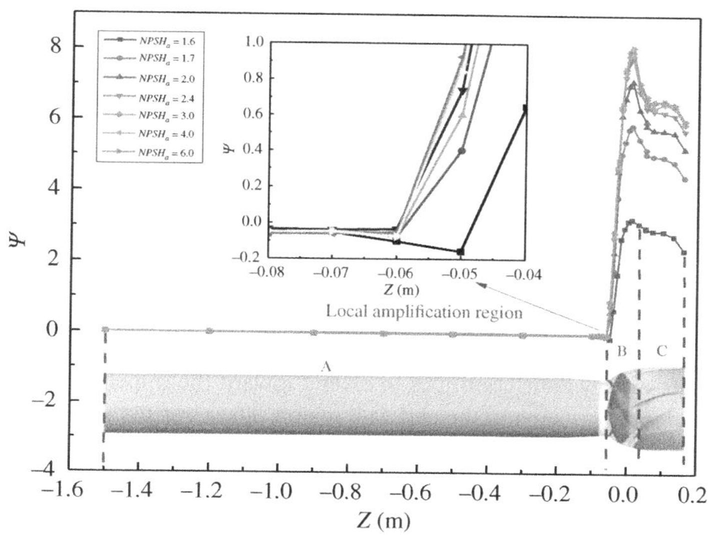


**Figure 10.8** Variation of head coefficient Ψ along the axial direction for different cavitation conditions.

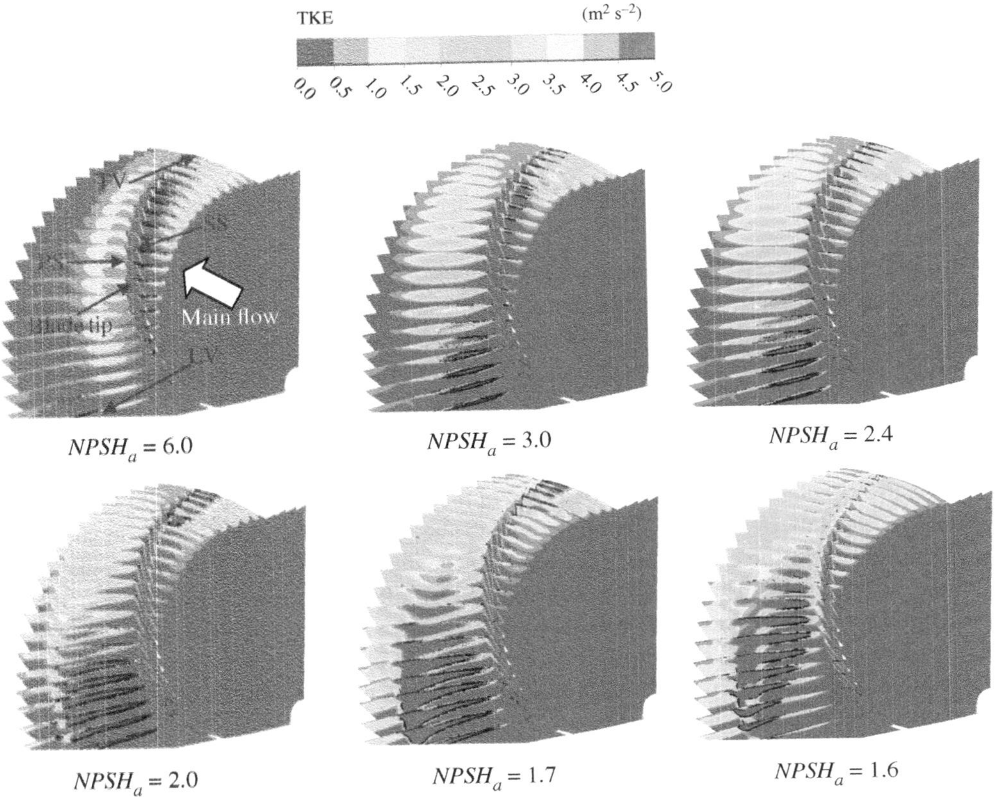


**Figure 10.9** Turbulent kinetic energy distribution in the impeller for various $NPSH_a$ values.

not discussed. The turbulent kinetic energy of the flow is an important indicator of the turbulent dissipation in the flow and the resulting fluid energy loss. Figure 10.9 compares the distribution of turbulent kinetic energy loss in the end wall region of the pump caused by the TLV-induced cavitation at various $NPSH_a$ values. When $NPSH_a$ is large, the high turbulent kinetic energy zone in the end wall region is affected by the cavitation induced by the TLV, and the high turbulent kinetic energy zone is almost the same as the location and area occupied by the TLV. On the other hand, with the decrease in $NPSH_a$, the high turbulent kinetic energy zone increases, and the position of TLV gradually shifts in the direction of the impeller outlet passage. When $NPSH_a = 6$ m, the high turbulent kinetic energy area mainly lies in the middle of the blade rim and near the TE of the blade, which is also the area affected by the cavitation due to TLV. When the $NPSH_a$ decreases to 3 m, the high turbulent kinetic energy caused by the TLV-induced cavitation gradually decreases, and the high turbulent kinetic energy area appears in the middle of the impeller runner and is caused by the separated flow in the end wall region. As the $NPSH_a$ continues to decrease, the high turbulent kinetic energy zone continues to extend toward the impeller outlet passage due to increase in the separated flow area in the end wall region. These results show that the cavitation induced by the TLV will lead to the formation of a high turbulent kinetic energy zone in the end wall area of the impeller. With decrease in $NPSH_a$, the high turbulent kinetic energy zone in the tip clearance region caused by the TLV-induced cavitation gradually decreases. Instead, the high turbulent kinetic energy zone in the rim region is caused by the separated flow in the end wall region. The high turbulent kinetic energy zone in the end wall region expands with the intensification/increase of cavitation and moves toward the impeller outlet, deteriorating the flow in the impeller.

### 10.3.3 Effect of Cavitation on the Output Power of the Pump

To have a deeper understanding of the internal energy characteristics of the pump, the impeller is divided into 10 domains along the streamline by using 11 sections, as shown in Fig. 10.10

The output power of the pump is equal to the moment imparted by the impeller multiplied by the rotational angular velocity of the pump shaft. The moment has two components: one is the moment generated by the pressure on the impeller surface, and the other is the moment generated by the viscous force on the non-slip wall [3]. Since the viscous force is small, therefore it can be neglected. The formula for calculation of power can be written as:

$$P = P_{\mathrm{p}} = \int (F_{\mathrm{p}} \times L)\mathrm{d}s \times \omega \tag{10.11}$$

In Eq. (10.11), $P$ is the pump output power in W; $P_{\mathrm{p}}$ is the power generated by the pressure on the impeller walls in W; $F_{\mathrm{p}}$ is the force due to the pressure on the impeller wall in N; $L$ is the distance from the axis of rotation of the pump to the point of resultant force in m; and $\omega$ is the angular speed of rotation of the pump in rad/s.

Figure 10.11 shows the output power of the impeller for different cavitation conditions. From this figure, it can be noted that different cavitation conditions can have significant effect on the output power of the impeller, and the output power shows a decreasing trend with decrease in $NPSH_a$. The stronger the cavitation, the lower the output power of the impeller. The results of numerical calculation show that the output power at $NPSH_a = 2$ m is only about 58% of that at $NPSH_a = 6$ m, and the output power at $NPSH_a = 1.6$ m is only about 35% of that at $NPSH_a = 6$ m. To further analyze the effect of cavitation on the output power of the impeller, the output power of each domain of the impeller is presented in the following paragraph.

Figure 10.12 shows the output power distribution in each region of the impeller under different cavitation conditions. When the $NPSH_a$ values are 6.0 and 3.0 m, respectively, the overall distribution of the output power in each region of the impeller under these two cavitation conditions is almost the same, except that the output power in the latter is substantially reduced in region three relative to the former. When $NPSH_a = 2.4$ m, the output power in domain 9 and domain 10 (impeller outlet) is significantly reduced due to the increase in cavitation. This may be caused by many dislodged cavitation bubbles at the impeller outlet, resulting in blockage of the flow passage. When the $NPSH_a$ value is reduced to 1.6, the output power of domain 1 and domain 2 is reduced to

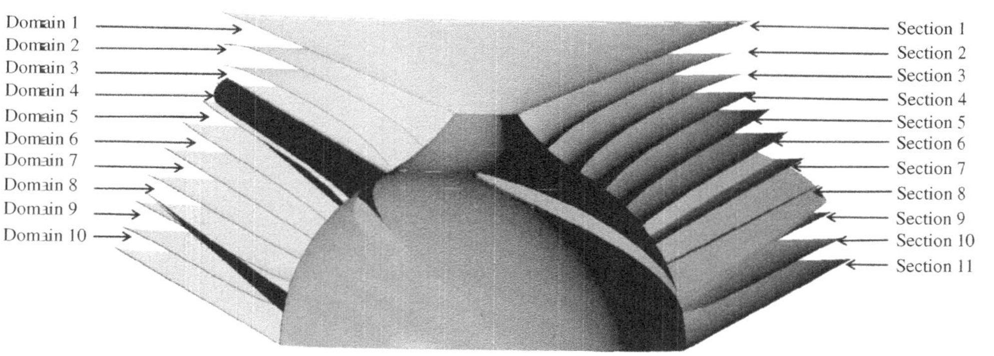


**Figure 10.10** Domain and section divisions.

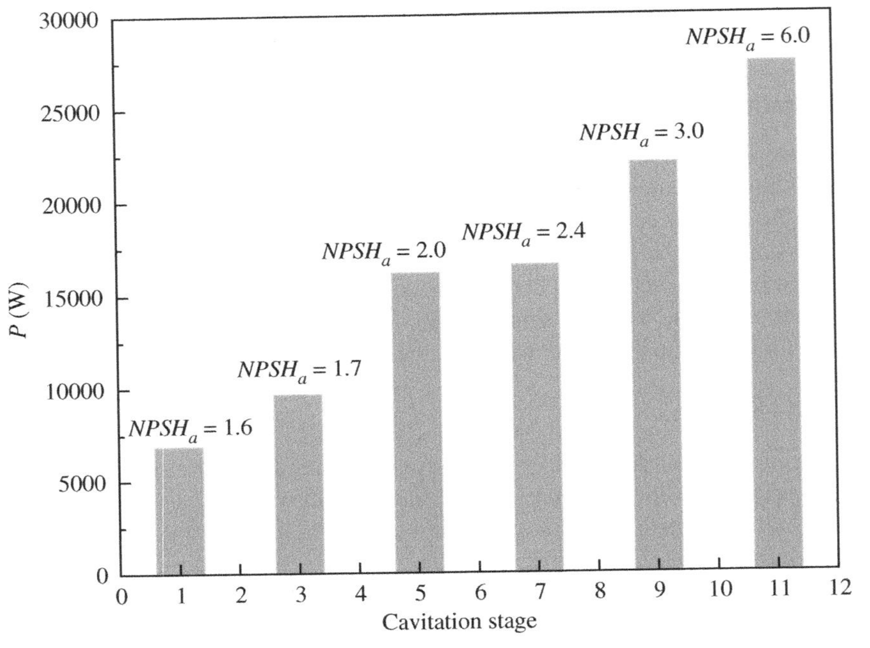


**Figure 10.11** Output power of the impeller for various $NPSH_a$ values.

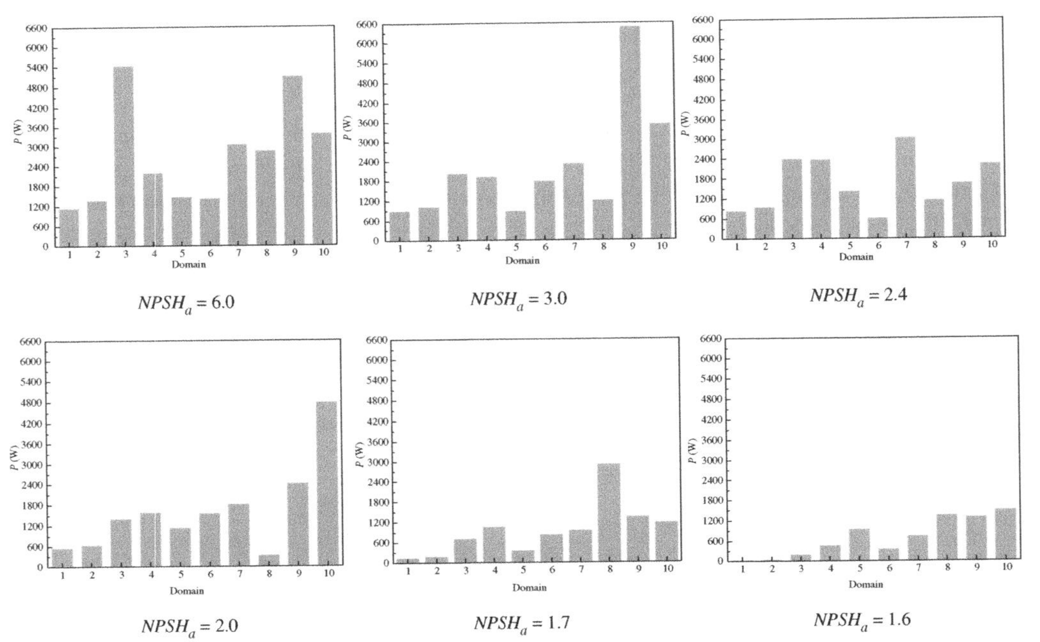


**Figure 10.12** Output power of each area of the impeller for various $NPSH_a$ values.

almost 0, and the output power of other domains is reduced to different degrees compared to that at $NPSH_a = 1.7$ m. In general, the output power of each region of the impeller decreases to different degree with the decrease in $NPSH_{a,}$ especially when the amount of cavitation is significant. This is due to the pressure gradient on the blade surface, resulting in the uneven distribution of blade load.

### 10.3.4 Effect of Cavitation on Energy Losses in the Pump

The output power of a mixed-flow pump can be divided into two parts: one is the net energy from the fluid, and the other is the energy lost by the flow, where the energy lost by the flow consists of the turbulent dissipation losses and the friction losses of the pump [4], which are defined as follows.

The energy loss due to turbulent dissipation can be written as:

$$P_{\text{tur}} = \int \rho \varepsilon \mathrm{d}V \tag{10.12}$$

where $\rho$ is the density of the two-phase fluid and $\varepsilon$ is the turbulent dissipation rate.

The energy loss due to friction can be written as:

$$P_{\text{Fri}} = \int F_{\text{fri}} \cdot v dS \tag{10.13}$$

where $F_{\text{fri}}$ is force due to the wall friction and $\boldsymbol{v}$ is the relative velocity between the fluid and the wall.

Figure 10.13 shows the turbulent dissipation losses in each domain of the impeller under different cavitation conditions. From this figure, it can be observed that the trend in loss due to

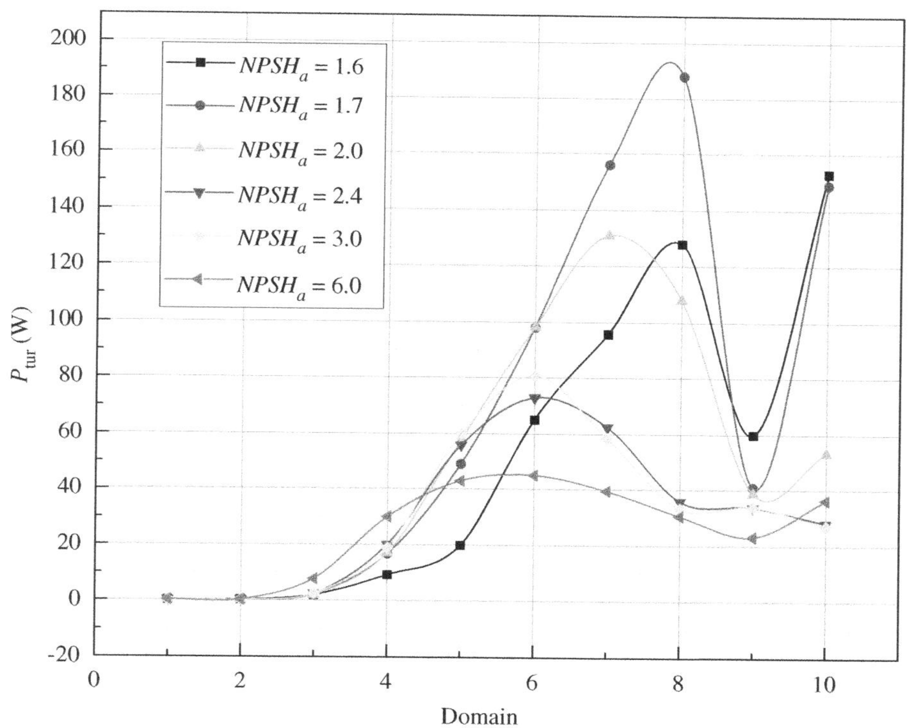


**Figure 10.13** Turbulent dissipation loss in each region of the impeller for various $NPSH_a$ values.

turbulent dissipation in each domain under different cavitation conditions is basically the same. When $NPSH_a = 6.0$ m, the maximum turbulent dissipation loss in the impeller is in domain 6. When the value of $NPSH_a$ decreases to 2.4, the maximum turbulent dissipation loss shifts from domain 6 to domain 7. As the value of $NPSH_a$ continues to decrease, the maximum turbulent dissipation loss continues to move to the next domain. When the value of $NPSH_a$ decreases to 1.7, the maximum turbulent dissipation is in domain 8 and the location of the maximum turbulent dissipation loss region does not continue to change with further decrease in the value of $NPSH_a$. At $NPSH_a = 1.6$ m, the maximum turbulent dissipation loss decreases slightly relative to that at $NPSH_a = 1.7$ m. In domain 10, the turbulent dissipation loss increases slightly relative to the upper region at lower $NPSH_a$ values, and the loss increases significantly at lower $NPSH_a$ values.

Figure 10.14 shows the friction losses in each section of the impeller under different cavitation conditions, and the energy loss due to friction is relatively larger compared to the turbulent dissipation loss. Overall, the change in the trend of friction loss is almost the same for each cavitation condition, and the friction loss shows an increasing trend from Sections 1–11. In Section 5, except for the valley value of friction loss at $NPSH_a = 1.6$ m, the first peak value of the friction loss appears at each cavitation condition, and the peak value decreases with the decrease in the $NPSH_a$ value. In Section 9, a valley value of friction loss occurs at $NPSH_a = 6.0$ m, and a second peak value of friction loss occurs at all other cavitation conditions, and the friction loss increases with increase in cavitation. Comparing Figs. 10.13 and 10.14, the region of large turbulent dissipation loss has the same location as the cross-section of the large friction loss, indicating that there is a larger turbulent dissipation loss in the region of large friction loss.

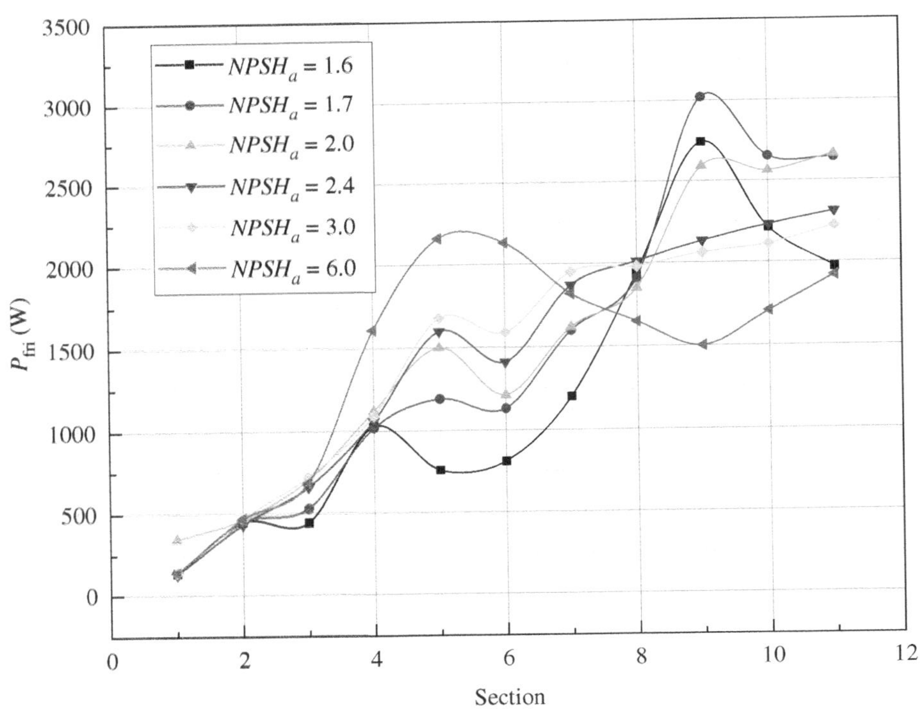


**Figure 10.14** Friction loss of each section of the impeller for various $NPSH_a$ values.

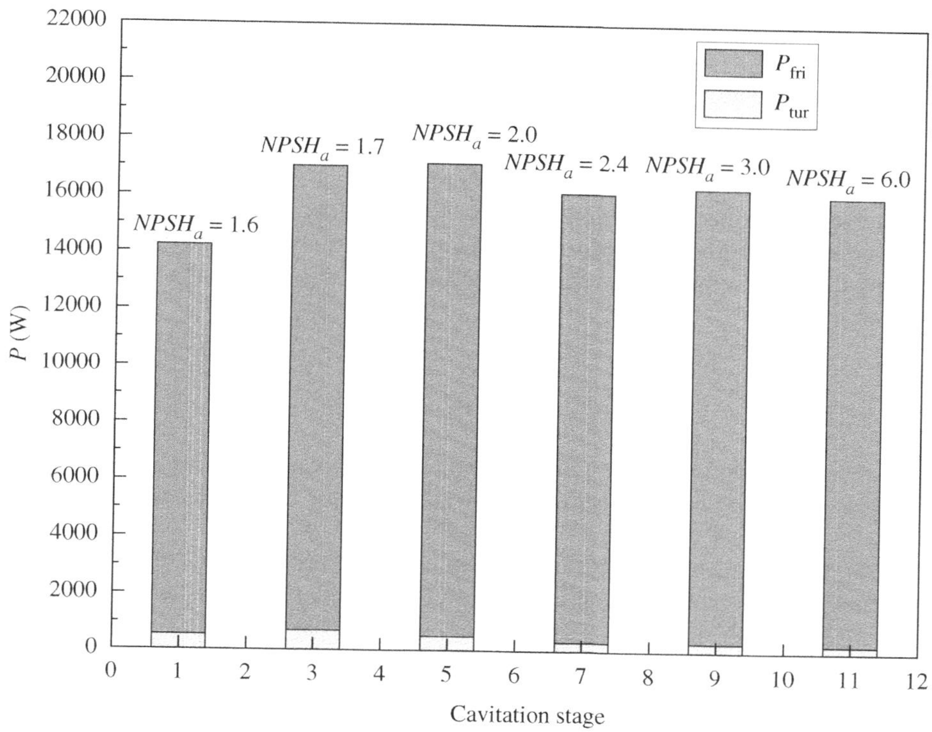


**Figure 10.15** Energy loss in the impeller area for various $NPSH_a$ values.

Figure 10.15 shows the total energy loss in the impeller region in different cavitation stages, including the turbulent dissipation loss and the friction loss components. It should be noted that the energy loss in the mixed-flow pump is mainly caused by the friction between the fluid and the wall, and the turbulent dissipation loss is smaller. Compared to the mild cavitation stage, severe cavitation leads to increased frictional and turbulent dissipation losses and increased total energy loss resulting in a decrease in the effective energy obtained by the pump, leading to a decrease in the overall pump performance.

## 10.4 Summary

(1) The cavitation intensity seriously affects the load distribution on suction surface (SS) of the blade, and the variation pattern of the low-pressure area distribution on SS is related to the cavitation location.
(2) With the deterioration of cavitation, the flow field structure in the impeller becomes more complex. In addition, the STLV becomes more obvious when the cavitation deteriorates. There are two typical TLV structures in the tip region, including the primary tip leakage vortex (PTLV) and the STLV structure. The interaction between PTLV and STLV disturbs the mainstream flow, causing the pump head to drop.
(3) As cavitation deteriorates, the working capacity of the impeller is weakened and the guide vane is also affected, resulting in a sharp drop in the energy performance of the pump.

Severe cavitation becomes the main reason for the decreasing head and unstable operation of the mixed-flow pump.

(4) The turbulent kinetic energy distribution characteristics are related to the degree of cavitation. The high turbulent kinetic energy zone in the end wall area expands with the intensification of cavitation and moves toward the impeller outlet, deteriorating the flow state in the impeller.

(5) Cavitation deterioration reduces the output power of the impeller, and the power distribution characteristics are affected by the degree of cavitation. The severe cavitation leads to increased frictional and turbulent dissipation losses and increased total energy loss, resulting in a decrease in the effective energy obtained, leading to a decrease in the overall pump performance.

## References

1 Zwart, P.J., Gerber, A.G., and Belamri, T. (2004). A two-phase flow model for predicting cavitation dynamics. In: *Fifth International Conference on Multiphase Flow*, Yokohama, Japan, vol. 152.

2 Ji, L., Li, W., Shi, W., and Agarwal, R. (2021). Transient characteristics of internal flow fields of mixed-flow pump with different tip clearances under stall condition. *Proceedings of the Institution of Mechanical Engineers Part A: Journal of Power and Energy* 235: 700–717.

3 Long, Y., An, C., Zhu, R., and Chen, J. (2021). Research on hydrodynamics of high velocity regions in a water-jet pump based on experimental and numerical calculations at different cavitation conditions. *Physics of Fluids* 33 (4): 045124. (1–16).

4 Shi, G., Wang, S., Xiao, Y. et al. (2021). Effect of cavitation on energy conversion characteristics of a multiphase pump. *Renewable Energy* 177: 1308–1320.

# 11

# Analysis of the Vortex Dynamics Characteristics in the Tip Region of the Mixed-flow Pump Under Cavitation

## 11.1 The Tip Leakage Flow Characteristics Under Cavitation

### 11.1.1 Tip Leakage Flow Characteristics

Figure 11.1 shows the characteristics of the tip leakage flow (TLF) on the blade along the flow direction by dividing the tip clearance region into multiple units along the flow direction. The leakage flow rate is expressed by the leakage flow intensity (the mass flow per unit area) of each unit. The leakage flow intensity can be expressed as:

$$Q_i = \frac{\int \rho_l v \cdot \boldsymbol{n} dA_i}{\int dA_i} \tag{11.1}$$

where $A_i$ is the area of the unit section in m$^2$, $v$ is the relative velocity of the fluid through the unit section in m/s and $\boldsymbol{n}$ is the normal vector on the surface of the unit section.

Figure 11.2 shows the leakage flow intensity distribution in the tip clearance region from the leading edge (LE) to the trailing edge (TE) of the blade at various cavitation conditions $NPSH_a$. It can be noted that the trend in the leakage flow intensity at the four cavitation conditions ($NPSH_a = 10.7$, 3.91, 2.48, and 1.98) is similar, and there are two peak intensity regions A and B, as shown in this figure. In the peak intensity region A which is close to the LE, the leakage flow intensity decreases with a decrease in $NPSH_a$ due to the development of tip leakage vortex and the cavitation in the tip clearance area. With the increase in cavitation, clearance cavitation gradually covers the top surface of the blade and becomes thicker, resulting in a decrease in the effective leakage area. Meanwhile, the development of the tip leakage vortex also inhibits the movement of the leakage flow from the top of the blade suction surface. In the peak intensity region B, the intensity distribution range of the leakage flow varies due to the influence of the leakage vortex cavitation, and the peak intensity region extends to the blade LE. The intensity of the leakage flow is enhanced by the cavitation, which almost coincides with the high-entropy production area. Table 11.1 quantitatively describes the flow rates of single-blade TLF for various values of the $NPSH_a$ and the relative leakage values of the total leakage at the flanges of the four blades at the design flow rate. It can be noted from this Table that the tip leakage of a single blade is not significantly affected by the tip leakage vortex cavitation (the variation is small), indicating that the energy loss in the pump is not due to the increased tip leakage rate caused by the cavitation.

According to Du et al. [1], the leakage flow can be divided into various parts along the flow direction of the impeller tip clearance, and each part plays a different role in the mechanism of inducing the tip flow instability. The leakage flow at different chord lengths of the blade tip is divided as follows. The first part of the leakage overflow area is in the range $0 \leq \lambda \leq 0.05$ (where

*Mixed-flow Pumps: Modeling, Simulation, and Measurements*, First Edition.
Wei Li, Leilei Ji, Ramesh Agarwal, Weidong Shi, and Ling Zhou.

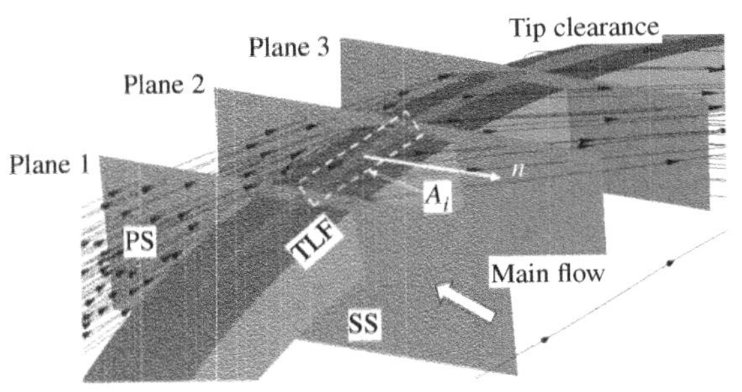


**Figure 11.1** Schematic diagram of the partition of the tip leakage flow region and nomenclature.

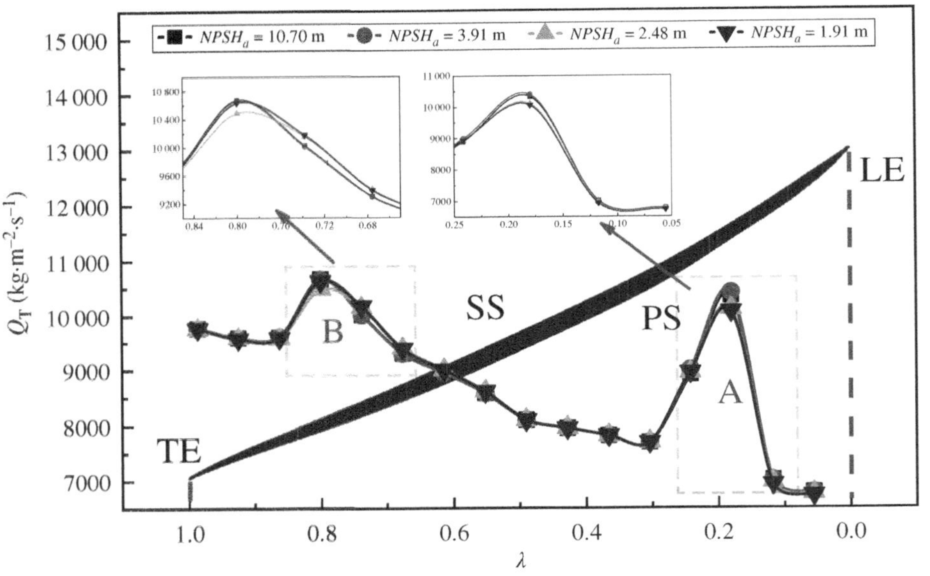


**Figure 11.2** Distribution of the tip leakage flow intensity in various cavitation conditions.

**Table 11.1** Comparison of tip leakage flow for various $NPSH_a$ values.

| $NPSH_a$ (m) | Single-blade tip clearance leakage flow (kg/s) | Relative leakage flow (%) |
|---|---|---|
| 10.70 | 0.9302 | 3.536 |
| 3.91 | 0.9305 | 3.537 |
| 2.48 | 0.9282 | 3.528 |
| 1.98 | 0.9296 | 3.533 |

$\lambda$ is the normalized value with $\lambda = 0$ being the LE and $\lambda = 1$ the TE) defined as the LE TLF; the second part of the leakage overflow area is in the range $0.5 \leq \lambda \leq 0.6$ defined as the leakage flow in the middle of the tip; and the third part of the leakage overflow area is in the range $0.8 \leq \lambda \leq 0.9$, defined as the TE TLF.

Figure 11.3(a) shows the three-dimensional streamline trajectories of the TLF at the tip LE at the design condition for various cavitation conditions $NPSH_a$. At $NPSH_a = 10.70$ and 3.91 m, the leakage streamlines cluster is concentrated and shows a higher flow velocity from the beginning of the LE to the TE. At $NPSH_a = 2.48$ m, the streamlines cluster gradually closes at the suction surface of the blade and begins to diverge from the beginning of the LE to the TE. At $\lambda = 0.7$, a distinctive local divergence zone of the streamlines cluster is formed. At this time, the tip leakage vortices break off, and a low-speed area is formed, which is affected by the end wall area to form a local separation region. When $NPSH_a = 1.98$ m, the streamlines cluster at $\lambda = 0.7$ is completely divergent, and the appearance of low-speed zones and the flow separation region is more evident. Figure 11.3(b) shows the three-dimensional streamlines' trajectories of the leakage flow in the middle of the tip at various cavitation conditions at the design flow rate. The streamlines' trajectories indicate that the leakage flow in the middle of the tip is related to the formation of the secondary leakage flow. More severe the cavitation, the more obvious the secondary leakage flow. Figure 11.3(c) shows the three-dimensional streamlines' trajectories of the tip TE leakage

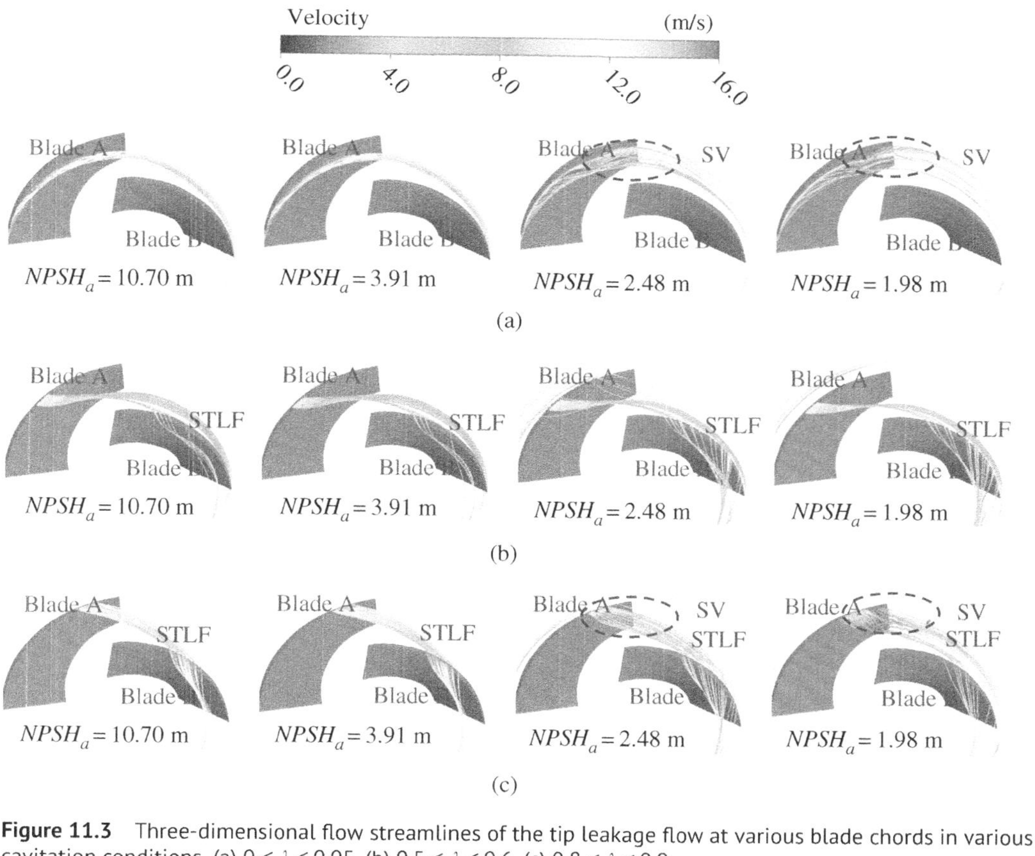


**Figure 11.3** Three-dimensional flow streamlines of the tip leakage flow at various blade chords in various cavitation conditions. (a) $0 \leq \lambda \leq 0.05$, (b) $0.5 \leq \lambda \leq 0.6$, (c) $0.8 \leq \lambda \leq 0.9$.

flow in various cavitation conditions at the design flow rate. At $NPSH_a = 10.70$ and 3.91 m, the tip TE leakage flow causes the secondary leakage flow in the middle of the adjacent blades. With a decrease in the $NPSH_a$ value, the streamlines cluster begins to diverge and forms a low-speed zone, and the secondary leakage area of adjacent blades extends to the blade TE. When $NPSH_a = 1.98$ m, the effect of the low-speed entrainment area is obvious; the flow separation occurs, and there is greater influence of the separation vortices. In addition, the leakage flow from the LE and the TE interacts in the range of $0.7 \leq \lambda \leq 1$, which disrupts the flow stability. The separation vortex and the low-speed zones formed in the end wall area result in the blockage of the flow passage, leading to a significant decline in the pump performance.

### 11.1.2 Evolution of the Vorticity and Its Intensity

Several studies reported in the literature have shown that the energy loss caused by turbulent dissipation is the main source of energy loss in the impeller. When there is a vortex in the velocity field, the uneven distribution of the velocity can lead to changes in the shape and direction of the cluster of vortices, and the strong fluctuations in the flow caused by the vortices are a major source of turbulence in the flow.

Figure 11.4 shows the vorticity distribution and the trajectory of the 3D streamlines on a series of radial sections in various cavitation conditions. It can be noted that at $NPSH_a = 10.70$ and 3.91 m,

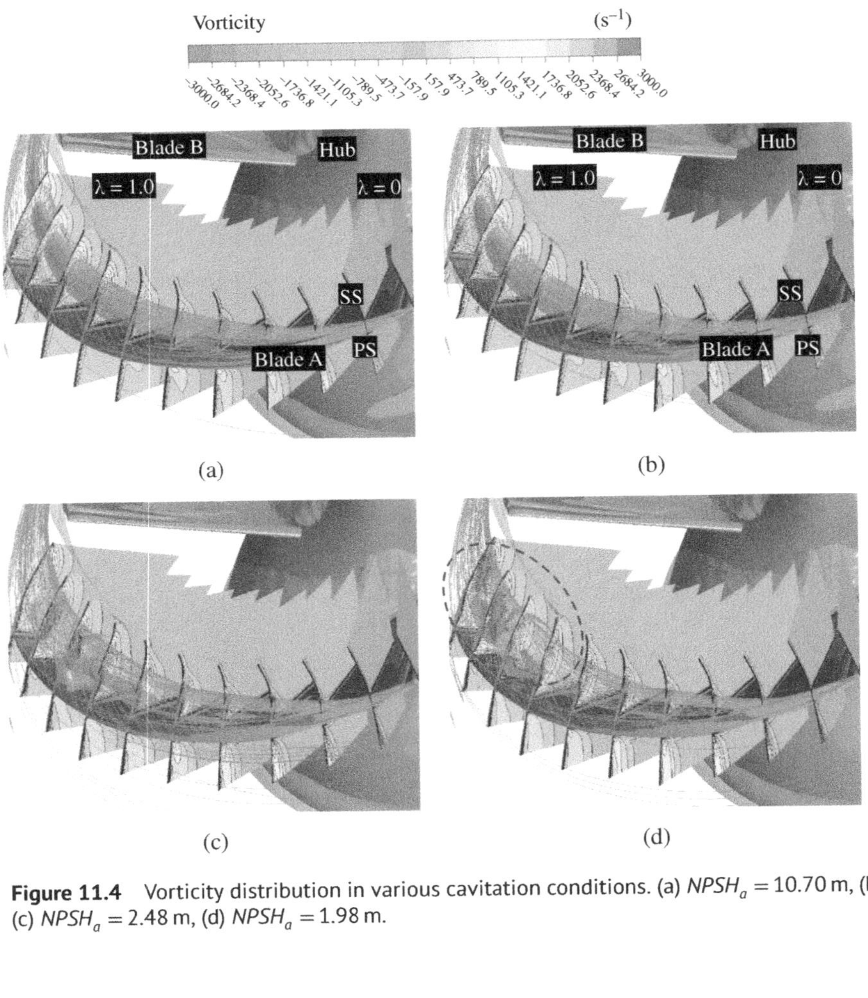


**Figure 11.4** Vorticity distribution in various cavitation conditions. (a) $NPSH_a = 10.70$ m, (b) $NPSH_a = 3.91$ m, (c) $NPSH_a = 2.48$ m, (d) $NPSH_a = 1.98$ m.

the tip LE leakage flow cluster almost coincides with the tip leakage vortex flow path, and the vorticity is concentrated in the leakage shear zone, and the turbulent kinetic energy in the tip leakage vortex core area is not large. When $NPSH_a = 2.48$ m, the tip leakage vortex breaks down at $\lambda = 0.7$, and the streamlines cluster becomes divergent from the non-cavitation and initial cavitation conditions, inducing the flow separation. In the range $0.7 \leq \lambda \leq 0.9$, the vorticity in the vortex core increases significantly. The breakdown of the tip leakage vortices is related to the development of cavitation both in scale and intensity. The low-pressure environment causes the cavitation to cover most of the area of the suction surface, which weakens the propagation of the leakage flow and inhibits the formation and development of the tip leakage vortices. When $NPSH_a = 1.98$ m, the tip leakage vortices disintegrate and fracture at $\lambda = 0.6$. The divergence of the streamlines' cluster spreads to the LE and the vortices due to flow separation form in the end wall area. At this time, the region of vorticity generation is mainly located in the range $0.7 \leq \lambda \leq 1$, and the area of high vorticity region becomes larger. In conclusion, the formation of the tip leakage vortices is inhibited when the $NPSH_a$ value is low.

The vorticity generation in the tip area of the blade can be divided into two parts: one part is the vorticity caused by the leakage flow and the other part is related to the vortices formed due to separation in the end wall area.

The analysis of the vortex intensity can clarify the evolution of the vortices in different cavitation conditions, since the vortex intensity reflects the evolution process of the vortices along the chord length. The vortex intensity and the vortex intensity per unit area can be expressed as:

$$\Gamma = \iint_A \omega dA \tag{11.2}$$

$$\Delta\Gamma = \frac{\iint_A \omega dA}{\int dA} \tag{11.3}$$

Figure 11.5 shows the distribution of the vortex intensity along the chord length of the blade on a series of radial sections in various cavitation conditions. The vortex intensity increases along the chord length and reaches its maximum value near the TE. It can be noted that by comparison, at $NPSH_a = 10.70$ and 3.91 m, the trends in the variation in vorticity $\Gamma$ and its values are generally consistent, reflecting the small effect of cavitation on the vorticity in the range $0.7 \leq \lambda \leq 1$. In the range $0.1 \leq \lambda \leq 0.5$, there is a small decrease in the vortex intensity with decreasing $NPSH_a$,

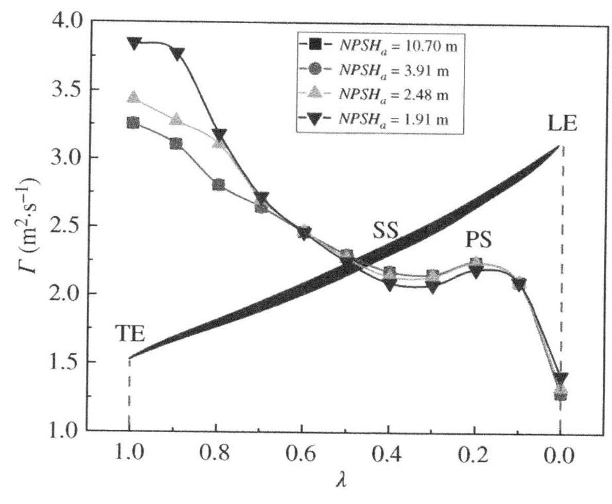


**Figure 11.5** Vortex strength distribution in various cavitation conditions.

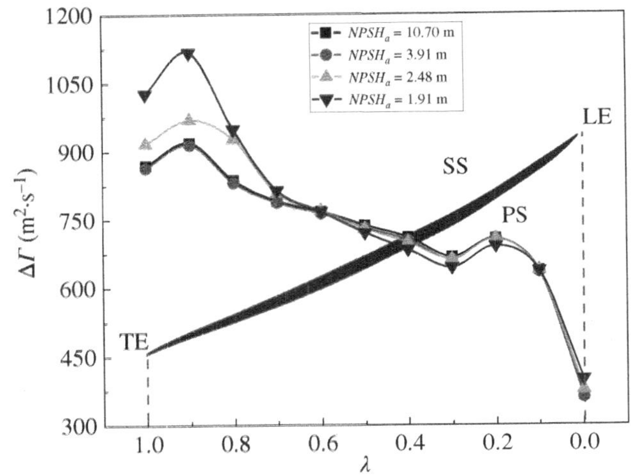


**Figure 11.6** Vortex strength distribution per unit area in various cavitation conditions.

which is associated with the scale of development of clearance cavitation and the tip leakage vortex cavitation. In the range $0.6 \le \lambda \le 1$, the intensity of the vortices at $NPSH_a = 2.48$ and 1.98 m increases significantly, showing a tendency to become stronger with the deterioration of cavitation. In addition, by dividing the vortex intensity of each section by the corresponding section area, the vortex intensity per unit area of each section $\triangle\Gamma$ can be obtained, as shown in Fig. 11.6. Considering the trend in the variation of the vortex intensity, there is almost no difference in the curves of vortex intensity per unit area at $NPSH_a = 10.70$ and 3.91 m. In case of severe cavitation, the vortex intensity per unit area $\triangle\Gamma$ increases in the range $0.6 \le \lambda \le 1$. It can be noted that the variation in vortex intensity per unit area $\triangle\Gamma$ characterizes the evolution of vorticity and clearly presents the process of vortex evolution under cavitation.

### 11.1.3 Turbulent Kinetic Energy Distribution in the Blade Tip Region

Figure 11.7 shows the turbulent kinetic energy distribution and the three-dimensional streamlines' trajectories on a series of radial cross-sections in various cavitation conditions. It can be noted that at $NPSH_a = 10.70$ and 3.91 m, the turbulent kinetic energy is concentrated in the leakage shear region. The turbulent kinetic energy in the area of the tip leakage vortex core is not large; however, it has a large value mainly due to the strong leakage shear in the leakage shear region and the large fluctuations in the velocity. With the decrease in $NPSH_a$ in the range $0 \le \lambda \le 0.7$, the turbulent kinetic energy generated by the leakage shear region decreases. Due to the influence of the low ambient pressure and the expansion and propagation of cavitation, the cavitation in the blade tip area develops fully and occupies the suction surface outlet, thereby restraining the propagation distance of the leakage flow. Due to the interaction of the tip LE leakage flow and the TE leakage flow, the flow separation occurs in the outlet area of the flow path, and the separation vortices form in the end wall area, resulting in the generation of high turbulent kinetic energy, as shown in Fig. 11.7(c,d).

Figures 11.8 and 11.9 show the curves of the turbulent kinetic energy and the turbulent kinetic energy per unit area along the chord length on the radial cross-sections in various cavitation conditions. Their distribution is like that of the vortex intensity and the vortex intensity per unit area,

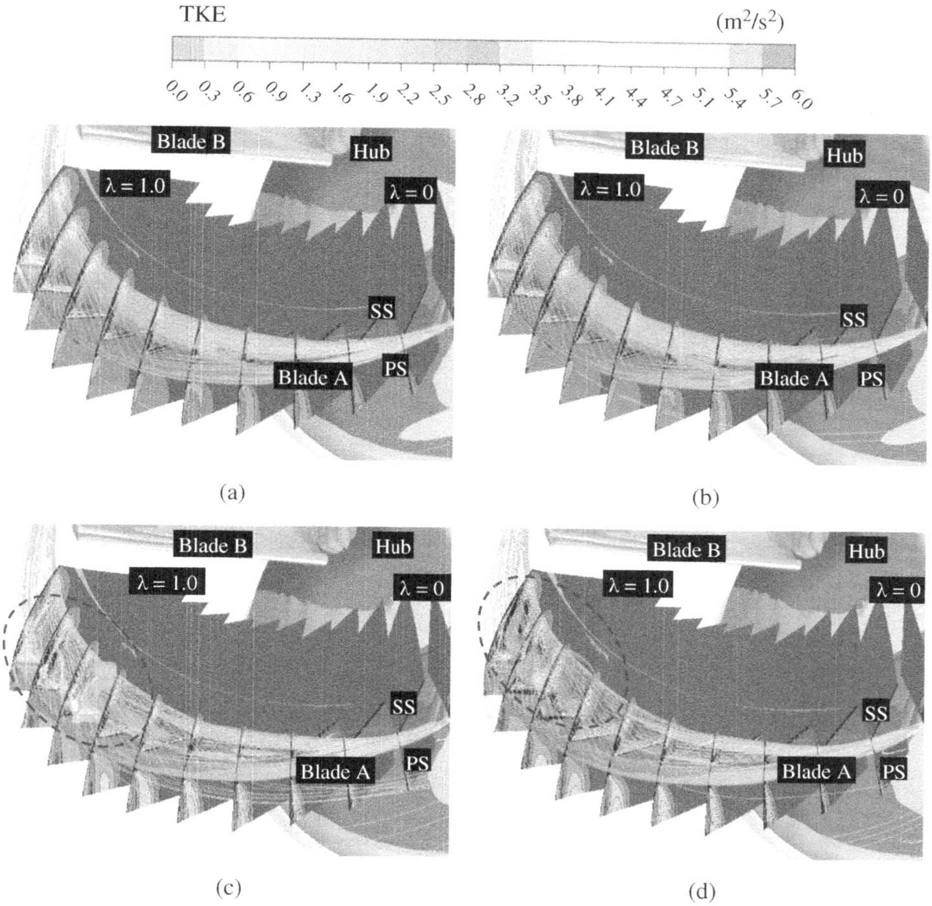


**Figure 11.7** Distribution of the turbulent kinetic energy in various cavitation conditions. (a) $NPSH_a = 10.70$ m, (b) $NPSH_a = 3.91$ m, (c) $NPSH_a = 2.48$ m, (d) $NPSH_a = 1.98$ m.

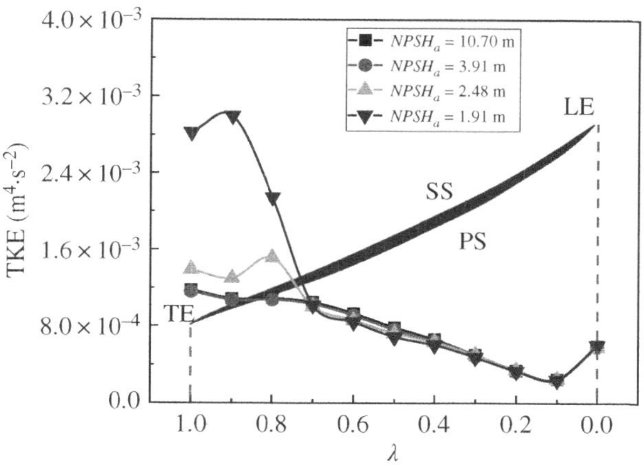


**Figure 11.8** Turbulent kinetic energy (TKE) distribution in various cavitation conditions.

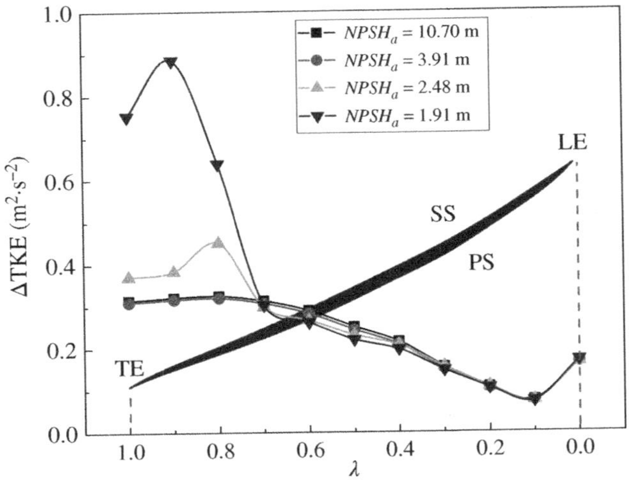


**Figure 11.9** Turbulent kinetic energy (TKE) distribution per unit area in various cavitation conditions.

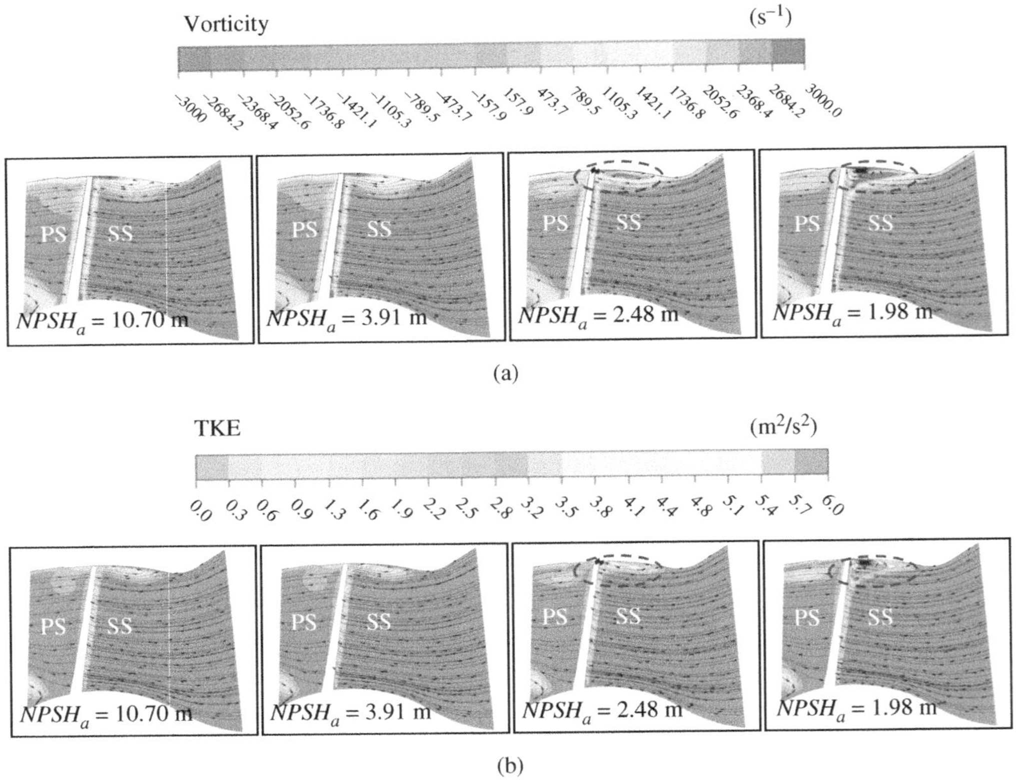


**Figure 11.10** Vorticity and turbulent kinetic energy distributions at $\lambda = 0.9$. (a) Vorticity distribution and (b) turbulent kinetic energy distribution.

which increases along the chord length and reaches the maximum value near the TE. In the range $0.1 \leq \lambda \leq 0.5$, the development of cavitation inhibits the formation and propagation of the leakage jet, and the turbulent kinetic energy intensity and the turbulent kinetic intensity per unit area decrease slightly due to cavitation. In the range $0.6 \leq \lambda \leq 1.0$, the flow separation occurs, and the vortices form as a result of separation at the outlet of the pump, which leads to the generation of high turbulence areas in severe cavitation conditions. Figure 11.10 shows the distribution of the vorticity and the turbulent kinetic energy in the radial cross-section at $\lambda = 0.9$. The figure shows that with the decrease in $NPSH_a$, the distribution range of high vorticity and high turbulent kinetic energy in the end wall area increases and develops along the radial and axial directions of the impeller. In conclusion, the distribution range of the high turbulent kinetic energy region is almost the same as that of the high vorticity region, and there is a positive correlation between the two intensities. Severe cavitation results in the radial and axial development of the high turbulent kinetic energy regions at the outlet of the flow passage of the impeller, which leads to an increase in the energy loss due to turbulent dissipation.

Figure 11.11 shows the distribution of the radial velocity $V_r$ and the axial velocity $V_a$ at $\lambda = 0.9$ in the cylindrical coordinates. It can be noted that in the outlet area of the flow passage, severe cavitation causes the axial velocity $V_a$ of the leakage flow to propagate toward the axial and radial direction in the impeller. The propagation intensity increases with a decrease in $NPSH_a$, and the development of high turbulent kinetic energy region is like that of the axial velocity $V_a$. A large pressure gradient exists in the velocity boundary layer, which controls the distribution of the high vorticity areas and results in large losses due to turbulent dissipation. In the case of the propagation

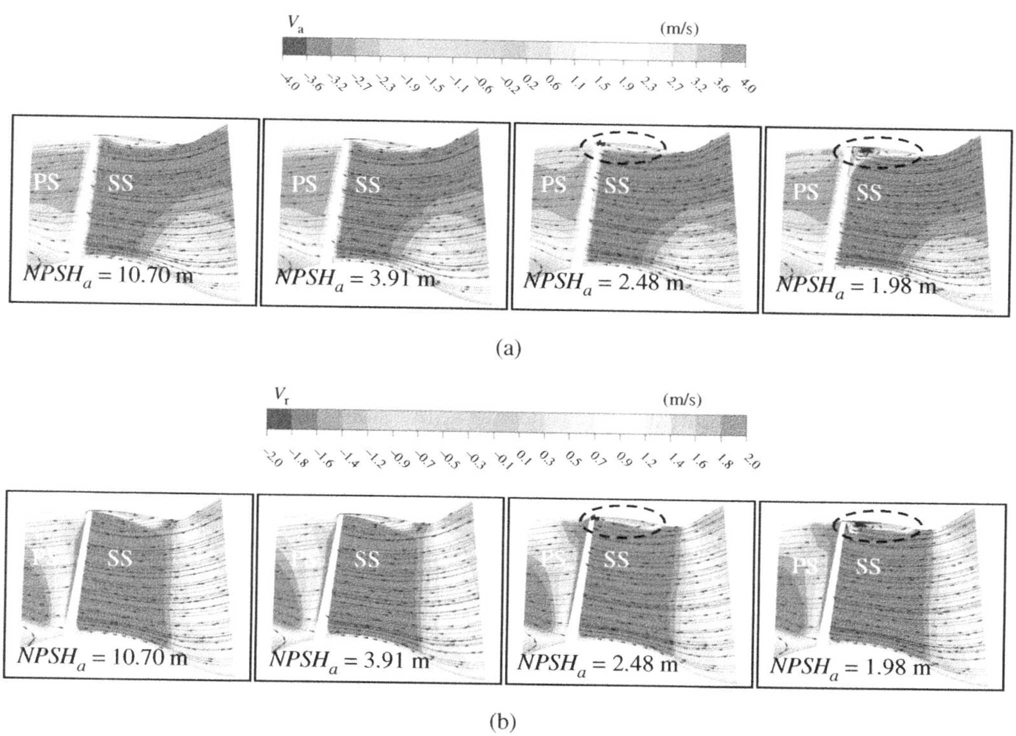


**Figure 11.11** Axial velocity $V_a$ and the radial velocity $V_r$ distributions at $\lambda = 0.9$. (a) Vorticity distribution and (b) turbulent kinetic energy distribution.

of the radial velocity $V_r$, the axial propagation intensity of the leakage flow increases, but the range of propagation area decreases compared to that in the non-cavitation operating conditions. In addition, the leakage flow has relatively little effect on the radial propagation of the radial velocity $V_r$. Based on the above analysis, the high turbulent kinetic energy region generated during the severe cavitation is found to be related to the propagation intensity and range of the axial velocity $V_a$ of the leakage flow. The gradient of the axial velocity $V_a$ of the leakage flow induces the generation of a high vorticity region, which is the main cause of the turbulent dissipation.

## 11.2 Analysis of the Vorticity Transport Characteristics in the Tip Region

The circumferential rotation of the impeller results in very different types of cavitation flows in the mixed-flow pump compared to those appear on the surface of an airfoil. The evolution and dissipation mechanisms of the cavitation vortex are described in this section by employing the relative vorticity transport equation in the cylindrical coordinates. The transport equation of relative vorticity in the vector notations for any coordinate system can be expressed as

$$\frac{D\vec{\omega}_R}{Dt} = (\vec{\omega}_{Rr} \cdot \nabla)\vec{V} - \vec{\omega}_R(\nabla \cdot \vec{V}) - \frac{\nabla \rho_m + \nabla p}{\rho_m^2} + (\nu_m + \nu_t)\nabla^2\vec{\omega}_R - 2\nabla \times (\vec{\Omega} \times \vec{V}) \tag{11.4}$$

where $\vec{V}$ is the relative velocity in m/s and $\vec{\Omega}$ is the angular velocity in rad/s. In the following discussion, cylindrical coordinate system is employed.

The leftmost term in Eq. (11.4), $D\vec{\omega}_R/Dt$ represents the relative vorticity variation rate in $s^{-2}$; on the right side of Eq. (11.4), the first term $(\vec{\omega}_R \cdot \nabla)\vec{V}$ represents the relative vorticity stretching in $s^{-2}$; the second term $-\vec{\omega}_R(\nabla \cdot \vec{V})$ represents the relative vorticity expansion/contraction in $s^{-2}$; the third term $-(\nabla \rho_m + \nabla p)/\rho_m^2$ represents the relative baroclinic torque in $s^{-2}$; the fourth term $(\nu_m + \nu_t)\nabla^2\vec{\omega}_R$ represents the relative viscous dissipation in $s^{-2}$; and the fifth term $-2\nabla \times (\vec{\Omega} \times \vec{V})$ represents the relative Coriolis force in $s^{-2}$. The Coriolis force term describes the interaction between the rotational motion due to the vorticity and the velocity gradient of the fluid particle. The viscous dissipation term has little influence at high Reynolds number [2], therefore, it is not considered in the analysis presented in this section. The transport of fluid involves the radial, circumferential, and axial motions when the impeller rotates to do the work on the fluid. In the previous studies reported in the literature, the circumferential, and axial vorticity in the pump and their transport terms in the vorticity transport Eq. (11.4) have been mainly studied, while the variation in the vorticity caused by the radial movement of the fluid has been ignored [3–5]. Previous studies have shown that the axial velocity $V_a$ of the leakage flow moves toward the impeller axially and radially during severe cavitation, which induces the generation of vortices. Therefore, in order to analyze the vorticity transport in the blade tip region, the radial, circumferential, and axial transport of the vorticity stretching term, the expansion/contraction term, the baroclinic torque term, and the Coriolis force term in the vorticity transport equation need to be understood, which are described in Section 11.2.1.

### 11.2.1 Radial Vorticity Transport in the Cavitation Region Near the Blade Tip

The radial vorticity distribution in the blade tip region in various cavitation conditions is shown in Fig. 11.12. At $NPSH_a = 10.70$ and 3.91 m, there is a large positive radial velocity distribution at the blade tip in the boundary layer on the suction surface of the blade, while a weak negative radial velocity distribution dominates the clearance jet area. With decrease in $NPSH_a$, the tip leakage vortices break up, resulting in the flow separation and the vortices due to separated flow in the

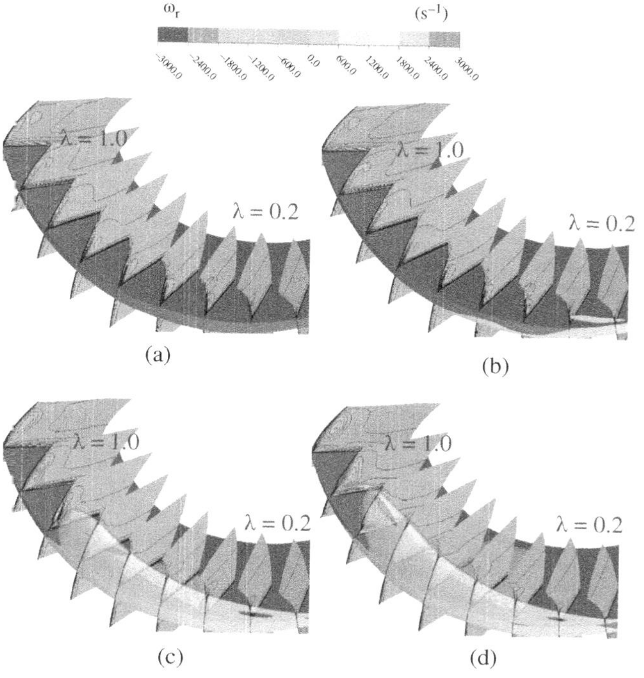


**Figure 11.12** Radial vorticity $\omega_r$ distribution in various cavitation conditions. (a) $NPSH_a = 10.70$ m, (b) $NPSH_a = 3.91$ m, (c) $NPSH_a = 2.48$ m, (d) $NPSH_a = 1.98$ m.

end wall area. In the range $0.6 \leq \lambda \leq 1.0$, the positive radial vorticity at the tip clearance develops toward the clearance jet region, and the radial vorticity distribution region in the boundary layer on the suction surface enlarges, and its intensity increases. At the same time, the area of the negative radial vorticity distribution and its intensity increases. The negative radial vorticity distribution area almost coincides with the tip LE leakage flow trajectory. Therefore, when the cavitation is severe, the entropy production of the large radial vorticity area at the suction surface of the boundary layer becomes large, which can be related to the distribution of the radial vorticity, thereby proving that the radial distribution of the vorticity is not negligible.

Figure 11.13 shows the radial distributions of each of the vorticity transport terms on a series of radial sections of the impeller in various cavitation conditions. The vortex stretching is caused by the velocity gradient, and its main function is to cause the vorticity tube to stretch thinner and twist to change its direction, which indicates its effect on changing the moment of inertia of the fluid element. The distribution of the radial vorticity stretching term is shown in Fig. 11.13(a). At $NPSH_a = 3.91$ m, the cavitation intensity in the blade tip area is weak, and the radial vorticity stretching term is not significantly different from that at $NPSH_a = 10.70$ m in intensity and scale. At this stage, the radial vorticity stretching term on the trajectory of the tip LE shows a negative distribution, and the absolute vorticity decreases from the LE to the TE, and the vorticity in the end wall area is mainly negative while the vorticity in the interaction area is positive. With the severity of cavitation, there is flow separation, and the vortices formed due to the flow separation induce the dissipation of the positive vorticity in the interaction region, where the end wall region is almost covered by a large negative radial vorticity stretching term. The vorticity expansion/contraction term is proportional to the divergence of the velocity, which indicates the effect of the fluid compression and expansion on the change in the moment of inertia of the fluid clusters. The distribution of the radial vorticity expansion/contraction term is shown in Fig. 11.13(b). At $NPSH_a = 10.70$ and 3.91 m, there

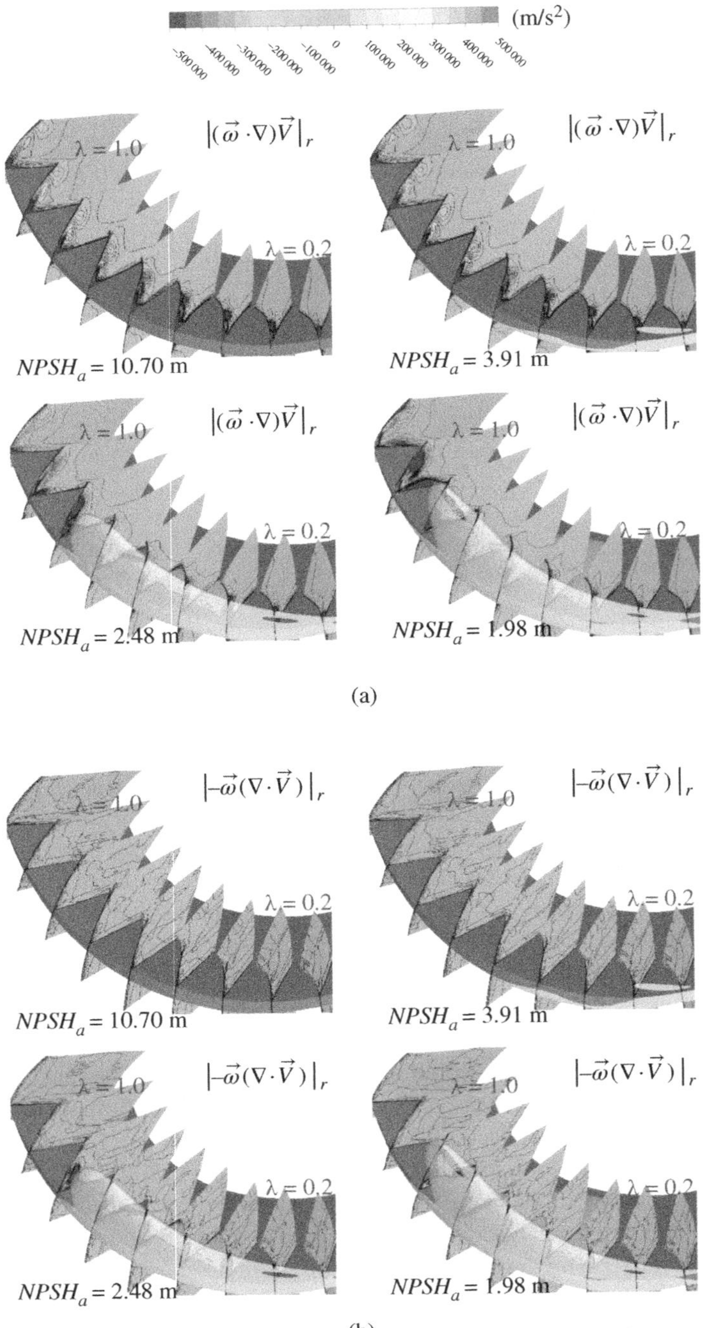


**Figure 11.13** Axial distribution of each transport term in the vorticity transport equation in various cavitation conditions. (a) Radial vortex stretching term, (b) Radial relative volumetric expansion/contraction term, (c) Radial baroclinic torque term, (d) Radial Coriolis force term.

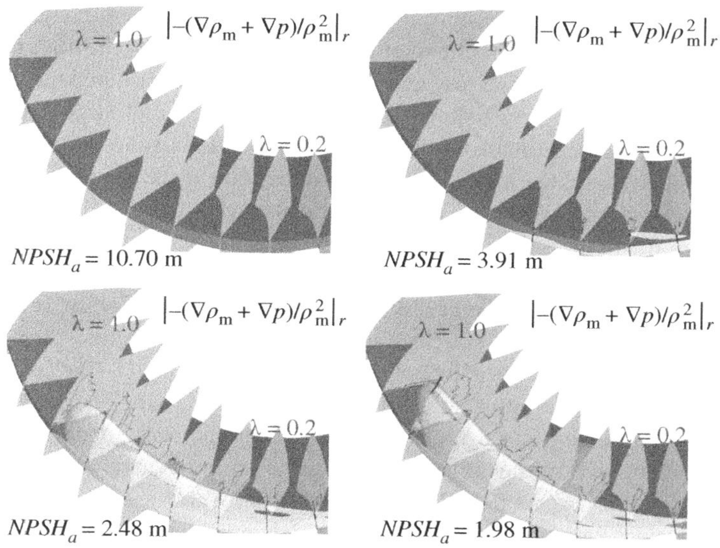


(c)

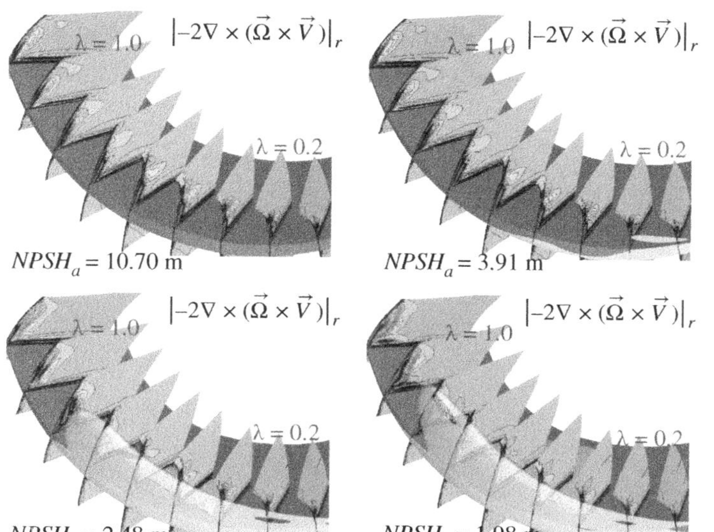


(d)

**Figure 11.13** *(Continued)*

is almost no vorticity distribution because the divergence of the velocity in the contraction term of the vorticity expansion is related to the velocity of the mass transfer between the phases. In the case of severe cavitation, the radial vorticity expansion/contraction term is distributed near the boundary layer of the clearance jet area and the suction surface. Significant gas-liquid exchange exists in the area of the tip leakage vortex cavitation, and the compressibility of the air bubbles results in a strong rotational effect, which is also the reason why the vorticity expansion/contraction term is mainly distributed inside the bubbles. The baroclinic torque term is generated by the difference between the bubble density gradient and the pressure gradient and is mainly distributed near the bubble surface. The radial baroclinic torque term is shown in Fig. 11.13(c). The radial baroclinic torque term is weaker than the other three terms in the vorticity transport equation. It can be noted from this figure that the distribution range of the radial baroclinic torque term increases with an increase in the cavitation intensity and its scale in the area on top of the blade. Long et al. [6] found that the baroclinic torque term can maintain the shape of the cavitation bubble, and the bubble collapse plays an important role in the generation of the vortices. The Coriolis force is an inertial force generated by the movement of the rotating fluid particles in a rotating coordinate system. It is perpendicular to the direction of motion of the fluid particles and only affects their direction without changing their velocity. The radial Coriolis force distribution is shown in the Fig. 11.13(d). At $NPSH_a = 10.70$ and 3.91 m, large positive and negative radial Coriolis forces occur in the boundary layer of the clearance jet area and of the suction surface, and the absolute vorticity value decreases and dissipates during the development toward the TE. With the deterioration of cavitation, the scale and magnitude of the radial Coriolis force near the blade TE significantly increase due to the influence of flow separation and vortices formed due to separation, which indicates that the radial Coriolis force cannot be neglected in analysis.

### 11.2.2 Circumferential Transport of Vorticity

The distribution of circumferential vorticity in the tip cavitation area in different cavitation conditions is shown in Fig. 11.14. Unlike the radial transport of vorticity, the circumferential transport

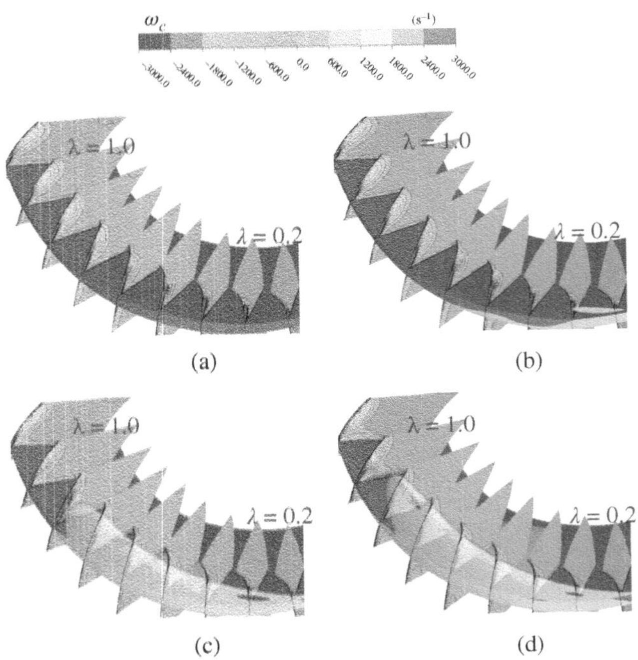


**Figure 11.14** Circumferential vorticity $\omega_c$ distribution in various cavitation conditions. (a) $NPSH_a = 10.70$ m, (b) $NPSH_a = 3.91$ m, (c) $NPSH_a = 2.48$ m, (d) $NPSH_a = 1.98$ m.

of vorticity is mainly reflected by the changes in the vorticity caused by the tip LE leakage flow and the clearance jet. At $NPSH_a = 10.70$ and 3.91 m, a large positive axial vorticity distribution appears in the clearance jet area, and the vorticity decreases during the development of the positive circumferential vorticity at the TE due to the tip LE leakage, indicating that the circumferential vorticity gradually dissipates. With the decrease in the $NPSH_a$ value, the scale and intensity of the cavitation in the blade tip area cause the streamlines cluster to diverge, and the circumferential vorticity in the tip LE leakage region is significantly reduced. The main source of circumferential vorticity at this time is the vortices formed due to the separation in the end wall area and the vortices formed by the interaction between the clearance jet and the tip LE leakage flow.

Figure 11.15 shows the circumferential distribution of each transport term in the vorticity transport equation on various radial sections of the impeller in various cavitation conditions. Unlike the distribution of radial vorticity stretching term, at $NPSH_a = 10.70$ and 3.91 m, there is large scale distortion in the flow due to positive circumferential vorticity stretching in the clearance jet area, while its distribution in the boundary layer on the suction surface is mainly negative. The distribution on the tip LE leakage flow is also negative, and the absolute vorticity decreases toward the TE. With the severity of cavitation, large positive and negative vorticity distributions occur simultaneously near the TE of the blade, which are related to the flow separation and formation of vortices due to separated flow as shown in Fig. 11.15(a). Considering the expansion/ contraction term for the circumferential vorticity, there is almost no distribution in the non-cavitation region. At $NPSH_a = 2.48$ and 1.98 m however, large negative distribution of the circumferential vorticity are dominant in the bubble interior and there are positive circumferential vorticity distributions near the bubble-liquid interface. This is due to the opposite condensation and evaporation processes between the dominant phases inside the bubble and in the boundary layer. The circumferential vorticity/expansion contraction term is shown in Fig. 11.15(b). The distribution of the circumferential baroclinic torque term is shown in Fig. 11.15(c), which is mainly distributed at the vapor-liquid interface of the cavitation bubbles and has the least impact compared to the other three terms in the vorticity transport equation and therefore is not discussed here in detail. When the impeller works in the axial rotation, the fluid mainly flows in the circumferential direction and the axial direction. In terms of the magnitude and distribution range, the circumferential Coriolis force is significantly stronger than the force in the axial direction, as shown in Fig. 11.15(d). It can be noted from this figure that at $NPSH_a = 10.70$ and 3.91 m, large positive circumferential Coriolis force term is distributed on the tip LE leakage flow region, and the absolute vorticity value decreases on the TE, while both the positive and negative values of vorticity are distributed in the clearance jet area. With the severity of cavitation, the positive circumferential Coriolis force distribution dominates near the TE, while the negative circumferential Coriolis force distribution dominates inside the bubble, which indicates that the cavitation bubble destroys the balance in the distribution of the circumferential Coriolis force.

### 11.2.3 Transport of the Axial Vorticity

The distribution of the axial vorticity in the tip cavitation region in various cavitation conditions is shown in Fig. 11.16. Unlike the distribution of the radial vorticity and the circumferential vorticity, the tip LE leakage has little influence on the distribution of the axial vorticity in the pump. At $NPSH_a = 10.70$ and 3.91 m, there are both positive and negative axial distributions of vorticity in the clearance jet area, and there is positive axial vorticity in the suction boundary layer, which may be related to the high-entropy production in the boundary layer. With the deterioration of cavitation, the negative axial vorticity distribution is dominant in the clearance jet area at the TE,

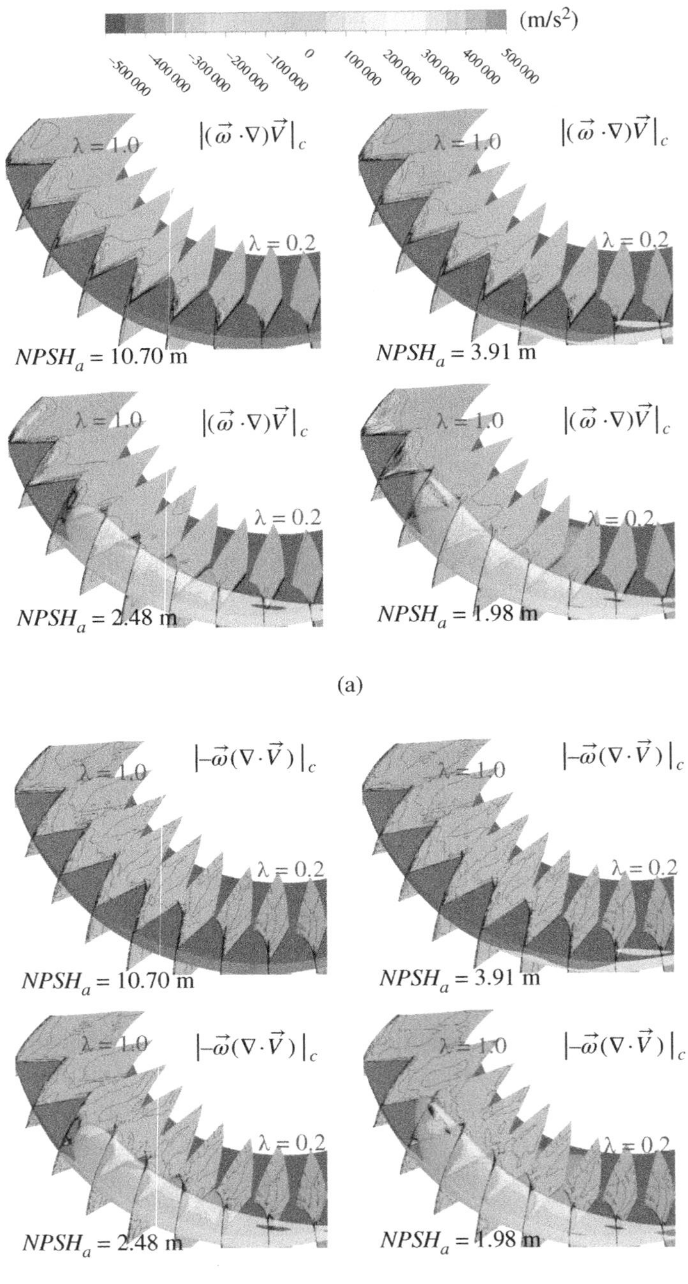


**Figure 11.15** Circumferential distribution of each transport term in the vorticity transport equation in various cavitation conditions. (a) Circumferential vortex stretching term, (b) Circumferential volumetric expansion/contraction term, (c) Circumferential baroclinic torque term, (d) Circumferential Coriolis force term.

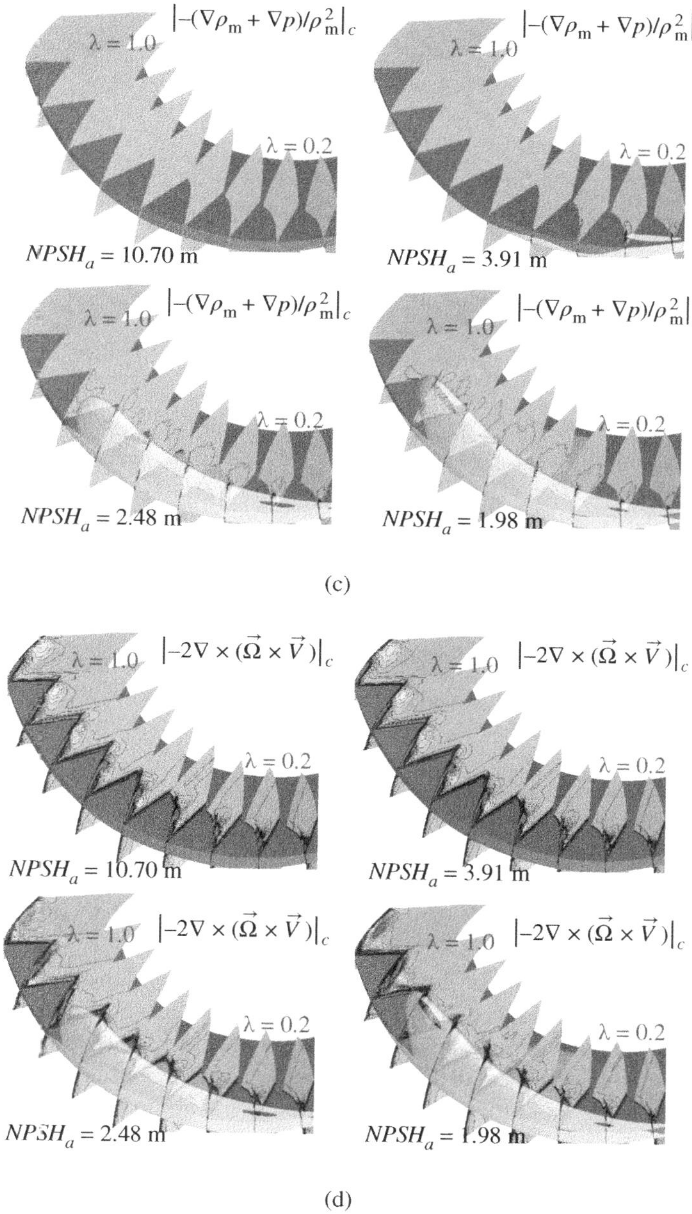


**Figure 11.15** *(Continued)*

and it can be related to the axial and radial propagation of the axial velocity $V_a$ of the leakage flow. In addition, the area of the higher magnitude of the positive axial vorticity distribution is related to the separated flow region and the formation of vortices due to separated flow. The more severe is the cavitation, larger is the positive axial vorticity distribution area in the TE region and higher is its magnitude.

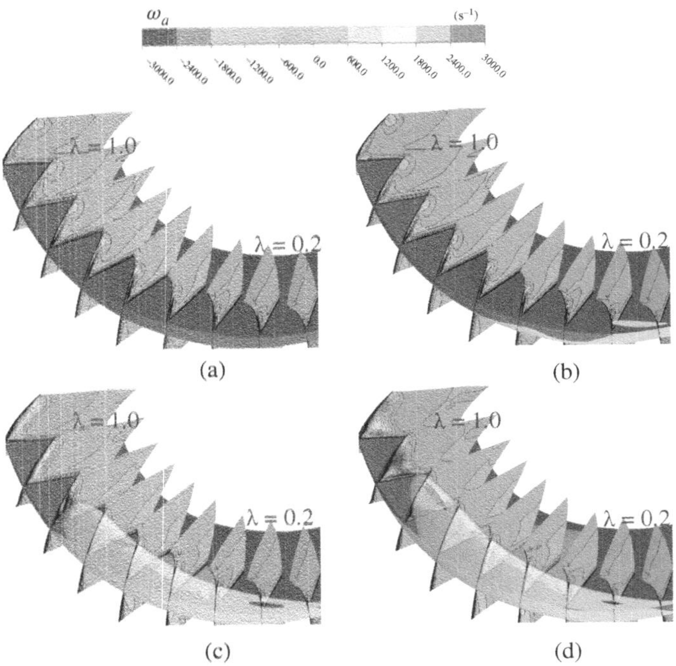


**Figure 11.16** Axial vorticity $\omega_a$ distribution in various cavitation conditions. (a) $NPSH_a = 10.70$ m, (b) $NPSH_a = 3.91$ m, (c) $NPSH_a = 2.48$ m, (d) $NPSH_a = 1.98$ m.

Figure 11.17 shows the axial distributions of each transport term on a series of radial sections of the impeller in various cavitation conditions. Distribution of the axial stretching term in each cavitation condition is shown in Fig. 11.17(a). When $NPSH_a = 10.70$ and 3.91 m, there is negative axial stretching distribution on the tip LE leakage flow trajectory, and the absolute vorticity value decreases toward the TE, reflecting the dissipation of vortices. Both sides of the negative vorticity area are wrapped by a higher magnitude of positive axial stretching term, one side of which is caused by a leakage jet. With the deterioration of the cavitation, the flow separation and the formation of vortices due to separation result in a low-speed flow area near the blade TE, which results in a significant increase in the distribution range and magnitude of the axial stretching term. The axial vorticity expansion/contraction term is identical to the circumferential distribution as shown in Fig. 11.17(b). Axial baroclinic torque term has little effect and therefore is not discussed further. Since there is no component of the Coriolis force in the axial direction, it is not discussed.

### 11.2.4 Quantitative Evaluation of the Vorticity Transport Terms

The radial, circumferential, and axial distributions of the vorticity stretching term, expansion/ contraction term, baroclinic torque term, and the Coriolis force term in the vorticity transport equation have been discussed in Sections 11.2.1–11.2.3. It was concluded that the characteristics of the radial vorticity transport in the cavitation flow should be considered, which have not been considered in previous studies in the literature. In addition, previous studies mainly focused on the qualitative analysis but did not carry out the quantitative analysis of the various components of the vorticity transport equation. Therefore, the intensity of the vorticity transport for each term of the vorticity transport equation is defined here as follows:

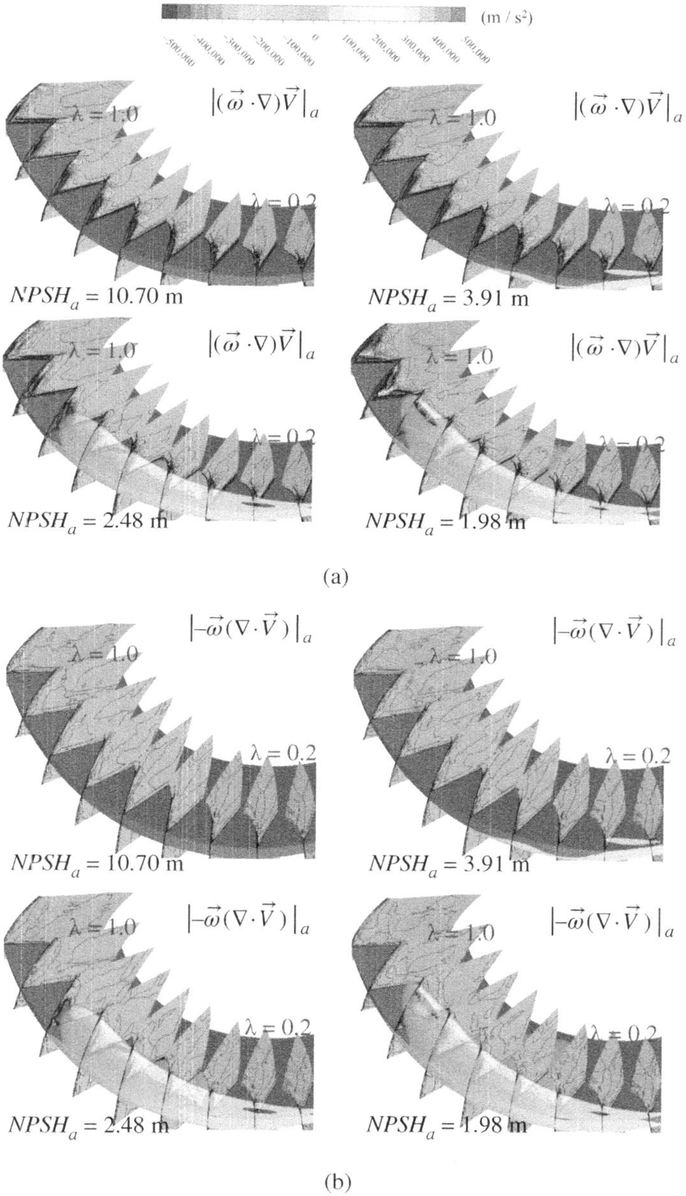


**Figure 11.17** Axial distribution of each transport term in the vorticity transport equation in various cavitation conditions. (a) Axial vortex stretching term, (b) Axial volumetric expansion/contraction term, (c) Axial baroclinic torque term.

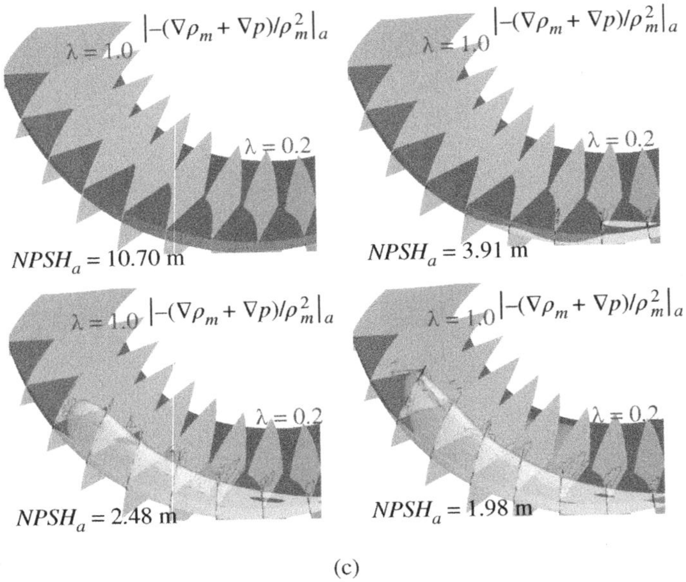


**Figure 11.17** *(Continued)*

The intensity of the vorticity stretching distortion term is defined as:

$$\Gamma_{\mathrm{VSD}i} = \frac{\int (|(\vec{\omega}_{\mathrm{Rr}} \cdot \nabla)\vec{V}|)_i dA}{\int dA} \tag{11.5}$$

The intensity of the vorticity expansion/contraction term is defined as:

$$\Gamma_{\mathrm{VEC}i} = \frac{\int (|-\vec{\omega}_{\mathrm{R}}(\nabla \cdot \vec{V})|)_i dA}{\int dA} \tag{11.6}$$

The intensity of the vorticity baroclinic torque term is defined as:

$$\Gamma_{\mathrm{BT}i} = \frac{\int \left(-\frac{\nabla\rho_{\mathrm{m}} + \nabla p}{\rho_{\mathrm{m}}^2}\right)_i dA}{\int dA} \tag{11.7}$$

and the intensity of the vorticity Coriolis force term is defined as:

$$\Gamma_{\mathrm{CORE}i} = \frac{\int (|-2\nabla \times (\vec{\Omega} \times \vec{V})|)_i dA}{\int dA} \tag{11.8}$$

where $i$ represents the component in the radial, circumferential, and axial direction in a rectangular coordinate system, and $A$ is the area of each radial section in $m^2$.

Figure 11.18 shows the vorticity transport intensity distributions on a series of radial sections in various cavitation conditions for the stretching distortion, expansion/contraction, baroclinic torque, and the Coriolis force terms in the vorticity transport equation. As shown in this figure, at $\lambda = 0$, the vorticity transport intensity at the LE of the blade is mainly influenced by the mainstream flow, and almost no leakage occurs; therefore, this case is not discussed here. In the range $0.1 \leq \lambda \leq 0.7$, the intensity of the vorticity stretching term decreases with an increase in the cavitation, which is related to the scale and intensity of cavitation. The serious cavitation in the blade

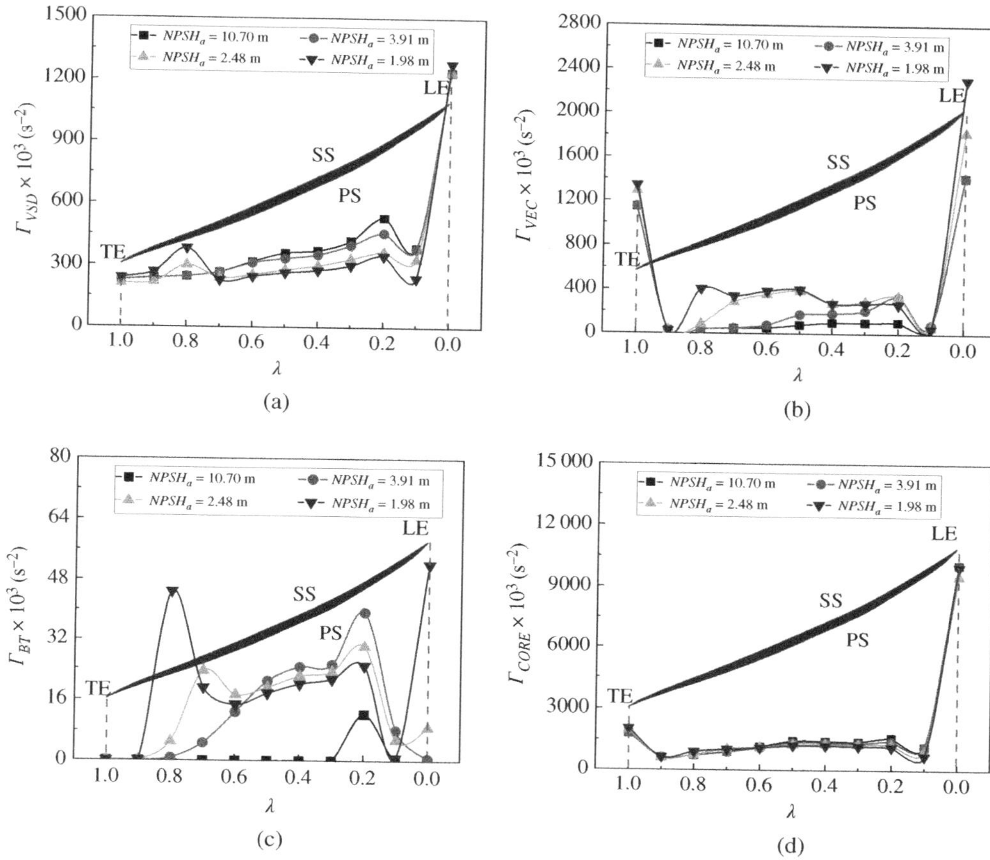


**Figure 11.18** Vorticity transport intensity distribution curves in various cavitation conditions. (a) $\Gamma_{VSD}$, (b) $\Gamma_{VEC}$, (c) $\Gamma_{BT}$, (d) $\Gamma_{CORE}$.

tip area weakens the vorticity intensity in the cavitation area, which agrees with the results shown in Fig. 11.6. In the range $0.7 \leq \lambda \leq 1.0$, the flow separation and the formation of vortices due to separation increase the distortion of the vortices due to stretching as well as the intensity of the vorticity transport. Since the mass transfer rate between the vapor and the liquid is proportional to the velocity divergence ($\nabla \cdot \vec{V}$), the variation in vorticity expansion/contraction intensity occurs mainly in the cavitation area, as shown in Fig. 11.18(b). With the severity of cavitation, the expansion/contraction intensity increases, and the influence area develops toward the TE. The results show that the increase in the scale and intensity of cavitation leads to the change in the intensity of the vortex transport due to the expansion/contraction term. The baroclinic torque term is related to the distributions of the pressure and density gradients; therefore, it affects the gas-liquid interface area and shows similar vorticity intensity characteristics as the intensity due to the expansion/contraction; however, its vorticity intensity is relatively weak. Due to the work done by the circumferential rotation of the impeller, the Coriolis force term cannot be ignored. The results show that the Coriolis force term plays a dominant role in the vorticity transport; however, cavitation has relatively small influence on the intensity of the Coriolis force.

Figure 11.19 quantitatively evaluates the radial, circumferential, and axial vorticity intensity distributions due to vortex stretching, the expansion/contraction, the baroclinic torque, and the

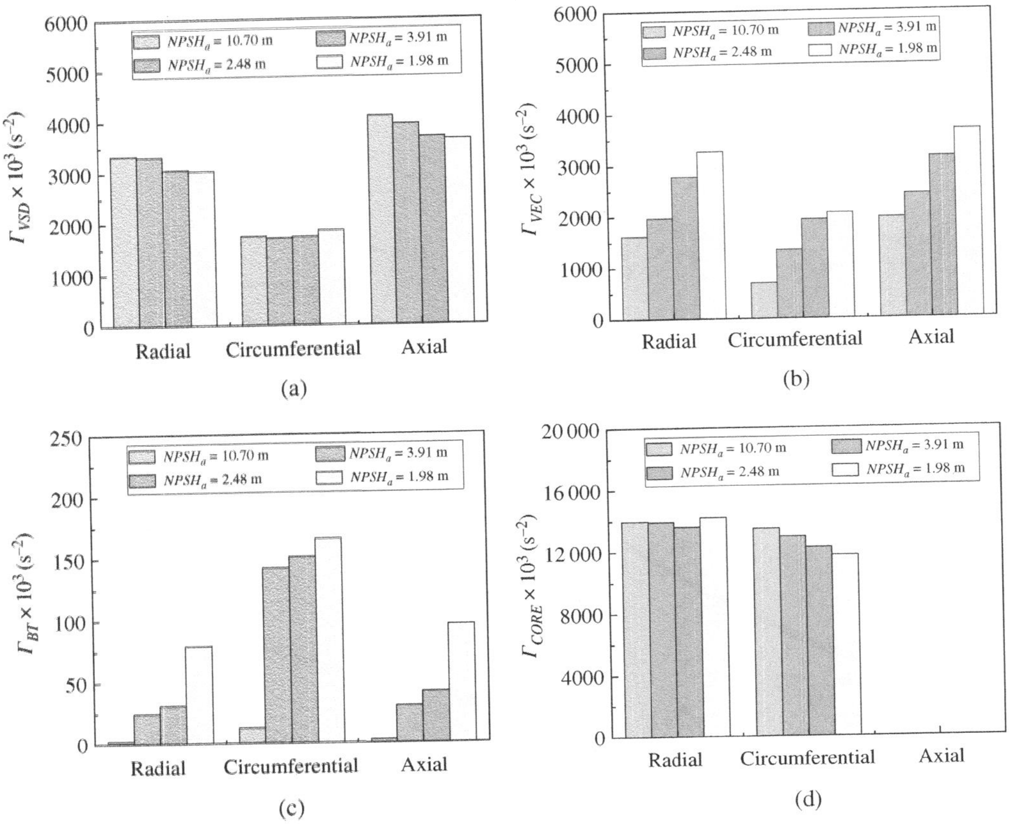


**Figure 11.19** Radial, circumferential, and axial distribution of different measures of vorticity in various cavitation conditions. (a) $\Gamma_{VSD}$, (b) $\Gamma_{VEC}$, (c) $\Gamma_{BT}$, (d) $\Gamma_{CORE}$.

Coriolis force terms in the vorticity transport equation in various cavitation conditions. It can be noted that the radial and axial intensity of the vorticity stretching term is stronger than that in the circumferential direction, and its intensity decreases with the decrease in the $NPSH_a$ value, and in turn the circumferential vorticity intensity increases. The vorticity expansion/contraction term is enhanced by the cavitation in the radial, circumferential, and axial directions and is equally important in all these directions in the vorticity transport process, which indicates that the components of the vorticity transport equation in the radial and axial directions are not negligible. The radial, circumferential, and axial vorticity intensity of the baroclinic torque term increases with an increase in the cavitation, while the total vorticity transport intensity is smaller than that of the vorticity expansion/contraction term. It can be noted from Fig. 11.19(d) that the Coriolis force term dominates the vorticity transport process, and the circumferential Coriolis force intensity tends to decrease under cavitation. Figure 11.20 quantitatively evaluates the intensity distribution of each transport term in the vorticity transport equation in various cavitation condition. It can be noted that the Coriolis force term plays a leading role in the vorticity transport process. The variation in the vorticity transport intensity caused by the cavitation is mainly reflected in the expansion/contraction term, while the baroclinic torque term has a small influence but plays an important role in maintaining the shape of the cavitation bubble.

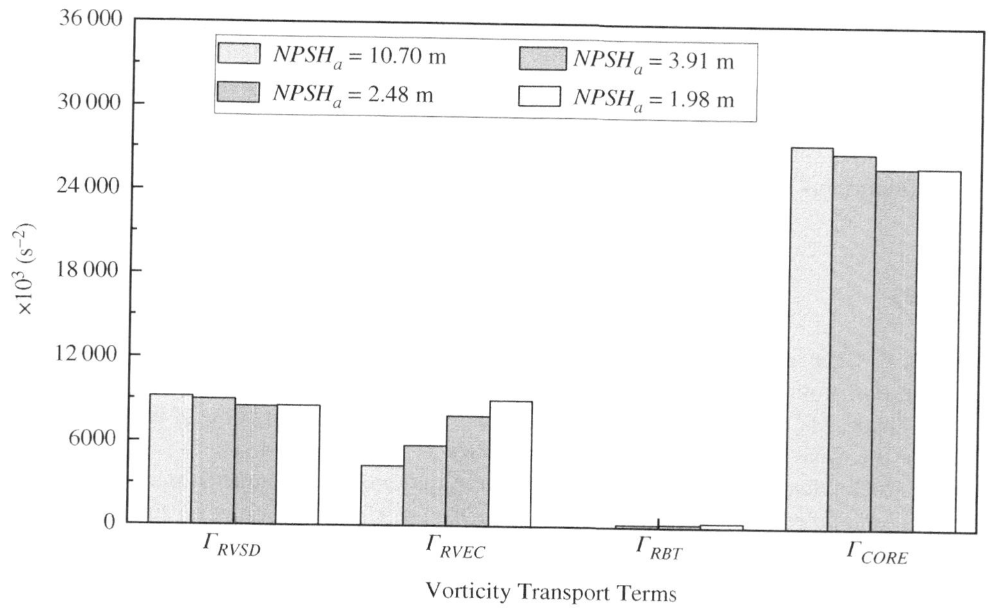


**Figure 11.20** Quantitative distribution of different measures of vorticity in different cavitation conditions

## 11.3 Summary

This chapter provided an analysis of the spatial evolution of the vorticity intensity and the turbulent kinetic energy intensity in the flow field of the mixed-flow pump in various cavitation conditions. Employing the vorticity transport equation, the contributions of each term in the vorticity transport equation to the vorticity generation process are evaluated qualitatively and quantitatively. The main conclusions can be summarized as follows:

(1) The blade tip leakage rate is less affected by the tip leakage vortex cavitation, which indicates that the energy loss in the pump is not caused by the increased tip leakage due to cavitation. The leakage flow at the blade LE gradually approaches the suction surface due to cavitation, and the tip leakage vortices break down. The leakage flow in the middle of the blade tip and the blade TE are responsible for the formation of the secondary leakage flow. The increase in cavitation leads to an increased secondary leakage flow. The interaction between the tip LE leakage flow and the TE leakage flow destroys the flow stability, resulting in the separated flow and the vortices due to the separated flow. The low-speed zones are formed in the end wall area, causing the blockage of the flow passage.

(2) The breakdown and fracture of the tip leakage vortices are related to the scale and intensity of cavitation in the flow. Low pressure causes cavitation to cover most of the area of the suction surface, which weakens the movement of the leakage flow and inhibits the formation and development of the tip leakage vortices. The distribution range of high turbulent kinetic energy region is almost the same as that of the high vorticity region, and there is a positive correlation between the two intensities. Severe cavitation results in the radial and axial development of the high turbulent kinetic energy region at the outlet of the flow passage toward the impeller, which leads to an increase in the energy loss due to turbulent dissipation.

(3) The contribution of each component of the vorticity transport equation to vorticity generation in cavitation flow is evaluated both qualitatively and quantitatively. It is found that the

variation in the vorticity transport intensity caused by the cavitation is mainly reflected in the expansion/ contraction term of the vorticity transport equation. The radial, circumferential, and axial vorticity intensity is enhanced by the cavitation, which indicates that the components of the vorticity transport equation in the radial and axial directions are not negligible. The Coriolis force term dominates the vorticity transport process, while the baroclinic torque term has a small influence but plays an important role in maintaining the shape of the cavitation bubble.

## References


1 Du, J., Lin, F., Chen, J.N., and Christoph, B. (2013). Flow structures in the tip region for a transonic compressor rotor, ASME. *Journal of Turbomachinery* 135 (3).

2 Wang, Y., Zhang, M., Chen, T., and Huang, B. (2018). Unsteady cavitating flow of liquid hydrogen around the ogive body. *Journal of Aerospace Power* 33 (8): 1845–1854.

3 Zhang, H. (2022). *Study on vortex evolution mechanism and instability characteristics of tip leakage flow in axial-flow pump*. PhD Thesis, Zhenjiang, Jiangsu University, viewed 6 June 2022, Research Direct database.

4 Shen, X. (2021). *Research on the mechanism of inception and evolution of trailing edge suction-side-perpendicular cavitating vortices induced by complicated cavitating flow in the tip region of an axial flow pump*. PhD Thesis, Zhenjiang, Jiangsu University, viewed 13 October 2021, Research Direct database.

5 Xu, S. (2020). *Investigation on the mechanism between vortex and cavitation in propelling pump tip clearance with front guide vanes*, PhD Thesis, Hubei: Wuhan University, viewed 4 August 2023, Research Direct database.

6 Long, X.P., Cheng, H.Y., Ji, B. et al. (2018). Large eddy simulation and Euler–Lagrangian coupling investigation of the transient cavitating turbulent flow around a twisted hydrofoil. *International Journal of Multiphase Flow* 100: 41–56.

# 12

# Multiphase Flow Simulations of Sediment Particles in Mixed-flow Pumps

## 12.1 Introduction

The pumps are energy conversion devices whose main goal is to convert the mechanical or rotational energy of the impeller into the fluid energy. Thus, the understanding of the various sources of energy loss and its reduction has consistently been a focal point of research to improve their efficiency and overall performance as described in Chapters 7–12 [1]. The energy loss in a pump can be categorized into the hydraulic loss, the volumetric loss, and the mechanical loss. The hydraulic loss encompasses the impact loss, the loss due to the vortex resistance, and the loss due to the friction. There have been many studies reported in the literature on the impact loss and the loss due to vortex resistance in the pumps operating with clean water as a working fluid. For example, Shi et al. [2] studied the impact loss at the impeller inlet for varying number of guide vane blades to determine an optimal number of blades to minimize the impact loss. Nie et al. [3] employed the velocity triangles concept to investigate the impeller impact loss, elucidating the impact of inlet angle and flow velocity on the loss. Li et al. [4–6] studied the energy loss caused by the internal vortex flow in the mixed-flow pumps. Ji et al. [7–9] explored the impact of the leakage vortex flow between the tips of the impeller blades on the internal flow field and hydraulic loss in the mixed-flow pumps. They analyzed the energy loss in the mixed-flow pump using the entropy generation theory and conducted the diagnostics. However, pumps are not always under clean water conditions. In practical engineering applications, pumps generally operate in a two-phase flow consisting of the solid sediment particles with water, where the loss due to friction becomes significant. The substantial friction losses occur due to interactions between the particles and the pump walls, among the particles themselves, and between the particles and the moving fluid.

Currently, there is a paucity of research on the frictional resistance loss in pumps, and the researchers studying the solid–liquid two-phase flow have long observed the anomalous results in predicting the hydraulic performance of the pump, prompting investigations into the causes of these anomalies in the results [10, 11]. The hydraulic performance of pumps is intrinsically linked to the energy, and understanding the causes of energy loss in the pumps holds the promise to discover solutions to improve their hydraulic efficiency. Zhao et al. [12] addressed the energy loss in pumps operating in solid–liquid two-phase flow and devised a formula to obtain the energy loss due to a unit weight of a solid–liquid mixture. Gandhi [13] introduced a method for determining the energy loss in the pump utilizing an analytical model grounded in the pump geometry and the characteristics of the suspended solid particles to estimate the energy head losses in the pumps. His objective was to predict the performance of the centrifugal slurry pumps handling the solid–liquid mixtures. Li et al. [14] employed the multiphase flow mixture model with the extended version of the k-ε turbulence model to analyze the pressure and velocity changes inside a slurry pump.

*Mixed-flow Pumps: Modeling, Simulation, and Measurements*, First Edition.
Wei Li, Leilei Ji, Ramesh Agarwal, Weidong Shi, and Ling Zhou.

They conducted an energy loss analysis on the critical overcurrent components in the pump. Sun et al. [15] utilized a combined energy analysis method to scrutinize the primary loss areas in the typical pipe pumps. Their analysis covered the power loss characteristics in various channel components of the pump, considering the total pressure change, entropy generation, and the energy balance equations. Luo et al. [16] studied the discontinuous online mixing pumps from an energy perspective utilizing a combined analysis method grounded in entropy generation theory and the Relative Energy Rotor Enthalpy (Rothalpy). They discerned the flow and energy conversion characteristics of the alternating media under various conditions. Shi et al. [17] investigated the flow field of various impellers of a centrifugal pump in the rotating frame employing a two-phase liquid–solid mixture and particle image velocimetry system. Their study revealed that the impellers with long and short blades significantly enhanced the "jet wake" structure, thereby mitigating the energy loss and improving the pump efficiency. Wu et al. [18] discussed the efficiency loss due to the key components of the pump and their conveying systems in the solid–liquid two-phase flow. They considered the energy-saving technologies for conveying systems and proposed effective measures to enhance the pump efficiency and its operational economics. Tang et al. [19], with the objective of improving the efficiency of pumps with impurity in the working fluid, formulated an optimization model with minimum energy loss as the objective function. The above-mentioned investigations shed some light on the energy loss in various components of different types of pumps, offering valuable insights for enhancing the performance of solid–liquid two-phase flow pumps and improving the efficiency of their fluid conveying systems.

With the increasing depth of research on solid–liquid two-phase flows, investigations have progressed in recent years beyond the simplistic examination of the energy loss solely due to the solid–liquid two-phase flow conditions inside the pump. Researchers have delved into understanding the intricate relationship between the varying solid-phase volume fractions and the energy loss. Shen et al. [20] employed a discrete phase model based on the Euler-Lagrange method to simulate the internal flow field of a screw centrifugal pump used for transporting the large-particle solid–liquid two-phase flow. Their analysis included the study of the energy changes, trajectory characteristics, and erosion damage due to the solid-phase particles in the computational domain for various solid volume fractions. Their findings revealed that the low-density fine particles exhibited longer trajectories, increased collisions with the flow components, higher energy loss, and relatively uniform erosion locations. Shi et al. [21] explored the external characteristics, the internal pressure field, the velocity field, and the solid distribution patterns in the pumps by varying the solid concentration and the particle size. Their research demonstrated that with an increase in the particle volume concentration, the energy exchange between the particles and the fluid media increased, leading to a corresponding rise in the hydraulic losses in the flow field. Wang et al. [22] conducted the performance tests on a centrifugal pump at low flow rates with varying particle concentrations. Their dimensionless analysis of the impact of the particle concentration on the pump head and efficiency indicated that the transport of the dense, fine particles could absorb a significant amount of energy, reducing the relative kinetic energy of the liquid phase and inhibiting the generation of eddies in the impeller channel. Liu [23] studied a guide vane centrifugal pump and conducted the numerical simulations of solid–liquid two-phase flow in the entire flow channel using the standard k-ε turbulent model and the multiphase mixture flow model. The study computed the energy loss for different solid-phase volume fractions, revealing that the energy loss gradually increased in the latter half of the pressure side of the blade. The energy of the fluid at the outlet of the pressure side decreased resulting in significant energy loss at the outlet. Wang [24] investigated the basic characteristics of a jet pump under the two-phase flow conditions simulating the coal powder volume concentration ranging from 0% to 50% in the production fluid from the geological formations. The results indicated that an increase in the coal powder volume concentration in the production fluid led to a reduction in the lifting capacity and efficiency of the jet pump. Moreover, it resulted in an elevated coal powder content carried by the pump, contributing to increased internal energy loss in various components of the jet pump. The amalgamation of varying solid-phase volume fractions and the energy loss in this example provides valuable insights for exploring the energy loss in mixed-flow pumps under conditions of multiphase flow of water with sediment particles.

To explore the influence of the sediment particles on the internal energy loss in a mixed-flow pump, numerical simulations conducted in the fluid domain (clean water) of the pump considering various solid phase (sediment particles) volume fractions are described in Sections 12.2–12.5. The results of the numerical simulations show the impact of different solid-phase volume fractions on the energy characteristics of the pump. The results also include the distribution patterns of solid particles in the impeller and the guide vanes along with an in-depth analysis of the associated energy losses. Finally, a comprehensive analysis is presented to elucidate the underlying mechanisms contributing to the energy loss in the channels of the impeller and the guide vanes.

## 12.2 Governing Equations of the Mixture Model of Multiphase Flow

The mixture model employs mass fractions to describe the distribution of various components such as fluids and solids in the multiphase mixture. It considers the variation in the mass fraction of each component and the diffusion among the components. The model exhibits strong coupling among the different phases, making it well-suited for simulation of the liquids containing the solid particles. Its demonstrated and proven capability in simulating multiphase flows in many industrial applications makes it suitable candidate for simulating the solid–liquid two-phase flow in the mixed-flow pump. In this chapter, the application of the mixture model is described for the multiphase flow simulation of the flow field in a mixed-flow pump consisting of clean water and sediment particles with various volume fractions. The governing equations of the multiphase flow of a mixture are described below [25]:

$$\frac{\partial \rho_{\mathrm{m}}}{\partial t} + \nabla \cdot \left(\rho_{\mathrm{m}} \vec{v}_{\mathrm{m}}\right) = 0 \tag{12.1}$$

In Eq. (12.1), the subscript m denotes the mixture of phases; $\rho_{\mathrm{m}}$ denotes the density of the mixture in kg/m$^3$; $t$ denotes time in s; and $\vec{v}_{\mathrm{m}}$ denotes the velocity of the mixed phase in m/s.

The momentum equation for the mixture is derived by summing up the individual momentum equations of all phases. The momentum equation is obtained as:

$$\begin{aligned}\frac{\partial}{\partial t}\left(\rho_{\mathrm{m}} \vec{v}_{\mathrm{m}}\right) + \nabla \cdot \left(\rho_{\mathrm{m}} \vec{v}_{\mathrm{m}} \vec{v}_{\mathrm{m}}\right) &= -\nabla p + \nabla \cdot \left[\mu_{\mathrm{m}}\left(\nabla \vec{v}_{\mathrm{m}} + \nabla \vec{v}_{\mathrm{m}}^{T}\right)\right] \\ &\quad + \rho_{\mathrm{m}} \vec{g} + \vec{F} - \nabla \cdot \left(\sum_{k=1}^{n} \alpha_k \rho_k \vec{v}_{\mathrm{dr},k} \vec{v}_{\mathrm{dr},k}\right)\end{aligned} \tag{12.2}$$

In Eq. (12.2), $\alpha_k$ denotes the volume fraction of phase $k$; $\vec{F}$ denotes the volume force in Newton; $\mu_{\mathrm{m}}$ denotes the viscosity of the mixture in Pa s; and $\vec{v}_{\mathrm{dr},k}$ denotes the drift velocity of phase $k$ in m/s.

The viscosity of the mixture is calculated by the equation:

$$\mu_m = \sum_{k=1}^{n} \alpha_k \mu_k \tag{12.3}$$

where $\mu_k$ is the viscosity of the phase $k$ in Pa s.

The drift velocity of phase $k$ in Eq. (12.2) is given by:

$$\vec{v}_{\mathrm{dr},k} = \vec{v}_k - \vec{v}_{\mathrm{m}} \tag{12.4}$$

The energy equation for the mixture model is given by:

$$\frac{\partial}{\partial t}\sum_k \left(\alpha_k \rho_k E_k\right) + \nabla \cdot \sum_k \left(\alpha_k \vec{v}_k \left(\rho_k E_k + p\right)\right) = \nabla \cdot \left( k_{\mathrm{eff}} \nabla T - \sum_k \sum_j h_{j,k} \vec{J}_{j,k} + \left(\bar{\tau}_{\mathrm{eff}} \cdot \vec{v}\right) \right) + S_{\mathrm{h}} \tag{12.5}$$

In Eq. (12.5), $k_{\mathrm{eff}}$ is the effective thermal conductivity in W/(m K); $E_k$ is the internal energy of the phase $k$ in KJ/mol; $\rho_k$ is the density of phase $k$ in kg/m$^3$; $\vec{v}_k$ is the velocity of the phase $k$ in m/s; $h_{j,k}$ is the enthalpy of substance $j$ in phase $k$ in KJ/mol; $\vec{J}_{j,k}$ is the diffusion flux of the substance $j$ in phase $k$ in kg/(m$^2$ s); $\bar{\tau}_{\mathrm{eff}}$ is the shear stress in Pa; and $S_{\mathrm{h}}$ is the heat source.

The effective thermal conductivity is given by the equation:

$$k_{\mathrm{eff}} = \sum \alpha_k \left(k_k + k_{\mathrm{t}}\right) \tag{12.6}$$

In Eq. (12.6). $k_k$ is the thermal conductivity of the phase $k$ in W/(m K), and $k_{\mathrm{t}}$ is the thermal conductivity of turbulent fluid in W/(m K).

The internal energy $E_k$ of phase $k$ in Eq. (12.5) is given by the equation:

$$E_k = h_k - \frac{p}{\rho_k} + \frac{v^2}{2} \tag{12.7}$$

In Eq. (12.7), $h_k$ is the specific enthalpy of phase $k$ in KJ/mol.

It should be noted that the equations described above are applicable to both the laminar flow and the fully turbulent flow. In case of turbulent flow, all the flow variables represent their time-averaged values, and the $\bar{\tau}_{\mathrm{eff}}$ is computed by employing a turbulence model. In the simulations presented in Sections 12.2–12.5, the standard $k$–$\varepsilon$ model described in Chapter 3 is employed. The governing transport equations for turbulent kinetic energy ($k$) and dissipation rate ($\varepsilon$) are given in Chapter 3, and its implementation details and performance for a mixed-flow model pump are described in Chapter 6. The multiphase governing equations and turbulence models are built-in in the computational fluid dynamics solvers Fluent and computational fluid dynamics X (CFX).

## 12.3 Pump Model and Mesh Generation

The mixed-flow model used in this chapter is the same as that used in the entire book; its geometric details and other relevant operational parameters are given in Chapter 3. Chapter 3 also describes the geometric modeling of the pump, the multi-block structured grid used inside the entire pump, and the special treatment near the wall to resolve the turbulent boundary layers. It also describes the mesh independence of the solution analysis and the number of nodes required for obtaining the mesh-independent solution. Here, the number of mesh nodes employed for obtaining the mesh-independent solution is given for the two main components of the pump, namely the impeller, which has 399 792 nodes, and the guide vane, which has 232 596 nodes. Other components of the pump have a smaller number of mesh nodes, as given in Chapter 3. Chapter 5 describes the experimental apparatus and measurements of the various flow quantities such as the velocity and the pressure including their fluctuations. The same experimental apparatus is used to obtain the data presented in Sections 12.4 and 12.5.

## 12.4 Two-phase Flow Characteristics of the Solid–Liquid Flow in Mixed-flow Pumps

### 12.4.1 Influence of the Solid Particles on the Energy Characteristics of the Mixed-flow Pumps

#### 12.4.1.1 Influence of the Solid Particles on the Pump Head and Efficiency in Various Flow Conditions

When conveying solid–liquid two-phase fluids, the introduction of solid particles significantly influences the head and the efficiency of the mixed-flow pumps, leading to changes in the pumps' energy characteristics. The inclusion of particles induces modifications in the composition of the fluid medium. When transporting the solid particles, the pump head is computed by Eq. (12.8) [26]:

$$H = \Delta Z + \frac{P_{\text{out}} - P_{\text{in}}}{\left(\rho_{\text{p}} \times C_{\text{V}} + \rho \times \left(1 - C_{\text{V}}\right)\right) g} + \frac{v_2^2 - v_1^2}{2g} \tag{12.8}$$

In Eq. (12.8), $H$ is the head in m, $\Delta Z$ is the position potential energy difference in m, $P_{\text{out}}$ is the outlet pressure in Pa, $P_{\text{in}}$ is the inlet pressure in Pa, $v_2$ is the average velocity at the pump outlet cross-section in m/s, $v_1$ is the average velocity at the pump inlet cross-section in m/s, $\rho_{\text{p}}$ is the density of the solid phase in kg/m$^3$, and $C_{\text{V}}$ is the volume fraction of the solid phase. $C_{\text{V}}$ is given by Eq. (12.9):

$$C_V = \frac{V_{\text{p}}}{V_{\text{p}} + V_{\text{f}}} \tag{12.9}$$

where $V_{\text{p}}$ is the volume of the solid phase in m$^3$ and $V_{\text{f}}$ is the volume of the liquid phase in m$^3$.

The efficiency of the pump is calculated by Eq. (12.10):

$$\eta = \frac{\left(\rho_{\text{p}} \times C_{\text{V}} + \rho \times \left(1 - C_{\text{V}}\right)\right) \times g \times Q \times H}{P} \times 100\% \tag{12.10}$$

where $g$ is the acceleration due to gravity in m/s$^2$, and $Q$ is the flow rate in m$^3$/h.

Figure 12.1 shows the energy performance of the mixed-flow pump in various flow conditions. As the flow rate increases, the head of the mixed-flow pump exhibits a declining trend. The head becomes lower when the solid particles are introduced compared to when only clean water is transported. Consequently, the inclusion of the solid particles results in a reduction in the pump's head. As to the efficiency, the mixed-flow pump demonstrates an increasing trend with the increasing flow rate until it reaches the design condition. Beyond the design condition, the efficiency experiences a decline. In general, in clean water conditions, the mixed-flow pump exhibits higher efficiency compared to the conditions of solid–liquid two-phase flow. It is evident that the addition of solid particles diminishes the efficiency of the mixed-flow pump.

#### 12.4.1.2 Influence of Different Volume Fractions of Solid Phase on the Pump Head and Efficiency

When conveying the solid–liquid two-phase fluids, the head and the efficiency of the mixed-flow pump undergo substantial changes due to the introduction of solid particles, which result in changes in the energy characteristics of the pump. The inclusion of solid particles modifies the composition of the fluid medium, in particular inducing notable changes in the fluid density $\rho_{\text{m}}$. The density of the two-phase flow medium is given by:

$$\rho_{\text{m}} = \left(1 - \frac{V_{\text{p}}}{V_{\text{m}}}\right) \rho_{\text{f}} + \frac{V_{\text{p}}}{V_{\text{m}}} \rho_{\text{P}} \tag{12.11}$$

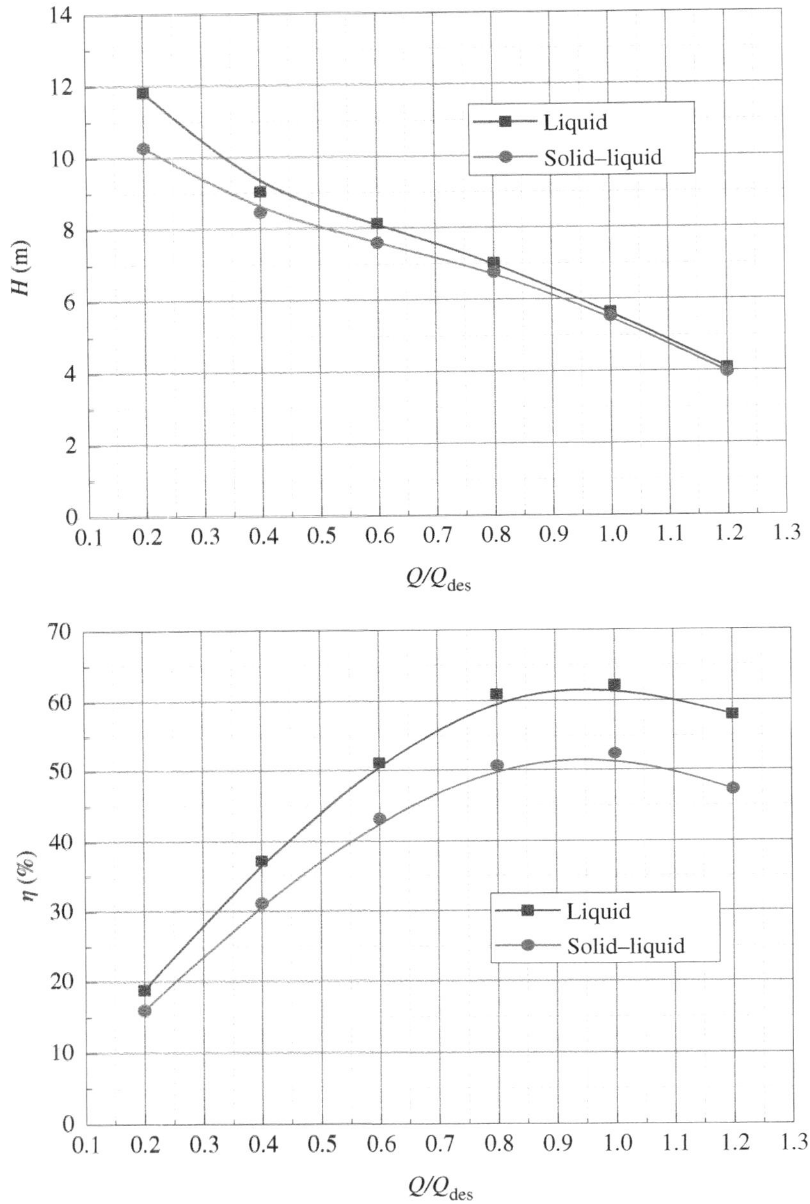


**Figure 12.1** Comparison of the energy characteristics between the clean water and the solid–liquid two-phase flow in the mixed-flow pump in various flow rates.

In Eq. (12.11), $\rho$ denotes the density in kg/m$^3$ and $V$ denotes the volume in m$^3$; the subscript $p$ denotes the particle phase, $f$ denotes the fluid phase, and $m$ denotes the mixed solid–liquid phase.

In what follows, the particles with a diameter of $dp = 0.5$ mm are chosen in the analysis for three different volume fractions, $C_v = 5\%$, $C_v = 7.5\%$, and $C_v = 10\%$. The pump head and efficiency for various particle inlet volume fractions are presented in Table 12.1. With an increase in the volume

**Table 12.1** Pump head and efficiency of the mixed-flow pump at various volume fractions of the solid particles.

| $C_v$ (%) | $H$ (m) | $\eta$ (%) |
|---|---|---|
| 5 | 5.703 | 58.023 |
| 7.5 | 5.670 | 56.432 |
| 10 | 5.526 | 52.358 |

fraction, particularly in the range from $C_V = 7.5\%$ to $C_V = 10\%$, a significant coupling effect between the particles and the flow field is observed. Moreover, the conveyance of the particles by the pump introduces additional energy losses, and the influence of the particles on the flow field becomes more pronounced, resulting in a declining trend in both the pump head and the efficiency. The introduction of the particles increases the density of the mixture compared to the clean water, which requires the impeller to exert additional work on the solid–liquid two-phase fluid flow compared to that on the single-phase clean water flow, requiring a larger shaft power. This implies that as the volume fraction of solid particles increases, the pump efficiency decreases.

The increase in particle volume fraction from $C_V = 5\%$ to $C_V = 7.5\%$ results in a 0.56% reduction in the head and a 2.74% decrease in the efficiency of the model mixed-flow pump. Further increasing the particle volume fraction from $C_V = 7.5\%$ to $C_V = 10\%$ decreases the pump head by 2.54% and causes a substantial drop of 7.22% in the efficiency, indicating significant energy losses. Clearly, the increase in the solid-phase particle concentration correlates with the increased energy losses in the pump.

### 12.4.2 Distribution of the Solid Phase Inside the Impeller of the Pump

The changes in the energy characteristics of the mixed-flow pumps are intricately linked to the energy conversion processes. The solid particles, as they move through the pump, are prone to collisions with each other and with various components of the pump, resulting in energy losses. Therefore, a key strategy for reducing energy losses involves a comprehensive understanding of the distribution patterns of the solid-phase particles in the pump. For a spherical particle of diameter $dp = 0.5$ mm, Fig. 12.2 shows the volume distribution of the solid-phase in the impeller at various volume fractions. Notably, distinct differences in the volume distribution of the solid-phase particles emerge in the impeller flow passage at various volume fractions. On the suction side of the blade, particles in the impeller flow passage predominantly accumulate in the outer edge area of the blade inlet, influenced by the liquid phase. A vortex concentration zone of the solid-phase particles is also evident in the front part of the wheel edge. At $C_V = 5\%$, fewer solid-phase particles are observed, and their movement is significantly influenced by the liquid phase. As they enter the impeller, these particles are swiftly carried by the liquid phase, rapidly traversing the blade passage. As the solid-phase volume fraction increases, the distribution of the particles in the inlet wheel edge area and in the vortex concentration zone gradually increases. On the pressure side of the blade, solid-phase particles concentrate primarily at the hub and move downstream from the impeller inlet to the outlet edge of the blade. With an increase in the solid-phase concentration, the influence of the liquid phase on the solid phase diminishes, and the distribution of the solid particles gradually expands from the hub to the blade edge and the outlet area of the pump. This expansion results in an increased distribution area of the solid-phase particles in the impeller flow passage.

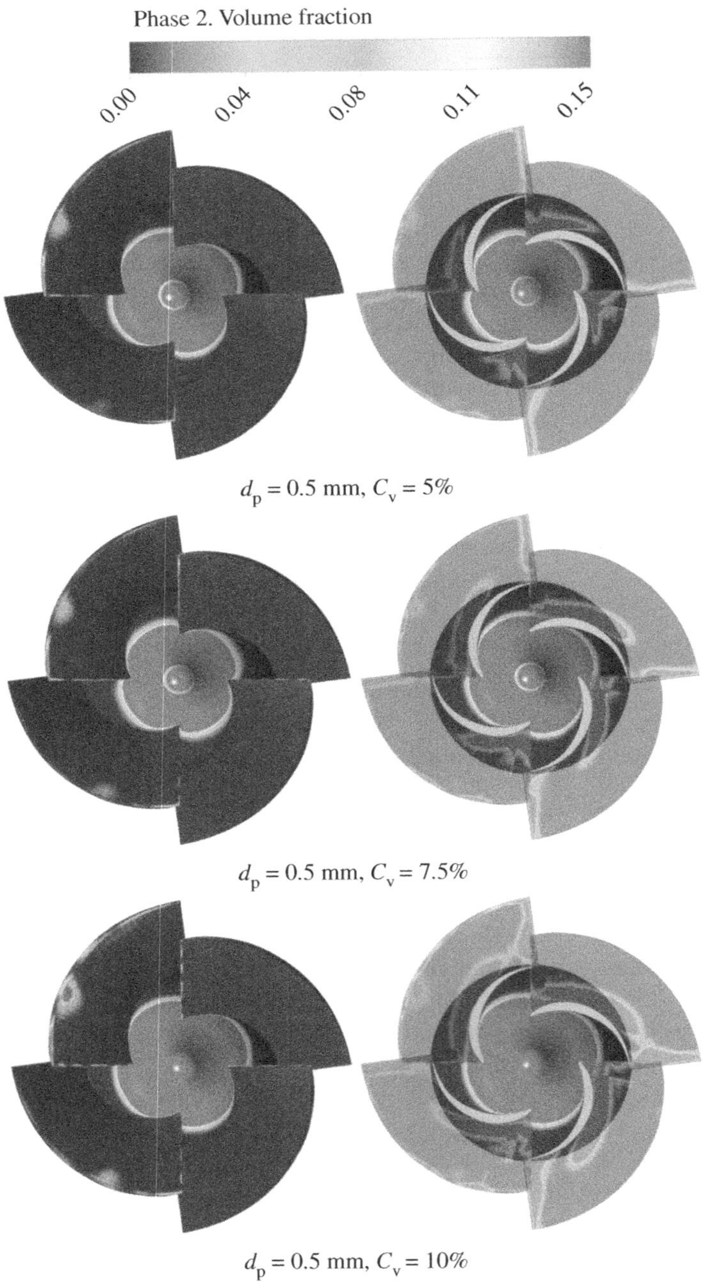


**Figure 12.2** Distribution of the solid phase at a cross-section in the impeller flow path for various volume fractions $C_v$.

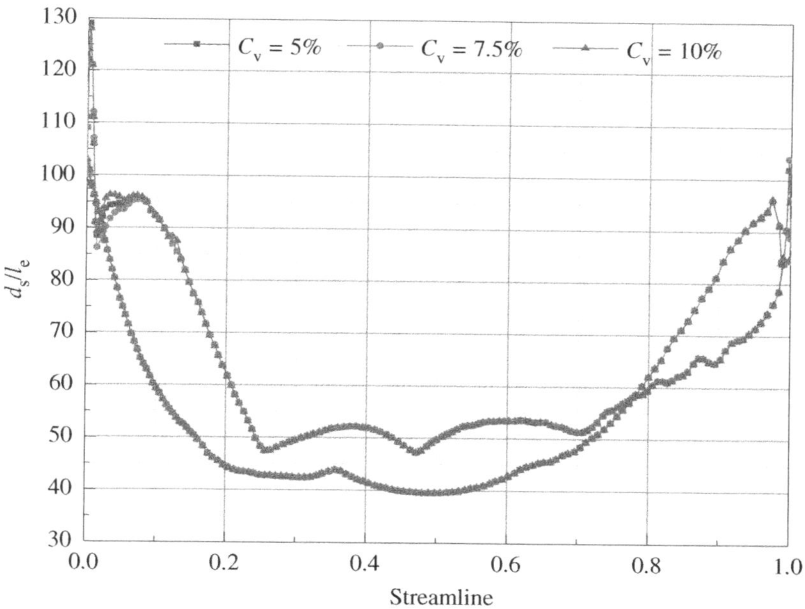


**Figure 12.3** Particle dynamic scale at the mid-streamline in the impeller.

In the investigation of the solid–liquid two-phase flow systems, the characterization of the solid particle diffusion is important, which involves considering the characteristic size of the energetic vortices. However, the sole reliance on the characteristic size of energetic vortices does not comprehensively determine particle diffusion. The critical determinant lies in the interaction between the particles and the energetic vortices. This interaction can be effectively evaluated by comparing the ratio of the particle diameter to the characteristic size of the energetic vortices, denoted as the dynamic scale of the particles $d_p/l_e$, where $d_p = 0.5$ mm is the particle diameter. The equation for the characteristic size of the energetic vortices is given by [27]:

$$l_e = C_\mu^{3/4} \frac{k_l^{3/2}}{\varepsilon_l} \tag{12.12}$$

In Eq. (12.12), $l_e$ is the characteristic size of the energetic vortices, $C_\mu = 0.09$, $k_l$ is the turbulent kinetic energy of the liquid phase, and $\varepsilon_l$ is the dissipation rate of the liquid phase.

As shown in Fig. 12.3, the region with a larger value of the particle dynamic scale is situated at both ends of the blade, while the particle dynamic scale is comparatively much smaller and is concentrated near the inlet and the outlet of the impeller. Notably, at the impeller inlet, the particle dynamic scale increases with the increase in the volume fraction of the solid phase. According to the literature [27], when the particle dynamic scale surpasses 0.1, it implies an intensified diffusion effect of the particles, consequently influencing the turbulent flow structures. It can be deduced that if the particle dynamic scale exceeds 30, it indicates a very high diffusion effect of the particles in the flow passage of the impeller.

### 12.4.3 Analysis of the Energy Loss in the Mixed-flow Pump

To provide a better understanding of the distribution of the energy losses in the pump resulting from the interaction between the flow structures and the solid particles at various solid-phase volume fractions, an analysis of the energy losses in the pump is presented in this section using the entropy production theory. The principle of entropy production (increase) in the flow is another way to express the second law of thermodynamics, which states that the incremental change in the entropy during an irreversible process is always greater than zero. The entropy generation in a system is the inevitable consequence of the irreversibility of a process, where the losses in the mechanical energy are converted into internal energy [16]. The addition of sediment particles in the clean liquid water in a mixed-flow pump enhances the irreversibility effect on the operation of the pump in the context of energy loss and entropy production. This effect includes the collisions between the sediment particles and the pump wall, as well as the collisions among the particles themselves. These interactions result in energy losses in the flow, leading to entropy production.

Figure 12.4 shows the percentage of the entropy production in different parts of the mixed-flow pump including the inlet, the impeller, the guide vane, and the outlet. It can be observed that the entropy production in the impeller and the guide vane accounts for the majority of the total energy loss in the pump. Furthermore, since the solid-phase volume fraction $C_V$ inside the pump increases from left to right in Fig. 12.4, it is obvious that the increase in $C_V$ increases the entropy production in the impeller and the guide vane and hence the energy loss. Comparing the entropy production for various solid-phase volume fractions, it is evident that as the flow rate increases, the entropy production in the impeller increases, while it decreases at the inlet and the outlet. In case of the guide vane, the variation in the entropy production for different solid-phase volume fractions is not large. It is possible that these conclusions may not be universal for all types of pumps, nevertheless they should provide a guideline for the level of energy loss in various components of a pump operating in clean water or water with some volume fraction of solid particles in two-phase flow.

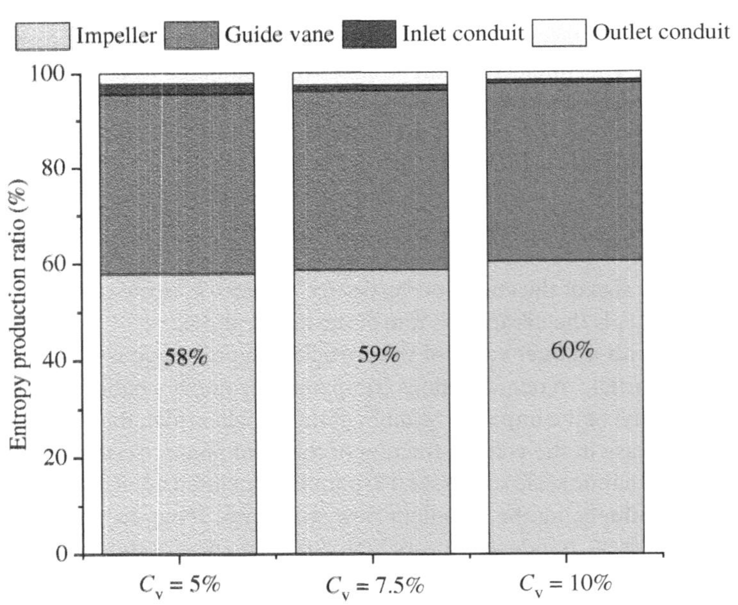


**Figure 12.4** Percentage of the entropy production in various fluid components of the mixed-flow pump.

### 12.4.4 Analysis of the Energy Loss Inside the Impeller

As shown in Fig. 12.5, monitoring line A is positioned on the blade edge at the impeller inlet, and monitoring line E is located on the blade edge at the impeller outlet. Other monitoring lines are evenly distributed between monitoring line A and monitoring line E on the blade edge. To compare the distribution patterns of the local entropy production in the impeller flow passage of the mixed-flow pump for various solid-phase volume fractions, the local entropy production at various cross-sections of the flow field is obtained, as shown in Fig. 12.6. This figure shows that the regions with significant energy losses are concentrated at the outer edge of the blade and near the hub, coinciding with the distribution of the solid-phase particles. This suggests that the distribution of the solid-phase particles influences the energy conversion efficiency of the impeller. Examining the distribution of entropy production for a specific solid-phase volume fraction, it is evident from the corresponding figure in Fig. 12.6 that the entropy production near the pressure side of the blade decreases from the impeller inlet to the impeller outlet. Analyzing the distribution of the entropy production at three different solid-phase volume fractions, as the volume fraction changes from $C_v = 5\%$ to $C_v = 7.5\%$, there is a slight reduction in the entropy production in the blade edge region but an increase in the entropy production near the hub. When the volume fraction changes from $C_v = 7.5\%$ to $C_v = 10\%$, there is not much change in the entropy production in the blade edge region, but there is an increase in the entropy production near the hub. This is because an increase in the solid-phase volume fraction results in a reduction in the liquid phase flow in the blade edge region, leading to a decrease in the entropy production in that area. The increase in entropy production near the hub can be attributed to the enhanced interaction between the liquid and the solid particles due to the increased solid-phase volume fraction.

Analysis of the entropy production values on different monitoring lines reveals a pattern where the solid particle distribution forms four peaks and four valleys corresponding to the number of blades. Notably, the entropy production values on monitoring lines B and C are considerably larger than those on the other monitoring lines. As the solid-phase volume fraction increases, it can be observed that there is a gradual decrease in the entropy production on the monitoring lines B and C. Meanwhile, the entropy production on the remaining monitoring lines remains relatively consistent.

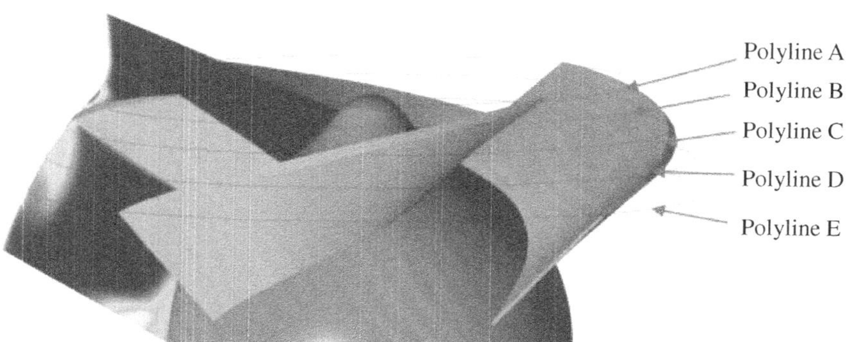


**Figure 12.5** The location of each monitoring line in the impeller and the cross-sectional position of the runner.

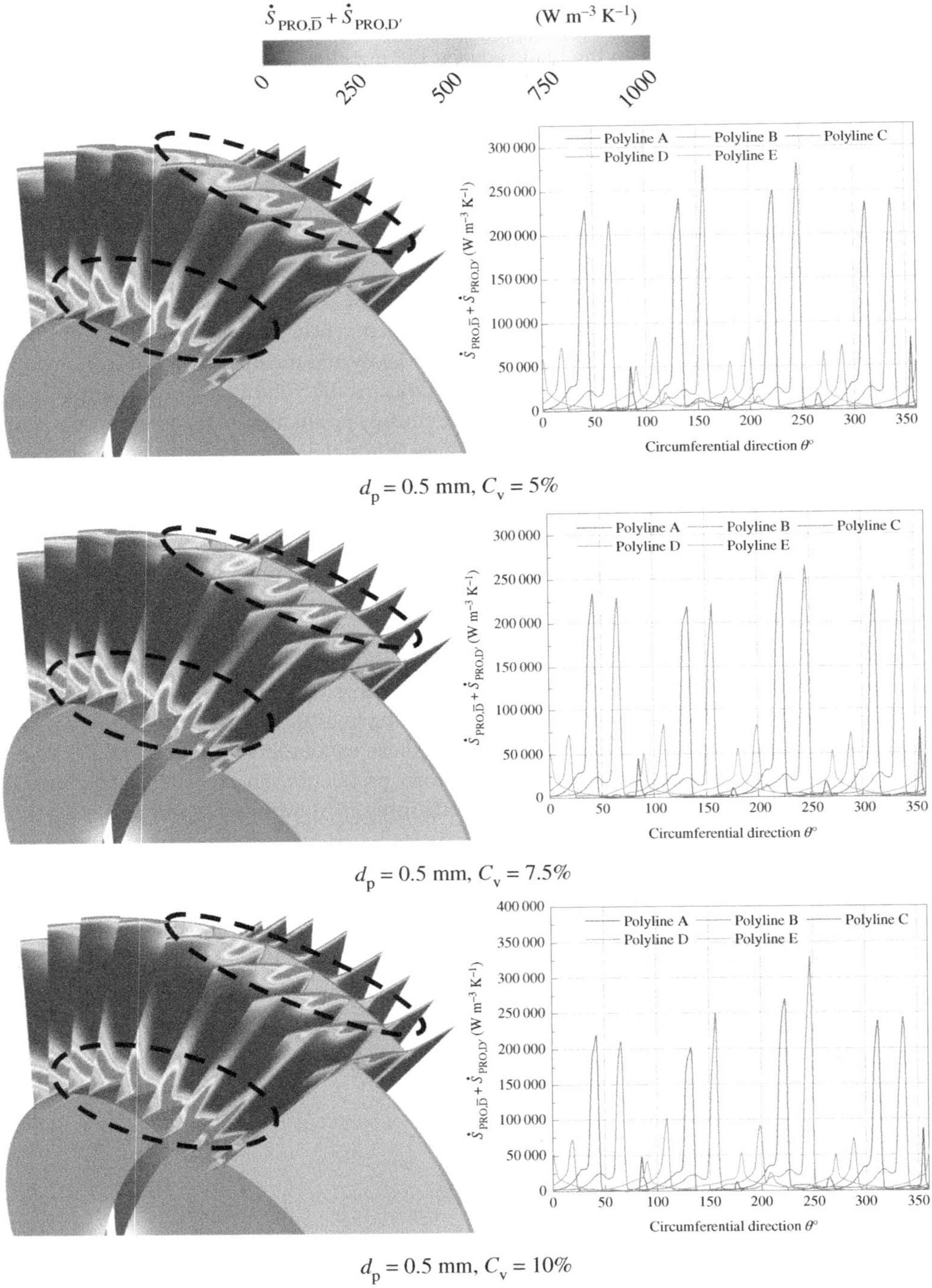


**Figure 12.6** Distribution of the local entropy production in various flow field sections of the impeller passage.

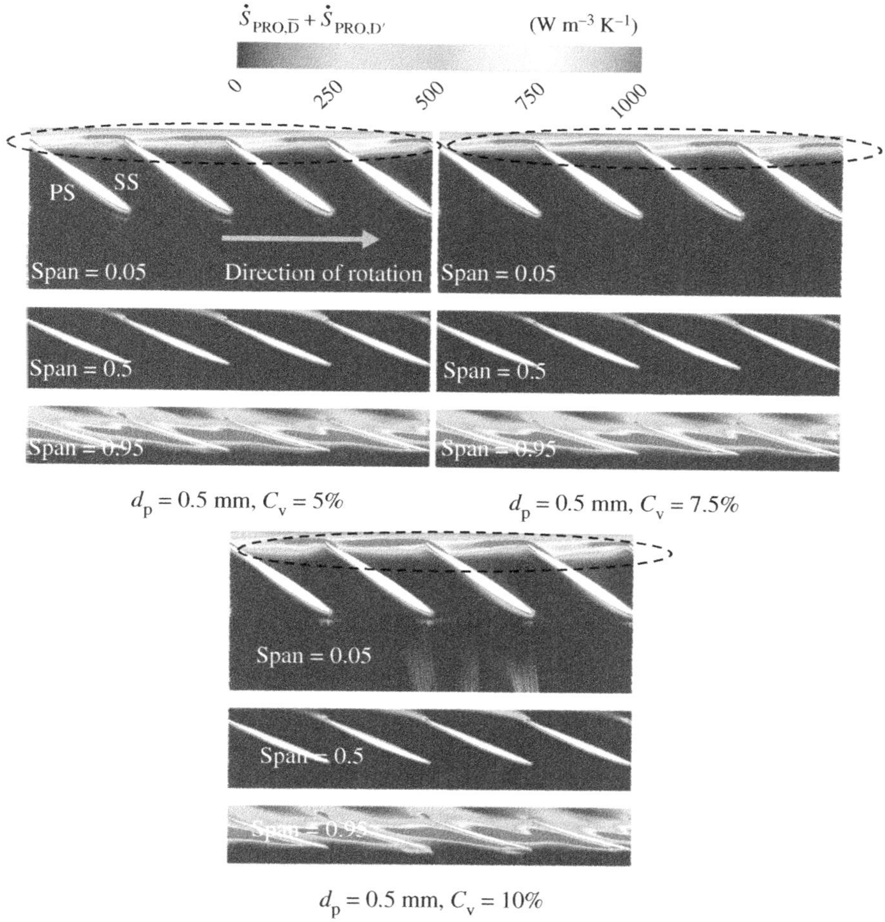


**Figure 12.7** Distribution of the local entropy production in different spanwise sections of the impeller for various volume fractions of the solid particles.

Now, consider the local entropy production distribution at different blade spans in the impeller to identify regions of high energy losses. For this purpose, the impeller domain is evenly divided along the spanwise direction. From Fig. 12.7, it is evident that at different solid-phase volume fractions, the local entropy production contour maps at various spanwise locations in the impeller show high-entropy production areas in the blade edge region at span = 0.95. This indicates that the outer edges of the blades of the mixed-flow pump impeller are the regions of significant energy losses. Furthermore, it can be noted that in the outlet region at blade span = 0.05 near the hub, there is a distribution of high-entropy production, which increases with an increase in the solid-phase volume fraction. Figure 12.8 shows the average entropy production distribution along the flow direction from the impeller inlet to the outlet. This figure shows that starting from the impeller inlet, the overall trend in the entropy production along the flow passage initially shows an increase and then a decrease. Moreover, with the increase in the solid-phase volume fraction, the entropy production in the impeller increases. This indicates that the regions with high-entropy production in the impeller flow passage are mainly concentrated in the middle region of the passage.

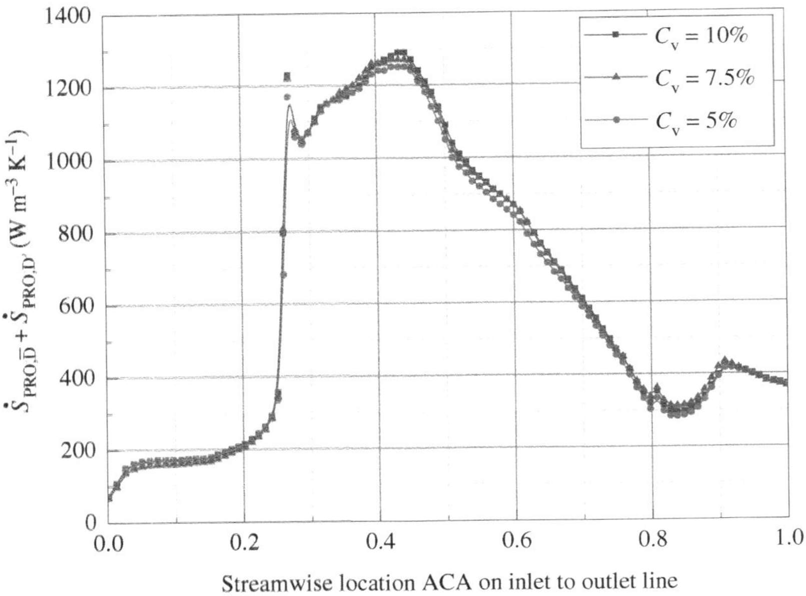


**Figure 12.8** Distribution of the average entropy production along the flow direction from the inlet to the outlet of the impeller.

The left diagram in Fig. 12.9 shows the distribution of turbulent kinetic energy dissipation in the middle section of the impeller flow passage, while the right diagram depicts the distribution of the turbulent kinetic energy dissipation on various monitoring lines in the impeller. In the middle section of the impeller flow passage, it can be observed that the turbulent kinetic energy dissipation shows a slight increase with an increase in the solid-phase volume fraction. It appears that the sole factor leading to an increase in the turbulent kinetic energy dissipation is the solid-phase volume fraction. Therefore, it can be inferred that an increase in the number of solid particles in the two-phase flow will result in higher turbulent kinetic energy dissipation.

Considering various monitoring lines in the impeller blade edge region, it can be observed that the turbulent kinetic energy dissipation on the monitoring line A at the impeller inlet blade edge is relatively low compared to that at other monitoring lines, indicating lower turbulent kinetic energy dissipation at the impeller inlet blade edge. On monitoring line E at the impeller outlet blade edge, the turbulent kinetic energy dissipation is higher than that on the monitoring line A but is lower compared to the other monitoring lines. Monitoring lines B and C exhibit large turbulent kinetic energy dissipation with similar levels of dissipation. Monitoring line D shows an intermediate level of turbulent kinetic energy dissipation, being higher than that at the impeller inlet and outlet blade edges but lower than that on monitoring lines B and C. Therefore, in the impeller blade edge region, the turbulent kinetic energy dissipation increases from the impeller inlet to the outlet and then decreases. The turbulent kinetic energy dissipation values mostly remain between 0 and 2 $m^2/s^2$, and with the increase in solid-phase volume fraction, the maximum turbulent kinetic energy dissipation value also increases.

Figure 12.10 shows the vorticity distribution in the mixed-flow pump at various volume fractions of the solid particles in the design flow condition. From this figure, it can be observed that at a volume fraction of $C_v = 5\%$, the vortices in the impeller are generated close to the edge of the blade suction surface. At $C_v = 7.5\%$, the vortices move away from the blade suction surface

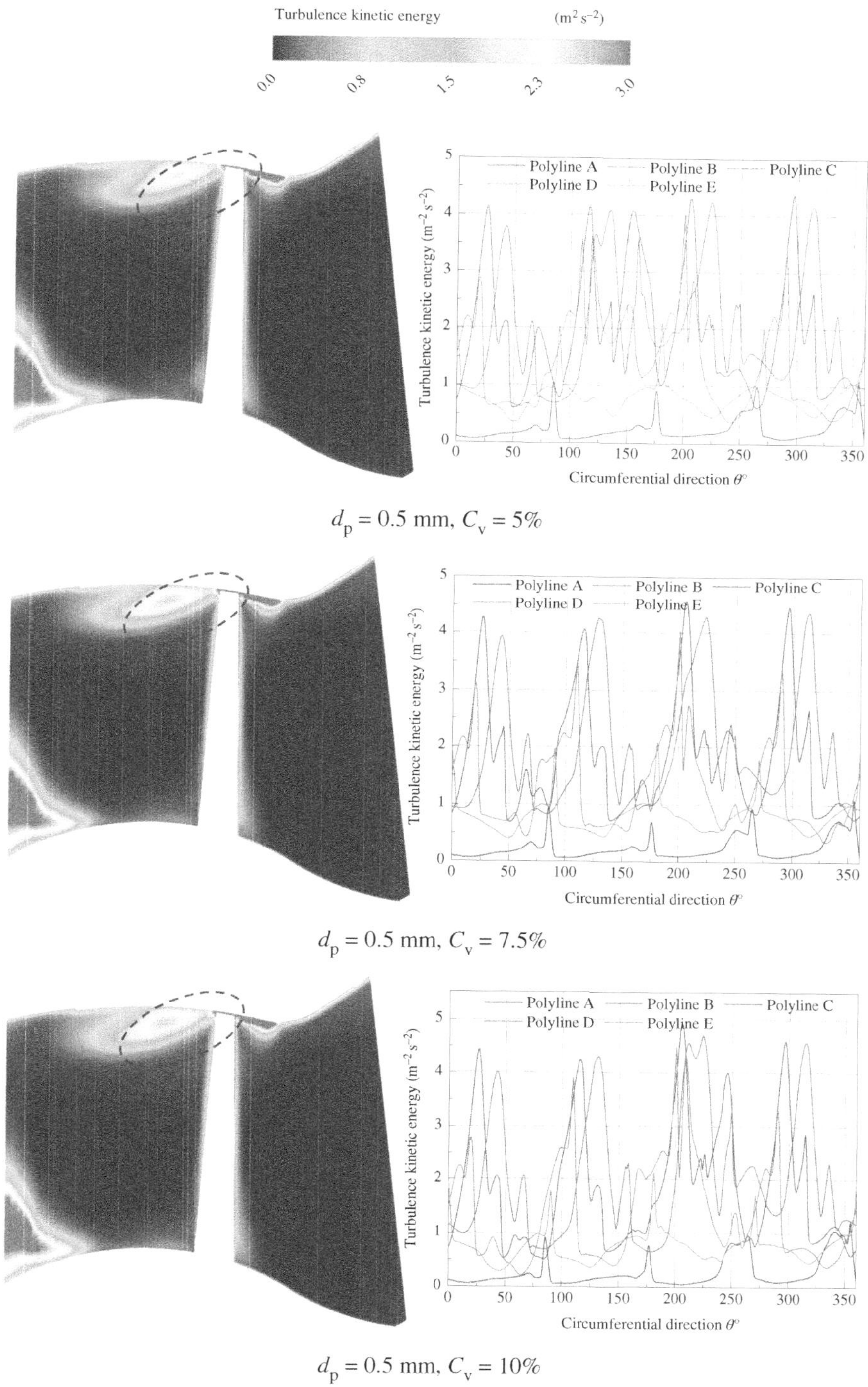


**Figure 12.9** Distribution of the turbulent kinetic energy for various volume fractions of the solid particles in the mid-section of the impeller passage.

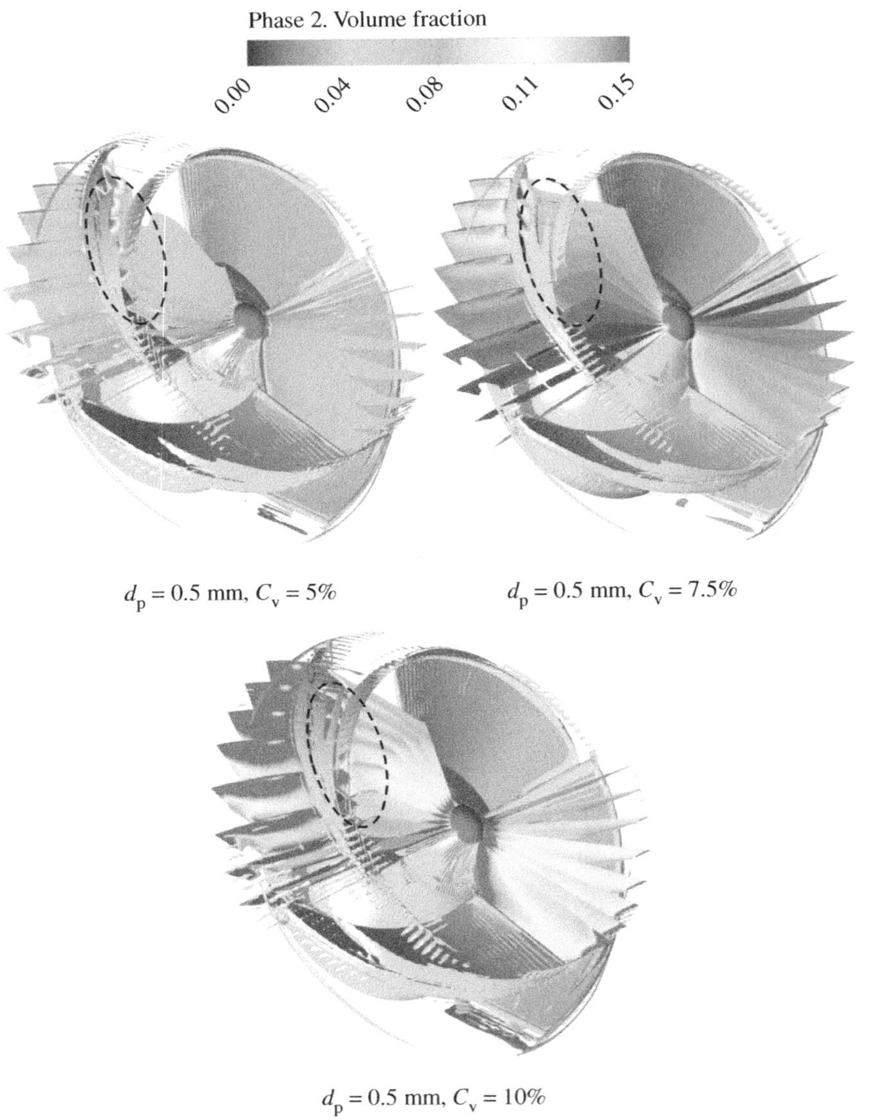


**Figure 12.10** Distribution of the vorticity and volume fractions of solid particles inside the impeller at the design flow rate.

when compared to $C_v = 5\%$, and the vorticity increases. At $C_v = 10\%$, the vorticity further increases, indicating that an increase in the solid-phase volume fraction leads to an increase in vorticity in the impeller. In the generation of the vortices, the rotational motion of the fluid in the vortices requires a certain amount of energy, leading to a decrease in the overall energy of the fluid, resulting in the energy loss.

## 12.5 Two-phase Flow Characteristics in the Guide Vane of Mixed-flow Pumps

### 12.5.1 Distribution of Solid Particles in the Guide Vane of a Mixed-flow Pump

Figure 12.11 shows the distribution of the solid-phase spherical particles of diameter $d_p = 0.5$ mm for various volume fractions in the two-phase flow at different cross-sections of the guide vane passage, as shown in the left graph of the figure. The sections in the guide vane passage from the inlet to the impeller exit are labeled as Planes 1–9. It can be observed that there is a significant difference in the distribution of the solid-phase particles at different volume fractions in the cross-sections of the guide vane passage. As mentioned earlier, when the solid-phase fraction $C_v = 5\%$ in the impeller, the solid particles are mainly concentrated in the exit region of the impeller and in the outer edge of the blade. Due to the low concentration of the solid-phase particles, as particles enter the guide vane immediately after leaving the impeller, they are entrained by the liquid phase. Since the guide vane does not perform rotational work, solid particles quickly pass through the guide vane passage along the outer edge of the guide vane. Therefore, for $C_v = 5\%$, the solid-phase distribution in the guide vane is mainly in the outer edge region.

For $C_v = 7.5\%$, the increase in the solid particles reduces the entrainment effect of the liquid phase on the solid phase. The distribution area of the solid particles in the guide vane passage expands with a tendency for the solid-phase distribution area to move closer to the hub of the guide vane. When the solid-phase volume fraction increases to $C_v = 10\%$, the distribution of solid particles in the guide vane passage shows a noticeable expansion from the outer edge of the guide vane toward the hub. The distribution area of the solid particles in the guide vane passage enlarges.

As the liquid flows in the guide vane passage, the enhanced turbulence intensity leads to changes in the velocity and direction of the liquid flow. Combined with the variations in the cross-sectional area of the guide vane passage and the guiding effect of the guide vane, the liquid may form rotating vortices during this process. The generation of the vortices undoubtedly affects the movement of the particles and the energy changes in the guide vane passage. The influence of vortices in the flow can be represented by the vorticity intensity with a higher vorticity intensity indicating greater turbulence in the flow, leading to higher energy losses. The equation for calculating the vorticity intensity of the vortices is given by:

$$\Gamma = \iint_A \left(\omega_y / \cos\theta - \omega_x / \sin\theta\right) dA \tag{12.13}$$

In Eq. (12.13), $\Gamma$ denotes the vorticity intensity; $\omega_x$ and $\omega_y$ denote the vorticity in the $x$ and $y$ direction, respectively; $\theta$ denotes the angle between the section and the $x$ direction; and $A$ denotes the area of the section.

The right graph in Fig. 12.11 compares the vorticity intensity at each cross-section. It can be observed from this graph that the vorticity intensity gradually decreases from the inlet to the outlet of the guide vane, and its sensitivity to the solid-phase volume fraction is relatively small. This indicates that, under the conditions of a two-phase flow with a solid–liquid mixture in the designed operating conditions, the impact of vortices in the guide vane passage is relatively small. The vorticity intensity tends to decrease along the flow direction, suggesting that the vorticity-induced energy losses in the guide vane passage are not as significant as in the impeller.

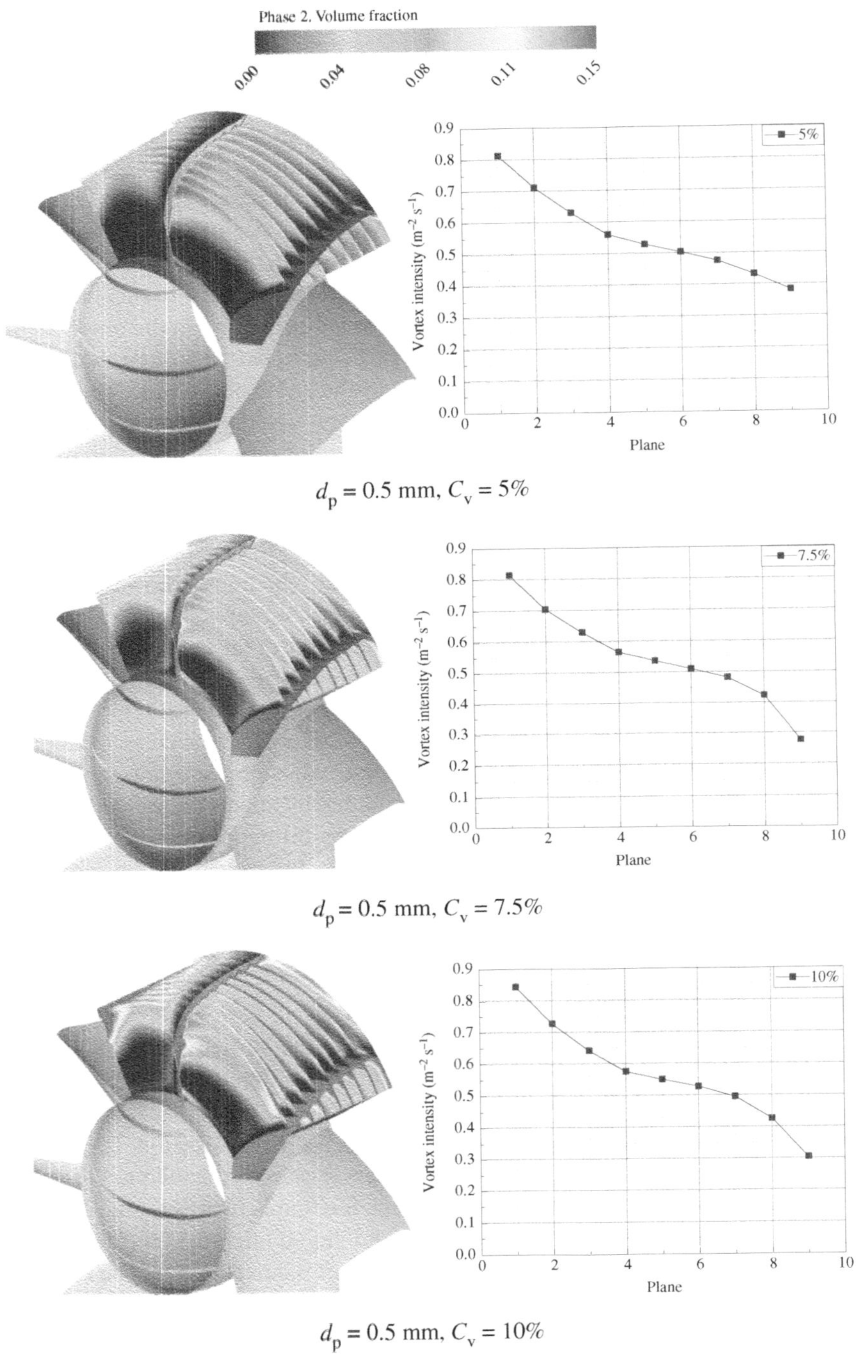


**Figure 12.11** Distribution of the solid particles in a plane section of the guide vane passage for various volume fractions of the solid particles.

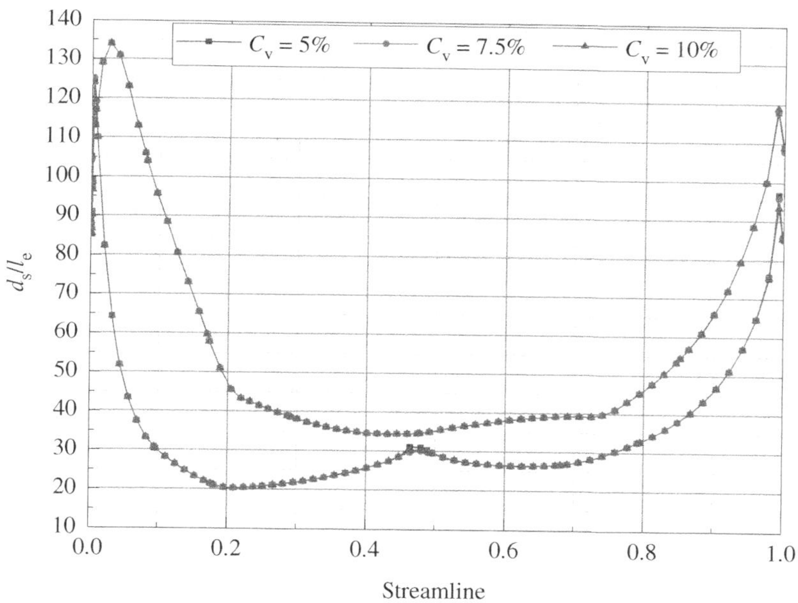


**Figure 12.12** Dynamic scale of particle motion along the midstream line in the guide vane.

The study of the particle diffusion in the guide vane passage reveals the distribution of the dynamic scale of particles along the midstream line of the guide vane, as shown in Fig. 12.12. Due to changes in the flow passage at the inlet and the outlet of the guide vane, fluid generates turbulence. Moreover, the presence of the solid particles enhances the turbulence intensity, resulting in a larger dynamic scale of the particles at the inlet and the outlet of the guide vane. The diffusion extent of particles in the guide vane passage remains similar under various solid-phase volume fractions.

## 12.5.2 Analysis of the Energy Loss in the Guide Vane

The entropy production analysis is employed to study the energy losses in the guide vane passage of the mixed-flow pump. Figure 12.13 shows the distribution of the local high-entropy production zones in the vertical axis sections in the guide vane passage. From this figure, it can be noted that in the design condition, the average entropy production at each section of the guide vane passage for the three different solid-phase volume fractions reaches its highest value in the outlet section of the guide vane. Simultaneously, based on the local entropy production contour maps at different guide vane passage sections, it can be observed that the high-entropy production zone in the middle of the guide vane passage develops downstream along the axial direction and gradually increases. The variations in the entropy production in the guide vane passage are similar for all solid-phase volume fractions, with the entropy production steadily increasing from the guide vane inlet to the outlet. Particularly at the guide vane outlet, the entropy production region occupies more than 30% of the passage, indicating significant energy loss.

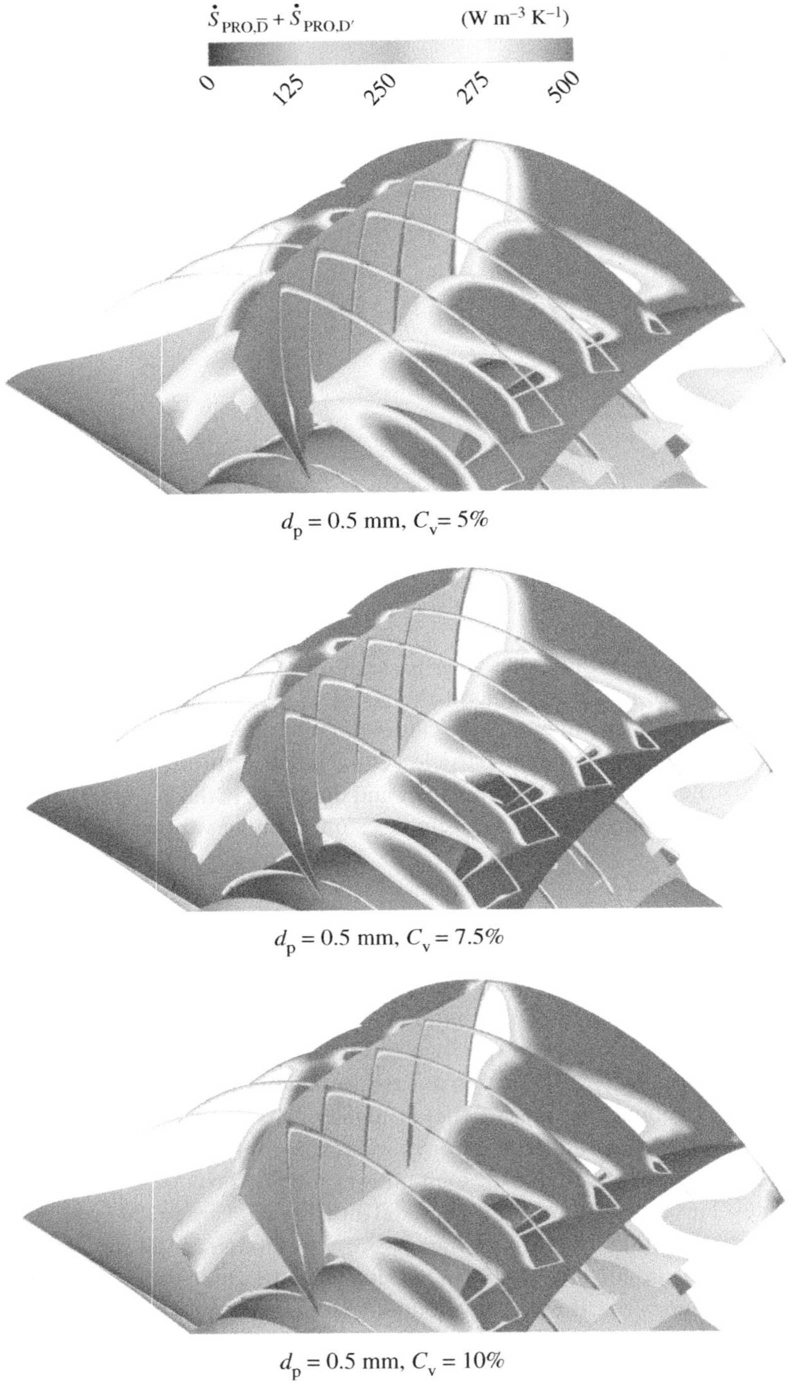


**Figure 12.13** Distribution of the high-entropy production regions in a cross-section of the guide vane passage.

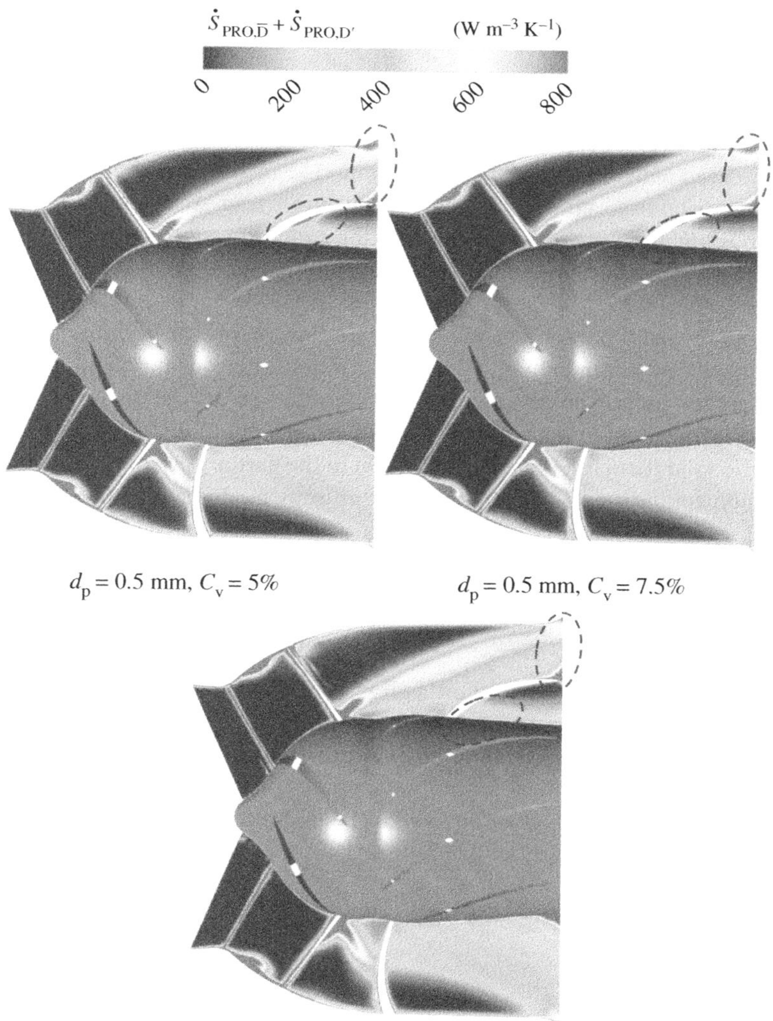


**Figure 12.14** Distribution of the entropy production in various cross-sections of the pump passage for various volume fractions of the solid particles.

The distribution of entropy production in various cross-sections of the mixed-flow pump passage under different solid-phase volume fractions is shown in Fig. 12.14. From this figure, it can be observed that the high-entropy production zone in the impeller is mainly concentrated in the region outside the blades and near the hub at the impeller outlet. The large entropy production region in the guide vane is mainly in the middle part of the guide vane passage, and the entropy production zone along the guide vane passage gradually expands. With an increase in the solid-phase volume fraction, the high-entropy production zone in the guide vane passage decreases slightly with some changes in its location.

## 12.6 Summary

Based on the material presented in this chapter, the following conclusions can be drawn:

(1) It has been shown in Chapters 7–11 that in various flow conditions, the mixed-flow pump exhibits a decrease in the head with an increase in the flow rate. The inclusion of the solid particles in the clean water contributes to a further decrease in the pump head. Also, the efficiency of the mixed-flow pump is higher in the clean water operating condition compared to that in the two-phase solid–liquid flow condition, which decreases with an increase in the volume fraction of the solid phase.
(2) As the volume fraction of the solid particles in the two-phase flow increases, both the head and efficiency of the mixed-flow pump exhibit a decreasing trend. The concentration of the solid-phase particles is predominantly observed in the impeller inlet, in the outer edge of the impeller, in the guide vane inlet, and in the outer edge of the guide vane. With an increase in the volume fraction of the solid particles, there is a reduction in the entropy production in the impeller blade edge region coupled with an increase in the entropy production near the hub area. It is observed that the regions with large entropy production and the turbulent kinetic energy dissipation in the impeller are situated in the middle of the impeller flow passage. The turbulent kinetic energy dissipation values in the impeller passage typically fall within the range of 0–2 $m^2/s^2$, and an increase in the solid-phase volume fraction results in higher turbulent kinetic energy dissipation.
(3) Solid-phase particles in the guide vane passage show a larger concentration in the outer edge region of the guide vane. Furthermore, an increase in the volume fraction of the solid phase correlates with a shift in the solid-phase distribution toward the hub area of the guide vane. Notably, the volume fraction of the solid phase has minimal impact on the entropy production and energy loss in the guide vane passage, with energy loss being predominantly localized in the middle region of the guide vane passage.

## References

1 Yuan, J.Q., Liu, Z., Huang, R. et al. (2023). investigation on energy conversion characteristics of vortex pump under bubble inflow. *CIESC Journal* 74 (09): 1–19.
2 Shi, G.T., Zhao, Z.W., Zhu, Z.H., and Zhou, M.H. (2022). Effect of guide vane number on energy loss in multiphase pump. *Machine Tool & Hydraulics* 50 (11): 90–93.
3 Nie, Q., Wang, Y.F., and Min, J.Q. (2016). Inlet and outlet impact losses analysis of impeller under off-design conditions. In: *Proceedings of The 2015 International Conference On Mechanics And Mechatronics (ICMM 2015)*, 425–432.
4 Li, W., Ji, L.L., Li, E.D. et al. (2021). Numerical investigation of energy loss mechanism of mixed-flow pump under stall condition. *Renewable Energy* 167: 740–760.
5 Li, W., Liu, M.J., Ji, L.L. et al. (2023). Study on the trajectory of tip leakage vortex and energy characteristics of mixed-flow pump under cavitation conditions. *Ocean Engineering* 267: 113225.
6 Li, W., Huang, Y.X., Ji, L.L. et al. (2023). Prediction model for energy conversion characteristics during transient processes in a mixed-flow pump. *Energy* 271: 127082.
7 Ji, L.L., Li, W., Shi, W.D. et al. (2020). Energy characteristics of mixed-flow pump under different tip clearances based on entropy production analysis. *Energy* 199: 117447.

8 Ji, L.L., Li, W., Shi, W.D. et al. (2021). Effect of blade thickness on rotating stall of mixed-flow pump using entropy generation analysis. *Energy* 236: 121381.

9 Ji, L., Li, W., Shi, W.D. et al. (2020). Diagnosis of internal energy characteristics of mixed-flow pump within stall region based on entropy production analysis model. *International Communications in Heat and Mass Transfer* 117: 104784.

10 Lu, L., Zhou, X.J., Wang, H.A., and Xu, Y.Q. (1995). Study on sand used in muddy water test of model water pump. *Pump Technology* 1995 (01): 22–26.

11 Li, Z.P., Huang, Z.Q., Zhang, A.X., and Xie, J.B. (2003). Research and design of hydraulic performance of multiphase flow pumps. *China Rural Water and Hydropower* 2003 (10): 61–63.

12 Zhao, Z.H., He, X.J., and Li, C.L. (1991). Calculation method of head and efficiency when centrifugal pump conveys solid-liquid mixture. *Fluid Machinery* 1991 (03): 20–24.

13 Gandhi, B.K., Singh, S.N., and Seshadri, V. (2000). Improvements in the prediction of performance of centrifugal slurry pumps handling slurries. *Proceedings of The Institution of Mechanical Engineers Part A-Journal of Power And Energy* 214 (A5): 473–486.

14 Li, W.H., Chen, Y., and Wang, R.X. (2009). Simulation and analysis of 100 Gz slurry pump based on liquid-solid two-phase fluids. *World Sci-Tech R&D* 31 (05): 852–854.

15 Sun, L.Y., Pan, Q., Zhang, D.S. et al. (2022). Numerical study of the energy loss in the bulb tubular pump system focusing on the off-design conditions based on combined energy analysis methods. *Energy* 258: 124794.

16 Luo, X., Cao, S.F., Pei, Y.J. et al. (2023). Energy conversion characteristics of an in-line mixing pump in solid-liquid two-phase dynamic mixing. *Arabian Journal for Science and Engineering* .

17 Shi, B.C., Zhou, K.L., Pan, J.P. et al. (2021). PIV test of the flow field of a centrifugal pump with four types of impeller blades. *Journal of Mechanics* 37: 192–204.

18 Wu, B., Zhang, J., and Xiang, Y. (2007). Study on energy saving technology of two-phase centrifugal pump transportation system. *Fluid Machinery* 2007 (07): 37–39.

19 Tang, L.F. and Tang, L.X. (2005). Optimal design of middle and low specific speed two-phase flow trash pump lmpeller geometric parameters. *Journal of Mechanical Transmission* 2005 (01): 41–43, 55–74.

20 Shen, Z.J., Li, R.N., Han, W. et al. The research on particle trajectory of solid-liquid two-phase flow and erosion predicting in screw centrifugal pump. In: *7th International Conference On Pumps And Fans (ICPF2015) 2016*, vol. 129, 012052.

21 Shi, H.X., Meng, J., and Li, Y. (2021). Numerical simulation of coarse particle two-phase flow in two-stage vortex pump. *International Journal of Simulation Modelling* 20 (4): 684–695.

22 Wang, Y.P., Li, W.Q., He, T.L. et al. (2021). The effect of solid particle size and concentrations on internal flow and external characteristics of the dense fine particles solid-liquid two-phase centrifugal pump under low flow condition. *AIP Advances* 11 (8): 085309.

23 Liu, H.J. (2018). *Analysis on Abrasion and Energy Loss Characteristics of Centrifugal Pump Blade in Solid–Liquid Two-phase Flow*, vol. 2018. Lanzhou University of Technology.

24 Wang, Z.B. (2023). *Research on Technology of Coal-bed Gas Well Jet Pump Carrying Coal Powder Drainage and Gas Production*, vol. 2023. China University of Petroleum (East China).

25 Li, Y. (2014). *The Research on Numerical Simulation and Abrasion Property of Solid-liquid Two-phase Flow Centrifugal Pump*, vol. 2014. Zhejiang Sci-Tech University.

26 Zhang, Z.C. (2017). *Investigation on Sediment Erosion Characteristics of Double Suction Centrifugal Pump*, vol. 2017. China Agricultural University.

27 Kock, F. and Herwig, H. (2004). Local entropy production in turbulent shear flows: a high-Reynolds number model with wall functions. *International Journal of Heat and Mass Transfer* 47: 2205–2215.

# Index



*Mixed-flow Pumps: Modeling, Simulation, and Measurements*, First Edition.
Wei Li, Leilei Ji, Ramesh Agarwal, Weidong Shi, and Ling Zhou.

Printed and bound by CPI Group (UK) Ltd, Croydon, CR0 4YY
17/09/2026
14964259-0002